DECOHERENCE AND ITS IMPLICATIONS IN QUANTUM COMPUTATION AND INFORMATION TRANSFER

NATO Science Series

A series presenting the results of scientific meetings supported under the NATO Science Programme.

The series is published by IOS Press and Kluwer Academic Publishers in conjunction with the NATO Scientific Affairs Division.

Sub-Series

I.	Life and Behavioural Sciences	IOS Press
II.	Mathematics, Physics and Chemistry	Kluwer Academic Publishers
III.	Computer and Systems Sciences	IOS Press
IV.	Earth and Environmental Sciences	Kluwer Academic Publishers
V.	Science and Technology Policy	IOS Press

The NATO Science Series continues the series of books published formerly as the NATO ASI Series.

The NATO Science Programme offers support for collaboration in civil science between scientists of countries of the Euro-Atlantic Partnership Council. The types of scientific meeting generally supported are "Advanced Study Institutes" and "Advanced Research Workshops", although other types of meeting are supported from time to time. The NATO Science Series collects together the results of these meetings. The meetings are co-organized by scientists from NATO countries and scientists from NATO's Partner countries – countries of the CIS and Central and Eastern Europe.

Advanced Study Institutes are high-level tutorial courses offering in-depth study of latest advances in a field.
Advanced Research Workshops are expert meetings aimed at critical assessment of a field, and identification of directions for future action.

As a consequence of the restructuring of the NATO Science Programme in 1999, the NATO Science Series has been re-organized and there are currently five sub-series as noted above. Please consult the following web sites for information on previous volumes published in the series, as well as details of earlier sub-series:

http://www.nato.int/science
http://www.wkap.nl
http://www.iospress.nl
http://www.wtv-books.de/nato_pco.htm

Series III: Computer and Systems Sciences - Vol. 182 ISSN: 1387-6694

Decoherence and its Implications in Quantum Computation and Information Transfer

Edited by

Tony Gonis

Lawrence Livermore National Laboratory, University of California, Livermore, CA, USA

and

Patrice E.A. Turchi

Lawrence Livermore National Laboratory, University of California, Livermore, CA, USA

IOS Press

Ohmsha

Amsterdam • Berlin • Oxford • Tokyo • Washington, DC

Published in cooperation with NATO Scientific Affairs Division

Proceedings of the NATO Advanced Research Workshop on
Decoherence and its Implications in Quantum Computation and Information Transfer
Mykonos, Greece
25–30 June, 2000

ISBN 1 58603 211 9 (IOS Press)
ISBN 4 274 90468 7 C3042 (Ohmsha)
Library of Congress Control Number: 2001094316

Publisher
IOS Press
Nieuwe Hemweg 6B
1013 BG Amsterdam
Netherlands
fax: +31 20 620 3419
e-mail: order@iospress.nl

Distributor in the UK and Ireland
IOS Press/Lavis Marketing
73 Lime Walk
Headington
Oxford OX3 7AD
England
fax: +44 1865 75 0079

Distributor in the USA and Canada
IOS Press, Inc.
5795-G Burke Centre Parkway
Burke, VA 22015
USA
fax: +1 703 323 3668
e-mail: iosbooks@iospress.com

Distributor in Germany, Austria and Switzerland
IOS Press/LSL.de
Gerichtsweg 28
D-04103 Leipzig
Germany
fax: +49 341 995 4255

Distributor in Japan
Ohmsha, Ltd.
3-1 Kanda Nishiki-cho
Chiyoda-ku, Tokyo 101
Japan
fax: +81 3 3233 2426

PRINTED IN THE NETHERLANDS

Preface

Decoherence is the physical process by which the classical world – the world of common sense – emerges from its quantum underpinnings. This physical process refers to the loss of phase coherence between the parts of a quantum system, because of the interaction of the system with the environment. The purpose of the Advanced Research Workshop (ARW) *Decoherence and its Implications in Quantum Computation and Information Transfer* was to provide an in-depth discussion of this physical process and of its relevance in understanding and interpreting quantum mechanics, with special emphasis given to its implications in implementing the practical aspects of quantum computation and information transfer. The ARW brought together scientists in diverse fields such as mathematical physics, quantum logic, and quantum computing for an intense, interdisciplinary discussion of decoherence and its practical implications in quantum computing.

Quantum computing represents possibly the epitome of the technological impact that quantum mechanics can be expected to have on technological developments in the foreseeable future. Its realization would allow the performance of tasks that are practically unthinkable in terms of classical physics. Such exotic occurrences as breaking hitherto unbreakable codes, searching enormous lists for specific entries, and communicating with messages that betray the presence of eavesdropping would fail within reach. These wonders are possible when one exploits the logical consequences of the laws that govern natural phenomena at the most fundamental level, namely, the laws of quantum mechanics.

Standing squarely in the way of these developments is decoherence. The aim of the workshop is to provide an in-depth discussion of the current state of affairs regarding the progress that is being made in understanding the effects of decoherence in quantum computation and information transfer and circumventing or alleviating the difficulties associated with it. The present is a particularly fitting time for discussions of an interdisciplinary nature, encompassing quantum logic, mathematics, and physics, addressing these issues. As we stand at the doorstep of the next millennium, we can ponder the possible progress that can be expected in this important area of applied physics. The invited speakers are among scientists who continue to make significant contributions in this domain of science and the resulting presentations and discussions were lively as well as deeply informative. We also hope that the participants at the conference and other scientists who peruse this proceedings will benefit from exposure to the issues under consideration here and will carry developments further as their own scientific career unfolds.

The subject matter of the ARW is divided along the three main categories mentioned above. After discussing the basic notion of interpretation of quantum mechanics and being eloquently reminded of Bohr's views and the Copenhagen interpretation, we proceed to more recent advances such as the logical interpretation. The EPR paradox is revisited, and the relevance of quantum logic to the understanding of decoherence is presented. The specific topics of quantum computing and information transfer are then presented, and their formal and experimental connection to the physical process of decoherenee is discussed. Special emphasis is given to recent developments such as non-locality, quantum error correction, and bypassing decoherence.

As the organizers and editors of this volume of proceedings, we wish to express our gratitude to the Office of Scientific and Environmental Affairs of the North Atlantic Treaty Organization (NATO) for their sponsorship of this ARW, and to the invited speakers and participants who worked diligently to bring this volume to fruition. We thank also the staff of IOS Press for their guidance, and their patience and understanding during the time that it took to collect the papers and assemble the present volume. It is our fervent hope that this volume will both inform as well as motivate further progress in the area of quantum computation and information transfer.

A. Gonis
P.E.A. Turchi

20 June, 2001

Contents

*) invited presentation

Foundations of Quantum Mechanics

Decoherence and its Implications in Quantum Computation and Information Transfer
T. Gonis and P.E.A. Turchi (Eds.)
IOS Press, 2001

Reading Bohr: Complementarity, Epistemology, Entanglement, and Decoherence

Arkady Plotnitsky
Department of English
Purdue University
West Lafayette, IN 47907
e-mail: aplotnit@sla.purdue.edu

... in quantum mechanics, we are not dealing with an arbitrary renunciation of a more detailed analysis of atomic phenomena, but with a recognition that such an analysis is in principle excluded.

—Niels Bohr

Abstract

On the basis of an extensive reading of Niels Bohr's works, quantum mechanics in Bohr's interpretation, known as complementarity, is argued to provide a consistent and complete description of the phenomena it considers, including quantum entanglement. This interpretation is also local, in the sense of fully conforming to the requirements of relativity theory. The argument is based on an epistemologically radical, and to some (Einstein among them) unpalatable, assumption that no physical properties of quantum objects or processes themselves are describable by means of this or, it appears, any theory; and it is further argued that no such properties are even ascribable at the quantum level. Instead, the formalism of quantum theory and its physical interpretation are seen as referring to the effects of the (quantum) interaction between quantum objects and the measuring instruments involved (described in terms of classical physics) upon the latter. The "efficacity" of the effects, understood as the (noncausal) power or agency producing them, defined by this interaction, is seen as indescribable and even inconceivable by any means, whether those of classical or quantum physics. We are only allowed an access to certain effects, such as those constituting the data of quantum physics, but not for the ultimate efficacity of these effects. This epistemology, however, is fully compatible with all the requirements of theoretical and experimental research in the field of quantum physics, as we know it now, and might be necessary in order to maintain some of these requirements. The implications of this epistemology for "decoherence" are also discussed.

1. Introduction

The current state of the debate about quantum mechanics is dominated by the arguments concerning the Einstein-Podolsky-Rosen (EPR) type experiments and the so-called (the term is due to Erwin Schrödinger) "quantum entanglement," the existence of a particular type of correlations between certain spatially separated quantum-mechanical events. These correlations appear to some to indicate the "nonlocality" of quantum mechanics (here understood strictly as the possibility of nonlocal influences that would be in conflict with relativity).[1] This particular way of thinking about the famously strange character of quantum physics emerged in the wake of Albert Einstein, Boris Podolsky, and Nathan Rosen's article, "Can Quantum-Mechanical Description of Physical Reality be Considered Complete?," to which Niels Bohr replied in an article under the same title, and Einstein's subsequent arguments on the subject [1] [2].[2] The latter refined the EPR argument by more sharply focusing it on the possible nonlocality of quantum mechanics (rather than on its incompleteness as a physical theory) and proposed an alternative between seeing quantum mechanics either as a complete but nonlocal or a local but incomplete theory. This alternative is suggested in the EPR article as well (albeit in a somewhat obscured form) and in Einstein's earlier arguments. By contrast, Bohr views the completeness of quantum mechanics as fully compatible with its locality. Or, at least Bohr offers an interpretation of quantum mechanics, complementarity, which, while preserving quantum entanglement, is local. It follows that only entanglement is inevitable in quantum mechanics and may, accordingly, be seen as part of the quantum-mechanical data, while nonlocality is not. By the same token, the role (perhaps irreducible in any event) of interpretation in quantum mechanics acquires further significance. Compelling reasons for Bohr's approach and epistemology emerged from the outset of quantum theory, in particular those pertaining to certain features of quantum-mechanical formalism and, correlatively, to the question concerning what physical events in space and time quantum-mechanical predictions may refer to. The requirements of locality provided further impetus and ultimately support to his views, even though Einstein's critique made Bohr refine, sometimes significantly, his argument. Bohr, however, already had in place a basic framework that enabled him to handle these additional complexities. Schrödinger's "cat-paradox" paper (1935) [3]; David Bohm's reformulation of the EPR type of experiments in terms of spin and then his (nonlocal) hidden variables theories; John S. Bell's theorem and his discontent with quantum mechanics [4]; Alain Aspect's experiments; and further refinements and extensions of Bell's theorem, as well as more recent investigations of decoherence and related questions, are all well-known developments and parts of the debate in question. The literature on the subject is massive, and it would not be possible to address it within the limits of this article. While, however, this literature confirms the significance of Bohr's thought for our understanding of quantum theory and its interpretation, it also reflects persistent problems of reading Bohr's reply to EPR and his writing in general, beginning with Einstein and Schrödinger, and extending to Bell and beyond.[3]

Bohr's writings may well present a special case, given the proportion of physics contained in verbal formulations rather than mathematical formulas, or formal logical deductions that have dominated the foundational work on quantum mechanics, including in some arguments concerning quantum nonlocality [5]. The peculiarity of these writings is in part due to the peculiar nature of quantum physics, and of Bohr's interpretation and epistemology of it. In commenting on the difficulties involved in "Discussion with Einstein on Epistemological Problems in Atomic Physics," Bohr said: "Rereading these passages, I am deeply aware of the inefficiency of expression which must have made it very difficult to appreciate the trend of the argumentation aiming to bring out the essential ambiguity involved in reference to physical attributes of objects when dealing with phenomena where no sharp separation can be made between the objects themselves and their interaction with the measuring instruments" ([6] 2, p.61).[4] Such separation and, hence, the description of (the properties of) quantum objects and processes themselves (as opposed to certain effects of their interaction with measuring instruments upon the latter) are impossible in Bohr's interpretation. This impossibility expresses the essence of Bohr's epistemology. As will be seen, however, this impossibility also opens the space of new possibilities of physics and knowledge in general.

On the other hand, some of the peculiarities in question are peculiarly Bohr's, especially insofar as Bohr's key terms, such as phenomena, individuality, atomicity, or complementarity, have unconventional and sometimes idiosyncratic meanings. In general, Bohr's writing appears to pose more substantial demands than customary in a scientific text as concerns paying special attention to particular formulations; carefully adhering to the particular meaning of his terms; understanding the philosophical (rather than only physical and mathematical) structure of his concepts; different languages involved (Bohr wrote and thought on the subject in several languages) and translations between them; and so forth. These demands are not always met by Bohr's readers, which leads to significant misunderstandings of his argumentation. Naturally, my point is not that one cannot disagree with Bohr's views or criticize his arguments, but the special conditions, often missed by Bohr's critics, that a meaningful reading or, if necessary, criticism would entail in his case.

Indeed, while this may be especially true in Bohr's case, I would argue more generally that the role of reading in physics is more significant than is commonly acknowledged, especially when dealing with quantum mechanics and its interpretation. Physics is also reading. It is the interpretation of texts, as well as of (and often jointly with) physical theories themselves. The history of quantum mechanics, from the work of founding figures to the most recent developments, offers remarkable examples of both, both in general and specifically as concerns our encounters with Bohr's ideas. Indeed this history appears to be indissociable from interpreting Bohr's ideas, from reading Bohr, as Roland Omnés has recently argued ([7], pp. 81-82, [8], pp. 49-52).

I shall pursue my argument by offering a macro-reading of "Discussion with Einstein." I shall also offer an example of, as it were, a micro-reading of a particular passage of Bohr's reply to EPR (famously invoking "the essential ambiguity" of their argument), where the question of the locality of quantum theory is especially germane. This micro-reading, however, is inevitably a macro-reading as well and involves an extensive traversing of Bohr's works. Finally, I shall comment on the implications of Bohr's epistemology for decoherence.

I would like to stress that the present article addresses primarily Bohr's <u>interpretation</u> of quantum mechanics, and even more specifically with the version presented in "Discussion with Einstein," which, at the very least, refines Bohr's earlier versions of complementarity. (These nuances are not always addressed or carefully sorted out to by Bohr's readers either.) Accordingly, most of my epistemological claims pertain to this interpretation rather than to the experimental data or mathematical formalism of quantum mechanics (if they can be seen as independent of an interpretation), or other interpretations of quantum mechanics, including those associated with "the Copenhagen interpretation." The latter term must be applied with great caution, given the differences between such interpretations and the thought of the different figures involved, even those who are considered, and consider themselves, close to Bohr (Heisenberg and Pauli, among them). These differences are much greater than it is usually argued and often outweigh the shared features, important as the latter may be. Indeed, I would argue that, once considered in all of its aspects, Bohr's interpretation (in the present reading or "interpretation") is unique and, I would argue, uniquely radical epistemologically. One several occasions, which I shall specify as I preceed, I shall advance arguments, both those arising from within Bohr's interpretation and relatively independent ones, that exceed the limits of Bohr's interpretation and lead to more general claims.

2. "The Peculiar Individuality of Quantum Effects": Complementarity and Epistemology

I begin with "Discussion with Einstein on Epistemological Problems in Atomic Physics," originally published in 1949 in the so-called "Schilpp volume," honoring Einstein [9]. Historically, it might be more appropriate to begin with his earlier (1935) reply to EPR, arguably the most decisive work in developing Bohr's ideas in their ultimate form, which is my main concern here. According to Bohr himself, however, "Discussion with Einstein" elucidates and refines Bohr's response to Einstein's arguments and "give[s] a clearer impression of the necessity of a radical revision of basic principles for physical explanation" ([6] 2, p.61). Bohr felt this necessity throughout the history of quantum theory, but it became especially pressing for him in the wake of the EPR argument. "Discussion with Einstein" becomes Bohr's most comprehensive and consistent exposition of his interpretation of quantum mechanics.[5] It also explicates and sometimes reinterprets his earlier works. Nevertheless, a proper understanding of the key ideas of the article requires taking into account these earlier works, especially, again, Bohr's reply to EPR and his so-called Como lecture of 1927, "The Quantum Postulate and the Recent Development of Atomic Theory" ([6] 1, pp. 26-51), which introduces complementarity. The latter appears there in the form of complementary features of physical descriptions, defined by Bohr as mutually exclusive, yet also necessary for the comprehensive analysis of the type of phenomena we encounter in quantum physics. Eventually it came to designate Bohr's overall interpretation of quantum mechanics. Although somewhat different in its approach (we may say that it is more "philosophical," while Bohr' post-EPR argumentation is more "empirical") and in some of its epistemological elements, the Como lecture remains germane to Bohr's thought and puts in place most key ingredients of Bohr's post-EPR version(s) of complementarity.

In particular, Bohr's argument, in "Discussion with Einstein" and other late essays, concerning "the peculiar individuality of quantum effects," crucial to his epistemology and my main concern here, is a refinement of his argument in the Como lecture. The latter presents its argument in terms of Planck's constant *h*—Planck's quantum of action—and "the quantum postulate" ([6] 1, p. 53), rather than in terms of "effects" and "phenomena," as in the latter works. By the time of "Discussion with Einstein," Bohr argues that "the peculiar individuality of quantum effects presents us, as regards the comprehension of well defined evidence, with a novel situation unforeseen in classical physics and irreconcilable with conventional ideas suited for our orientation and adjustment to ordinary experience [on which classical physics is based]. It is in this respect that quantum theory has called for a renewed revision of the foundations for the unambiguous use of elementary concepts" ([6] 2, p.62). Quantum entanglement and the question of locality of quantum mechanics further contribute to the necessity of these revisions.

What is this "individuality" and what makes it so peculiar, and why "effects" (a term that itself became essential for Bohr), rather than quantum objects themselves? It is, again, the Como lecture that introduces the subject of and the very term "individuality," by way of a hesitant parenthesis ("effects" will have to wait). Bohr says: "Notwithstanding the difficulties which ... are involved in the formulation of quantum theory it seems that its essence may be expressed in the so-called quantum postulate, which attributes to any atomic process an essential discontinuity, or rather individuality, completely foreign to classical physics and symbolized by the Planck's quantum of action" ([6] 1, p.53; emphasis added). This hesitant shift from "discontinuity" to "individuality" (eventually further supplemented by "indivisibility" and "wholeness") is the opening move in the development of one of Bohr's most radical and innovative concepts. It took him twenty years and much further development of his interpretation of quantum mechanics before this concept crystallized at the time of "Discussion with Einstein." By then it came to designate phenomena that appear only at the level of the interaction, irreducible in Bohr's interpretation, between quantum objects and measuring instruments, rather than anything applicable to quantum objects themselves. As will be seen, the very concept of phenomenon as applicable in quantum mechanics was by then redefined by Bohr in the manner correlative to his view of "individual quantum effects." Indeed both are equivalent.

I shall follow Bohr in presenting the subject primarily in terms of the double-slit experiment and the position-momentum complementarity, rather than in terms of spin, as is more common in current discussions, although, as will be seen, spin may in fact offer a more natural general model of the situation. The double-slit experiment, the archetypal quantum-mechanical instrument, may be argued to contain all of the key epistemological features of quantum mechanics. It may, accordingly, be useful to summarize the double-slit experiment in a way consistent with Bohr's understanding.

A sufficient number (say, a million) of elementary particles, such as electrons or photons, emitted from a source, is allowed to pass through the diaphragm with two slits and leave their traces on the screen. (It is convenient to consider, as Bohr does, an intermediate diaphragm with one slit between the source and the two-slit diaphragm.) If both slits are open and no arrangements are made that would allow us to check through which slits the particles pass, a wave-like interference pattern will emerge on the screen, in principle (there may be practical limitations as to what kind of set-up can be realized), regardless of the distance between slits, or the time-interval between the emissions of each particle. More accurately, one should speak of a pattern analogous to the traces that would be left by classical waves in a corresponding medium. This pattern is the actual manifestation and, according to most standard interpretations (there are alternative views), the only possible manifestation of the "wave" character of the quantum world. In Bohr's interpretation, one can speak of "wave propagation" or of any attributes of the classical-like phenomenon of wave-propagation only provisionally or by convention, or else symbolically. The same, however, is also true as concerns the attributes of classical particle motion, in particular trajectories. We see on the screen only classically manifest traces of quantum objects. The latter themselves are destroyed or in any event made unavailable in the process of this, in Bohr's terms, "irreversible amplification" of our encounters with quantum objects to the classical level, which, as will be seen, is a kind of "decoherence" process ([6] 2, p.51)

If, however, we could know through which slit particles pass (indeed merely setting up the apparatus in a way that such a knowledge would in principle be possible would suffice), the interference pattern would inevitably disappear.

The situation is correlative to the uncertainty relations and to the statistical character of quantum theory; and it can be given a well-known statistical interpretation.[6] Indeed, by virtue of this correlation, the statistical nature of quantum mechanics appears to be irreducible. In Bohr's and most other (standard) interpretations, quantum mechanics only predicts the probabilities of certain occurrences and only predicts such probabilities (as probable outcomes of certain measurements on the basis of previously performed measurements), rather than accounts for the motion of quantum objects themselves in the way classical mechanics does for classical objects. All actual information is obtained only by measurements, and, unlike in classical mechanics, a measurement gives us no information about the past of the system prior to it. Nor, by virtue of uncertainty relations, can sequentially repeating measurements with the same accuracy help us in predicting subsequent measurements in the way they do in classical mechanics. Moreover, the statistical character of quantum mechanics cannot be seen as arising merely from insufficient information about "the properties of mechanical systems of great structural complexity [as in the case of classical statistical physics]" ([6] 2, p.34). Instead, in Bohr's interpretation, it is itself an (enigmatic) aspect of "the peculiar individuality of quantum effects." In other words, we do not know and, in Bohr's view, ultimately cannot know why certain probability rules, say, those of Planck's law or Max Born's "square moduli" rule for calculating quantum-mechanical probabilities in fact apply, but they do. Now, whenever the interference pattern is found, one cannot assign probabilities to the two alternative "histories" of a "particle" passing through either one slit or another on its way to the screen. If we do, the quantum-mechanical probability sum law (based in adding amplitudes rather than probabilities) would not be obeyed and the conflict with the interference pattern will inevitably emerge, as Bohr stressed on many occasions ([2], pp. 146-47; [6], pp.46-47).[7] One can also put it as follows. In counting the probability involved, we must, as it were, take into account the possibility of a particle passing through both slits, when both are open to it, in calculating the probabilities of the outcomes of such experiments, otherwise the probability will not come out right. (Hence, the probability counting procedure in quantum physics is different from that of classical statistical physics.) We cannot, however, again, at least in Bohr's interpretation, assume that a particle can actually do so in space and time.

The behavior just described, sometimes also known as the quantum measurement paradox, is indeed remarkable. Other common locutions include puzzling, strange, and mysterious. The main reason for this reaction is that, if one speaks in terms of the attributes and behavior of quantum objects themselves (usually the source of the trouble), in the interference picture the behavior of each particle appears to be "influenced" by the location of the slits. Individually or (which would be even more troubling) collectively, particles appear somehow to "know" whether both slits are or are not open, or whether there are or are not counting devices installed. Any attempt to picture or conceive of this behavior in terms of physical attributes of quantum objects themselves appears to lead to a logical contradiction; or be incompatible with one aspect of experimental evidence or the other; or entail strange or mysterious behavior; or require more or less difficult assumptions, such as that of attributing volition or personification to nature in allowing particles individual or collective "choices" ([6], 2, pp.51, 73); or imply nonlocality. The latter alternative was first proposed by Einstein and is legitimate and, arguably, the most rational under this attribution. Or, if local, quantum theory would be incomplete in this view.

Bohr, by contrast, sees the situation as indicating the essential ambiguity in ascribing conventional (and ultimately any) physical attributes, such as wave-like or particle-like behavior (or kinematical and dynamical variables of classical physics), to quantum objects themselves or in referring to their independent behavior. He writes: "To my mind, there is no other alternative than to admit that, in this field of experience [rather than with properties of quantum objects], we are dealing with individual phenomena and that our possibilities of handling the measuring instruments allow us only to make a choice between the different complementary phenomena we want to study" ([6] 2, p.51). This inevitability would hold insofar as there is no local interpretation that is (epistemologically) essentially different from Bohr's, on which point I shall comment below. At the very least, Bohr's interpretation is both consistent with the quantum-mechanical predictions and local, in contrast, say, to Bohm's hidden variables theories. It is possible that the assessment of the relative merits of both interpretations would change were nonlocality to follow from the formalism of quantum theory or data themselves, as some argue [5] [10]. I would contend (for the reasons to be explained below) that such arguments are so far unconvincing [11].[8] The term phenomena in the above passage by Bohr must now be understood in Bohr's special sense (Bohr's statement makes no sense otherwise), which I shall properly explain below. Briefly, it refers to the effects of the interaction between quantum objects and measuring instruments, under specified experimental conditions, such as two different and ultimately mutually exclusive or complementary types of effects manifest in the double-slit experiment. The choice of observing either one or the other (both cannot be observed together) is the choice to which Bohr refers.

The undesirable features in question become removed without affecting the integrity of the data or formalism of quantum theory, once this type of interpretation is in place and reference to the properties of quantum objects is suspended. This is a reasonably trivial application of the standard logical deduction under the nontrivial and to some (Einstein among them) unacceptable, but far from irrational, epistemological assumptions entailed by Bohr's interpretation. More generally, Bohr's interpretation is defined by its deliberate logical consistency; and Bohr specifically denies that any departure from ordinary logic (for example, multivalued logic) is necessary, at least within his interpretation ([6] 3, pp.5-6; [12], p. 48). Along these lines (there are considerable differences otherwise), Bohr's interpretation may be related to Omnés's "logical interpretation of quantum mechanics" [7][8].

Crucially, in Bohr's interpretation, a reference to quantum objects or processes themselves would remain ambiguous even when one speaks of single such attributes, rather than, as would be more common, of a simultaneous attribution of joint properties, involved in various uncertainty relations, and even at the time when the measurement takes place. This stronger claim or, again, the corresponding interpretation appears to have emerged in Bohr's arguments in the wake of the EPR article. It is clear, however, that this is what Bohr has in mind, since in speaking of this ambiguity there he never qualifies it by a reference to either joint properties or the uncertainty relations. As he says in his reply to EPR, "In fact to measure the position of one of the particles can mean nothing else than to establish a correlation between its behavior [not its position!] and some instrument ... which defines the space frame of reference" ([2], pp. 148; emphasis added). In other words, in Bohr's interpretation we are not dealing with the "elements of reality" pertaining to quantum objects (here, a "position"), such as those that EPR want to attribute to them, but only with the (in terms of their physical description) classical properties pertaining to measuring instruments or the reference frame they establish. The former type of reference, or at least an assumption of its unambiguous possibility, is found in most of Einstein's arguments, and is responsible for the essential ambiguity of these arguments, under the conditions of Bohr's interpretation.

It is not of course that what we designate as "quantum objects" do not exist; quite the contrary, the existence of "quantum objects" of some sort and of their actions, such as those producing certain effects, is an inevitable inference given the data (i.e. "effects") in question. The point is that such attributes as "quantum" (in any sense, such as that of discontinuity or even complementarity) or even any attributes do not at all apply to the ultimate constituents of nature as we understand them now. We do not and cannot know or conceive of what sort of objects quantum objects may be, nor can we ultimately, in all rigor, think of them as "objects" (a relatively recent philosophical concept), to let alone attribute physical properties to them or their behavior by analogy with classical physics. (We can, again, easily do both in relation to certain effects subject to quantum-mechanical treatment.) Accordingly, when Bohr speaks of the "behavior of quantum objects," this phrase should be read in this general sense of the existence of something that is beyond the classical level of description and that (in the interaction with measuring instruments) produces observable effects, rather than implying an ascription of conventional physical or perhaps any other attributes to this behavior anymore than to quantum objects, ultimately the attribute of "behavior" itself included.

The quantum character of the world manifests itself only in and is, as I shall here term it, the "efficacity" of, the quantum-mechanical effects in question. I here use the term "efficacity" in its dictionary sense of power and agency producing effects, but, in this case, without the possibility of ascribing this agency causality in any sense hitherto available. The quantum-mechanical efficacity in this may be seen as part of "a rational generalization of the very ideal of causality," which Bohr associates with "complementarity" ([6] 2, p.41). The ultimate nature and character of this efficacity must, thus, be in turn seen as disallowing any unambiguous reference to it. Indeed it is now seen not only as inaccessible (as would be more common, also in interpreting Bohr) but also as inconceivable by any means that are or perhaps may ever be available to us. At a more immediate level of quantum measurement this efficacity may be seen in terms of the quantum interaction between quantum objects and measuring instruments, which (or rather certain of their components) are classically-described, but which are also capable of quantum interaction with quantum objects. The last qualification is crucial and is often missed or not adequately considered by Bohr's readers, critics and champions alike, on which I shall further comment below. In this case the quantum constitution of measuring instruments is, although, again, essential in principle, "not relevant for the problems of the adequacy and completeness of the quantum-mechanical description in its [epistemological] aspects here discussed" ([6] 2, p.51). That is, this constitution need not, although may be (as, for example, in von Neumann's framework) explicitly considered to ascertain the adequacy and completeness of quantum mechanics. Ultimately this efficacity would depend on and perhaps be defined by the quantum constitution of both quantum objects and measuring instruments, a complex subject that cannot be addressed here.[9]

Eventually Bohr came to see such effects and, hence, quantum-mechanical data in terms of "phenomena," the concept developed by him in the wake of the EPR argument around 1938 in the so-called Warsaw lecture [13]. Phenomena emerge and become indivisible as a result of the irreducible interaction between quantum objects and measuring instruments; that is, it becomes impossible to consider the quantum objects under investigation independently of their interaction with the measuring instruments and, hence, isolate them from phenomena. Phenomena are defined by the appearance of the particular, specifically complementary, effects recorded in certain parts of measuring instruments under rigorously specifiable experimental conditions. This specification must itself be seen as part of the phenomena ([6] 2, p.64). For, in Bohr's interpretation, the influence of these conditions can be never eliminated in considering the outcomes of quantum-mechanical experiments in the way it can, at least in principle, be done in classical physics. For example, if seen independently of the quantum-mechanical context of its appearance, each mark on the screen in the double-slit experiment would be perceived in the same way or as the same phenomena in the sense of the philosophical (say, Husserl's) phenomenology of consciousness. Such a mark would appear as the same regardless of the difference in the conditions and, hence, outcome ("interference" or "no interference") of the double-slit experiment. According to Bohr's definition, however, a mark is seen as a different individual phenomenon depending on these conditions, which are always mutually exclusive in the case, such as this, of complementary phenomena. As Bohr says:

> This point is of great logical consequence, since it is only the circumstance that we are presented with a choice of either tracing the path of a particle or observing interference effects [once we have a sufficient number of the particles hitting the screen], which allows us to escape from a paradoxical necessity of concluding that the behavior of an electron or a photon should depend on the presence of a slit in a diaphragm through which it could be proven not to pass. We have here to do with a typical example of how the complementary phenomena appear under mutually exclusive experimental arrangements and are just faced with the impossibility, in the analysis of quantum effects, of drawing any sharp separation between an independent behavior of atomic objects and their interaction with the measuring instruments which serve to define the conditions under which the phenomena appear. ([6] 2, pp. 46-47)

In contrast to the situation that obtains in classical physics, phenomena can never be subdivided so as to make it possible to disregard or compensate for the interactions between measuring instruments and quantum objects, and to allow one to isolate the latter or their behavior as subject to a classical-like or, again, any conceivable physical description in space and time. Any attempt at such subdivision will only introduce "new possibilities of the interaction between [quantum] objects and measuring instruments" which "reveal the ambiguity in ascribing customary physical attributes to atomic objects" ([6] 2, pp.40, 51). As will be seen, the very formalism of quantum theory must be adjusted accordingly. The situation leads Bohr to his concept of the indivisibility or wholeness of phenomena ([6] 2, pp.40, 72-73, [6] 3, p.4). This concept has nothing to do with David Bohm's concept of "wholeness" as the nonlocality of the quantum world, indebted as the latter may be to Bohr's ideas. Bohr's concept of the wholeness (indivisibility) of phenomena is coextensive with the locality of his interpretation.

Quantum individuality itself, originally seen in the wake of Planck's discovery as quantum discontinuity and, hence, atomicity at the level of quantum objects, is now understood as the individuality of phenomena, defined through the interactions between quantum objects and measuring instruments. These interactions are, again, irreducibly constitutive of the data ("effects") in question and are in each case unique. Thus, in the case of the double-slit experiment, we will be presented with two distinct multiplicities (defined by the effects of "particles" hitting the screen) of individual phenomena (each of which is indivisible), depending on two different sets of conditions of the experiment. One of these sets of conditions will lead to the emergence of the interference pattern, "built up by the accumulation of a large number of individual processes, each giving rise to a small spot on the photographic plate, and the distribution of these spots [following] a simple law derivable from the wave analysis" ([6] 2, pp.45-46). Each single spot, however, must be seen as a different individual phenomenon that depends on the conditions in which the event occurs. Quantum-mechanical predictions crucially depend on this distinction as well. Two different overall patterns, "interference" and "no-interference," thus, pertain to two (very large) sets of different phenomena.

The same type of considerations would apply to, along with a host (indeed all) of other standard quantum-mechanical phenomena, the measurements of either the position or momentum of a given particle, which are always mutually exclusive in view of the uncertainty relations. A given arrangement can only be seen as suitable either for one measurement or another, but never for both together, which is a crucial point of Bohr's argument with Einstein, specifically as concerns the EPR experiment. The essence of complementarity is this mutual incompatibility and exclusivity of two distinct experimental arrangements, which are always necessary, as opposed to dealing with a single experimental arrangement or an underlying (quantum) system, from which, in the language of the EPR article, different "elements of reality" could be arbitrarily extracted. What, according to Bohr, Einstein did not realize or at least did not fully appreciate and what "we must realize" (emphasis added) is "that in the problem in question [including the EPR experiment] we are not dealing with a single specified experimental arrangement" or phenomenon, from which we can extract one or another element of reality at will. Instead, "we are referring to two different, mutually exclusive [complementary] arrangements" or phenomena ([6] 2, p.57; [2], p. 149). Or, again, at least this interpretation is logically consistent and compatible with both the data in question and locality. In other words we are not, or need not be, in the freedom of choice with respect to a given physical situation or configuration of reality, as Einstein appears to have thought, which view of the situation may indeed entail nonlocality. Instead we may see the situation as offering us the freedom of choice in dealing with two different situations of physical reality at the macro-level, without referring to the attributes of quantum objects themselves ([2], p. 149). This view leads to an epistemology and an interpretation of quantum mechanics that allows for locality, even as it preserves quantum entanglement, along with all other standard quantum-mechanical data. In this interpretation, the concept of reality itself can only apply at the level of phenomena rather than quantum objects themselves, which may be real in the sense that they exist but appear to prevent us from conceiving of the way in which they exist.

"The problem," Bohr adds, "again emphasizes the necessity of considering the whole experimental arrangement [or, again, the phenomenon], the specification of which is imperative for any well defined application of the quantum-mechanical formalism" ([6] 2, p.57). The latter necessity deprives us of the possibility of either further subdividing or otherwise rearranging phenomena so as to avoid complementarity at that level. It further deprives us of the possibility of referring, even in principle, to the properties of quantum objects so as to see complementarity arising (however inevitably in practice) in measurement as an effect of something that is, in principle, free of complementarity (for example, in the manner of hidden variables). As Bohr writes:

> Within the scope of classical physics, all characteristic properties of a given object can in principle be ascertained by a single experimental arrangement, although in practice various arrangements are often convenient for the study of different aspects of the phenomena. In fact, data obtained in such a way simply supplement each other and can be combined into a consistent picture of the behavior of the object under investigation. In quantum physics, however, evidence about atomic objects obtained by different experimental arrangements exhibit a novel kind of complementary relationship. Indeed, it must be recognized that such evidence, which appears contradictory when combination into a single picture is attempted, exhausts all conceivable knowledge about the object. Far from restricting our efforts to put questions to nature in the form of experiments, the notion of complementarity simply characterizes the answers we can receive by such inquiry, whenever the interaction between the measuring instruments and the objects forms an integral part of the phenomena. ([6] 3, p.4)

This makes Bohr's use itself of the term "complementarity" peculiar, since the word complementary usually indicates parts of or adding up to a whole, which is precisely what is by definition impossible under the condition of complementarity in Bohr's sense. This difference is here captured by Bohr's reference to "different aspects of the (same) phenomena" in classical physics, as opposed to two different, mutually exclusive or complementary, phenomena in quantum physics. The latter phenomena are never reducible to or derivable from a single entity of any kind, which gives a very peculiar character to the efficacity behind the quantum-mechanical effects. But this is also what gives the concept a well-defined meaning and what enables a consistent and rigorous interpretation of quantum mechanics. If certain aspects of the situation seem inconceivable, this is because they in fact are, which point is crucial to Bohr's view. They certainly are not available to anything conceived on the model of classical physics, but more radically on any model that is or possibly ever will be available.

In Bohr's post-EPR works the complementarity of phenomena in Bohr's sense represents the physical meaning of the uncertainty relations. The variables themselves involved in actual measurements can refer only to the physical behavior of certain parts of the measuring instruments, rather than (unless symbolically) that of the quantum objects themselves. It is an important and often missed or unappreciated point that, in Bohr's interpretation, only certain parts of measuring instruments are subject to classical physics, while other parts are subject to quantum behavior, which enables their interaction with quantum objects resulting in the appearance of observable and measurable effects. (I shall return to this point presently.) From this perspective a more accurate explanation of the meaning of uncertainty relations is as follows. The data recorded in certain parts of our measuring instruments as a result of their interactions with quantum objects is of the same type as the data resulting from the measurement of classical objects in their interaction with the measuring instruments. (The latter behave in the same way.) In classical physics, however, we can, at least in principle, always measure both variables in question simultaneously, and disregard or compensate for the interaction between the objects in question and measuring instruments. By contrast, in quantum mechanics we can only measure or, in Bohr's view, unambiguously define either one or the other variable of that type, say, a change in the momentum of a certain part of the measuring instruments involved, but never both simultaneously. Hence, classical-like determinism is not possible even at the macro level of effects, each of which effects itself is a classically describable physical object. The effects themselves of the interactions between quantum objects and measuring instruments upon the latter can, however, be described in the realist manner, since any single variable of either type by itself can always be predicted with a probability equal to unity. In this view, such variables can only be seen as defining either the positional coordinates of the point registered in some part of the measuring instruments or, conversely, a change in the momentum of another such part, under the impact of its interaction with the object under investigation. They cannot be seen as referring to the attributes of quantum objects themselves. As Bohr writes in a rather striking sentence: "in this context, we are of course not concerned with a restriction as to the accuracy of measurement, but with a limitation of the well-defined application of space-time concepts and dynamical conservation laws, entailed by the necessary distinction between [classical] measuring instruments and atomic [quantum] objects" ([6] 3, p.5).[10] Even this application takes place in reference to measuring instruments rather than to quantum objects, to which, again, no classical-like physical attributes can be unambiguously applied and whose behavior, as will be seen, cannot be seen in terms of any formalism, (more obviously) classical or (less obviously, but more crucially) quantum. In terms of the corresponding variables of the measuring instruments involved, both variables can never be simultaneously defined. As referring to quantum objects, not even a single variable can be defined.

Quantum individuality itself is no longer seen in terms of quantum objects or processes themselves, as it was (specifically in terms of the discontinuity of radiation under certain conditions) immediately in the wake of Planck's discovery. Hence, Bohr argues, quantum physics reflects a new feature of atomicity, foreign to the mechanical (atomic) conception of nature, extending from Democritus's atomism, on which classical physics is based and which are sometimes, reciprocally, based on classical physics ([6] 2, pp. 33, 71; [6] 3, p.2). This physics would describe the behavior of its objects as such (in, at bottom, causal, even if practically statistical or chaos-theoretical, terms). By contrast, in Bohr's interpretation, the only ultimate "atomic" (i.e. indivisible) entities that can be rigorously described by quantum theory are certain indivisible configurations of experimental technology, phenomena in Bohr's sense.[11] Any attempt at further subdividing such phenomena can only produce other phenomena of the same indivisible nature. Obviously, these "atoms" are not physically atomic, since they are composed of more elementary (ultimately quantum) physical constituents, but in the sense that no different form of access to quantum objects is possible, regardless of how far we can divide and subdivide our experimental technology or matter itself. These circumstances prevent any possibility for quantum objects to appear or to be considered or even conceived of in any specific (such as "atomic") form independently, outside of the instrumental <u>enclosures</u> of specific experiments, of which Bohr also speaks as "closed phenomena" ([6] 2, p.71).

Thus, contrary to Einstein's view, quantum mechanics in Bohr's interpretation allows for an interpretation that does "offer an <u>exhaustive</u> description of the <u>individual</u> phenomena" rather than "merely ... the means of accounting for the average behavior of a large number of atomic systems [as statistical ensembles]" ([6] 2, p.61, emphasis added). But it redefines what "individual phenomena" are or, in Bohr's interpretation, can possibly be, given the data in question in quantum theory. Complementary phenomena can never be reduced to or derived from a single experimental arrangement, or to the possibility of extracting the values of such quantities from a single arrangement, especially as referring to the underlying (independent) quantum object(s). There is no single physical situation of which these two situations would be parts or aspects, or from which they could be derived, of which they would be effects. The efficacity of such (complementary) effects is as individual and as unique each time as are the effects themselves, as I shall explain presently (the proposition also requires qualification).

Einstein said (in 1936) that "to believe this [i.e. that quantum theory offers an exhaustive description of individual quantum processes] is logically possible without contradiction," and ultimately he thought that Bohr did most justice to the problem, in his reply in the Schilpp volume. The latter also contains an exposition of his own views on the subject [15]. However, he saw it "as so very contrary to [his] scientific instinct that [he could not] forgo the search for a more <u>complete</u> conception" ([14], [6] 2, p.61, emphasis added). By this point, Einstein saw the situation in terms of the alternative for quantum theory as that between being "complete and nonlocal" vs. "local but incomplete," and appears to (mis)read Bohr along these lines. Thus, he was searching for a local and complete conception of quantum physics. The latter, I argue here, becomes possible once one accepts Bohr's interpretation and epistemology, according to which "in quantum mechanics we are not dealing with an arbitrary renunciation of a more detailed analysis of atomic phenomena, but with a recognition that such an analysis is, <u>in principle</u>, excluded" ([6] 2, p.62; Bohr's emphasis).

I here maintain the above contextualization of this claim as pertaining to Bohr's interpretation and epistemology of quantum mechanics, even if Bohr had a stronger, less interpretation-dependent, claim in mind. He might have thought that this claim inevitably follows from the formalism of quantum mechanics and perhaps (along with the relevant aspects of the formalism itself) from the quantum-mechanical data, specifically the uncertainty relations

He might well have been right on this, but the subject requires a separate and difficult analysis, which cannot be undertaken here. In any event, a more limited, interpretation-dependent, claim suffices for my point at the moment and my main argument in this article, which primarily concerns the possibility of a local and complete interpretation of quantum mechanics or the data in question and the fact that Bohr's fulfills both requirements. From this perspective a counter-argument to Einstein's is that, once the epistemology in question is accepted, it becomes possible to interpret quantum mechanics as both complete and local. Of course, as Bohr noted, Einstein's "attitude, even if ... [it] might seem balanced in itself" still "implies a rejection of [this] argumentation" on epistemological grounds, even if Einstein saw the latter argumentation as compatible with locality, which is, again, not altogether clear (6 [2], p.62; emphasis added).

Bohr's interpretation makes a more detailed analysis of the quantum-mechanical situation excluded at, correlatively, two levels. First, it is impossible to further subdivide phenomena so as to avoid complementarity, which, again, disallows us to see (complementary) quantum phenomena in terms of a single configuration from which we could arbitrarily select (or derive) either one complementary part or another. Instead, we always deal with two mutually exclusive, un-summable (complementary) situations, which can never be assumed to be parts of the same overt (phenomenal) or hidden whole, or be derivable from a single overt or hidden source. Second, our analysis cannot in principle appeal to the physical properties of quantum objects or the nature of quantum processes themselves. It may, again, be added that the ultimate analysis of why statistical predictions are possible in quantum mechanics may be, in principle, excluded as well. Bohr defines the difference between classical statistical physics and quantum mechanics in epistemological terms: "In quantum physics, we are presented not with intricacies [we encounter in classical statistical physics], but with the inability of the classical frame of concepts to account for the peculiar feature of indivisibility, or 'individuality,' characterizing elementary processes" ([6] 2, p.34; [2], p.149). For Bohr quantum mechanics is as complete as a theory could be under these conditions, and his interpretation reflects the possibility of this claim. Indeed, from this viewpoint, it is not even clear in comparison to what such an account would be incomplete, physically speaking, assuming that no (epistemologically) essentially different local account is available. Would merely a lack of conformity with certain classical-like models, whose ultimate correspondence, however indirect and mediated (Einstein's position on this was complex), to "physical reality" has never been established, be enough to see Bohr's conception as lacking in completeness? Bohr certainly did not think so, which made his interpretation at least sufficient, if not inevitable.[12] No local alternatives were on the horizon at the time, and, as I shall suggest, in essence (there are significant nuances) not even now.

Bohr's interpretation, thus, entails a new form of epistemology. Hence, Bohr's persistent references to "the epistemological lesson of quantum mechanics." This epistemology is that of knowledge and conceptualization (only) in terms of effects, while it leaves the nature and character of the ultimate efficacity behind these effects outside the limit of all the available or even conceivable knowledge and conceptualization.[13] That is, we cannot assign to this efficacity any given or conceivable general character, say, by analogy with classical physics or in the manner of hidden variables theories, while assuming that some or even all specific elements involved are partially or even altogether unknown. In his later works, Bohr, interactively with "phenomena," speaks of the key features of quantum data in terms of quantum effects throughout, in particular, again, of "the peculiar individuality of the quantum effects" ([6] 2, pp.40, 62). Reciprocally, however, the existence of this irreducible unknowable efficacity can only be inferred from the peculiar character of its effects, for example, the individual character of all effects involved and the complementary character of some of them. A few specific consequences of this view are worth stressing.

First, while this efficacity manifests itself, always in mutually exclusive circumstances, through the complementary effects of both types, it cannot be thought of in terms of a single underlying (hidden) governing "wholeness." It would not make a difference whether such wholeness is seen as indivisible in itself or as "atomic." In the latter case it would be correlated with manifest individual effects (such as individual marks left on the silver bromide plate in the double-slit experiment), while remaining subject to an underlying coherent architecture (which may or may not evolve in time) that is not manifest itself in these effects. This efficacity is neither single in governing all of its effects, nor multiple in the sense that it could allow one to assign an unambiguous separate specific efficacity to each lawless individual effect. It may be seen as (irreducibly) multiple in the sense that the (inconceivable) efficacious processes that give rise to each individual phenomenon are each time different. While each time inconceivable, this efficacity is nevertheless different each time and, hence, is as unique as, and jointly with, the effects produced. This individuality of the (unknowable) efficacity of each individual effect is a structural, defining part of the <u>individuality</u> of phenomena in Bohr.

Secondly, an efficacity of that type cannot be seen in terms of independent properties, relations, or laws, which, while unavailable, would define a certain material entity that would exist in itself and by itself, and which would be responsible for giving rise to the (available) effects in question. Instead, it must be seen as reciprocal with and, again, indivisible from its effects: it can never be, in practice and in principle, conceived as isolated, separate from them. Nor, however, can it be seen as fully continuous with these effects either. This is not contradictory given the nature of this efficacity, in part, again, arising in view of the necessity of maintaining the features of description, which would be contradictory otherwise. All quantum-mechanical and specifically complementary effects appear only within the manifest strata of such indivisible configurations, that is, phenomena. The latter, however, also contain the inaccessible and even inconceivable stratum or (we may not be certain) strata. It is this stratum or strata that defines the level of quantum objects and processes, including those involving the interaction between quantum objects and measuring instruments, from which the phenomenal effects in question are accordingly discontinuously detached, once the experiments are performed. This efficacity is both irreducibly inaccessible and yet irreducibly indissociable from that part of the overall configuration which is accessible, that is, subject to Bohr's phenomena and hence open to representation, conception, knowledge, and so forth. As such, this efficacity is part of the <u>indivisibility</u> or <u>wholeness</u> of phenomena in Bohr's sense.

This conception is, definitionally, non-realist; it is of course also (and, again, definitionally) non-causal. At best we can say that this efficacity is "real" in the sense that it has actual physical existence, specifically in its capacity to produce certain particular quantum-mechanical effects, such as complementarity, particular statistical and correlational, or quantum-informational, data, entanglement, and so forth. But it allows for no concept of reality that has so far been or will ever be available to us (in particular those borrowed from classical physics) to be applied to it consistently with the effects in question in quantum physics (at least while maintaining the locality of quantum theory). It is not, again, that nothing exists that gives rise to the effects in question (including when we are not there to interact with it), but only that whatever produces these effects is inconceivable in classical and perhaps any possible terms. "Efficacity" and "existence" (or "nonexistence") are, too, among such terms, as are the possibility or impossibility of "conceiving" of it, or "possibility" or "impossibility," or "it" or "is."

Thus, in Bohr's interpretation, quantum mechanics not only does not offer an analysis of the ultimate efficacity of the observational effects it considers, but it tells us that such an analysis is in principle impossible, an epistemological cost that is exorbitant or even prohibitive to some, beginning with Einstein. This may be the ultimate meaning of Bohr's statement that "we are not dealing with an arbitrary renunciation of a more detailed analysis of atomic phenomena but with a recognition that such an analysis is in principle excluded" ([8] 2, p.62).

In this sense, the quantum objects may be even more radically inaccessible than the interior of black holes. While it must be applied with caution, the analogy itself is not out of place, insofar as the "naked" singularity, inevitably (i.e. by virtue of the equations that describe a black hole) found in the interior of a black hole, can never be seen. Black holes must, of course, in fact be seen as quantum objects in view of Stephen Hawking's famous theorems. On the other hand, in certain recent versions of string theory the ultimate constituents of matter are in fact configured as microscopic black holes. This idea is in turn partly due to Hawking, who, actually, suggests that this type of objects may introduce more radical levels of, at least, indeterminacy (but, it appears, also of noncausality and arealism) into physics [48].

It may also be added that this limit upon how far our knowledge can penetrate into black holes in no way appears to limit our advancement of knowledge concerning them. Nor, accordingly, should we be alarmed by the presence of similar limits in quantum theory. As Bohr stresses, "such argumentation does of course not imply that in atomic [quantum] physics, we have no more to learn as regards experimental evidence [defining these effects] and the mathematical tools appropriate for its comprehension" ([6] 3, p.6). Nor, one might add, would it imply that quantum-mechanical effects, such those of quantum entanglement, might not be used in practice, for example, in quantum computing. Hence, throughout this article, I only appeal to the ultimate impossibility of knowledge or conception as concerns the efficacity in question. "Indeed," Bohr adds "it seems likely that the introduction of still further abstraction into formalism will be required to account for the novel features revealed by the exploration of atomic processes of very high energy" ([6] 3, p.6). The history of quantum physics has demonstrated and continues to demonstrate just that, as all of its experimental and theoretical findings so far appear to be consistent with Bohr's epistemology. Such effects as the identity (in the sense of indistinguishability) of particles in quantum statistical ensembles; global quantum effects; quantum entanglement; "decoherence;" and many key findings of quantum field theories and other higher level theories all have momentous significance and implications—experimental, theoretical, and practical.

It does follow that, in Bohr's interpretation, the formalism of quantum theory refers only to such effects rather than to properties of quantum objects or features of quantum processes (in space and time) themselves. In other words, while certain ("effect"-constituting) parts of measuring instruments are described by the formalism of classical physics, there is no presupposition that the quantum-mechanical formalism in any way describes the ("undisturbed") quantum process before the measurement interference takes place, or between instances of such interference. Instead this formalism describes these interactions and their impact on the measuring instruments, rather than the properties, even single properties, or behavior of quantum objects. Hence, it would not be possible to speak in terms of the physical reality of, say, states in linear ("coherent") superposition in the sense that the corresponding formalism (such as Schrödinger's equation or the Hilbert-space formalism) describes the physical (in space and time) behavior of quantum objects themselves. The latter instead is, again, seen as describing particular effects pertaining to the impact of "quantum objects" on measuring instruments or, more generally, on the physical world to which we can relate directly through observation and measurement. Heisenberg's original (matrix) formulation of quantum theory appears to conform to this view more readily, and may be seen as shaping Bohr's view all along. For his "matrices" and the (formal) Hamiltonian equations (retained from classical physics but now applied to these matrices) in fact relate to the observable radiation effects, rather than to the motion of electrons (this was what he calls new "kinematics"). The issue would, however, require a separate discussion.

This view gives a new and more radical meaning to Bohr's argument, concerning the irreducible discrimination "between the objects under investigation and the measuring instruments which serve to define, in classical terms, the conditions under which the phenomena appear" ([6] 2, p.50). This discrimination is, we recall, consistent with the indivisibility of phenomena themselves. Indeed Bohr sees "this necessity of discriminating in each experimental arrangement between [them]" as "a principle distinction between classical and quantum-mechanical description of physical phenomena" ([2], p. 150). For, while "in classical physics the distinction between object and measuring agencies does not entail any difference in the character of the description of the phenomena concerned," in quantum physics it does ([2],1 p.50; [6] 2, p.50; [6] 3, p.3).

This and related statements by Bohr may suggest (and have suggested to some) that while parts of measuring instruments are described by means of classical physics, the behavior (in space and time) of quantum objects is described by means of quantum-mechanical formalism. Bohr obviously says the former, but he clearly does not ever say and does not mean the latter. Instead, the difference between the two descriptions is that classical theory describes the classical world, specifically certain parts of the measuring instruments, while quantum theory describes the interaction between the measuring instruments (described classically) and quantum objects, indescribable by means of either classical or quantum theory. The totality of possible individual quantum phenomena (or effects) and specifically the complementary character of some of them can be comprehensively and consistently accounted for by the formalism of quantum theory, while it cannot be unaccounted for by means of the classical theories. The two types of formalism themselves remain ultimately incompatible, even though they must coexist in Bohr's interpretation.

As Bohr points out, "it is true that the place within each measuring procedure where this discrimination is made is in both cases largely a matter of convenience" ([2], p. 150). The arbitrariness of this place, however, or the arbitrariness of the "cut," as it is also called, becomes a natural feature of this interpretation rather than a problem, as it would be, and often has been, when we assume that the quantum-mechanical formalism describes the space and time behavior of quantum objects themselves, as it is, for example, in most discussions of the Schrödinger cat paradox and decoherence (e.g. Ref [25]), on which I shall further comment below. At this point of his reply to EPR, Bohr brings into consideration the so-called transformation theorems of quantum mechanics. The latter mathematically ground the EPR argument and, according to Bohr, "perhaps more than any other feature of formalism contribute to secure its mathematical completeness and its correspondence with classical mechanics" ([2], p.145, Note). For "by securing its proper correspondence with the classical theory the [transformation] theorems exclude in particular any imaginable inconsistency in the quantum-mechanical description, connected with a change of the place where the discrimination is made between object and measuring agencies. In fact it is an obvious consequence of [Bohr's] argumentation that in each experimental arrangement and measuring procedure we have only a free choice of this place within a region where the quantum-mechanical description of the process concerned is effectively equivalent with the classical description" ([2], p. 150).

What makes this last point (which also conveys the deeper epistemological aspects of Bohr's correspondence principle) so crucial is that quantum objects are always on the other side of the "cut" and indeed may be rigorously defined accordingly. This point is, again, intimated by Heisenberg's original presentation of his matrix version of quantum mechanics. In Bohr's interpretation, the predictions of quantum mechanics refer strictly to classically manifest effects upon the measuring instruments, which can be rigorously correlated with specific configurations involving quantum objects. It is only this correlation that enables any possible identification of such a measurement with the position or momentum of a quantum object. The correlation is, according to Bohr, real, or rigorous, while the identification can only be, in his words, "symbolic." Bohr, as we have seen, does say that "to measure the position of [a] particle can mean nothing else than to establish a correlation between its behavior and some instrument rigidly fixed to the support which defines the space frame of reference [or some equivalent classical-like configuration]" ([2] pp. 149-48; in this order.) He is, however, careful to say "behavior," which is quite different from saying "position." The sentence also compels one to argue that in Bohr's scheme this "behavior" itself, too, cannot be described but only correlated with the behavior of the measuring instruments, the behavior describable by means of classical physics.

At one end, by virtue of their classical nature, the individual effects in question can be isolated materially and phenomenologically—we can perceive and analyze them as such—once an experiment is performed, although they cannot be isolated from their efficacity by even conceiving, let alone analyzing, their physical emergence. By contrast, at another end, quantum objects and processes can never be isolated, either materially (from the measurement process and measuring instruments) or perceptually or even conceptually (we cannot in principle conceive of what actually happens at that level or how it happens).

Reciprocally, however, Bohr's point just discussed bears crucially on the question of the classical description of the measuring instruments involved. It follows that, rather than to the constitution of measuring instruments as a whole (as is often assumed by Bohr's interpreters), this description applies only to the effect-constituting parts of these instruments. Their, as it were, efficacious components are quantum and, as such, are responsible for the effects in question and the classical behavior or those parts that, in their quantum interactions with quantum objects, produce these effects. Naturally, these components and these interactions are subject to quantum-mechanical treatment, including uncertainty relations, and Bohr's epistemology. In this respect, Bohr's interpretation may not be as different from von Neumann's and related interpretations, as it sometimes argued in view of the not suitably qualified (along the lines just indicated) claim that in Bohr's interpretation the apparatus is described classically. Von Neumann himself was quite aware of this proximity, as his view of the "cut" would appear to indicate.[14] It is true that in von Neumann's interpretation and other interpretations of that type the apparatus is treated as the quantum-mechanical, rather than classical, system.[15] However, on the one hand, as just explained, so is the interactive-efficacious part of the apparatus in Bohr. On the other hand, the motion of the "pointer" registering the measurement result, crucial to von Neumann's scheme, is subject to classical motion or, epistemologically, perception, and as such is in fact analogous, even if not identical, to the classically described parts of the apparatus in

Bohr. In other words, von Neumann's scheme, too, has a classical component at the level of the ultimate, perceptible or "phenomenal" (both in the usual and in Bohr's sense) effects of measurement. There are, of course, other differences between both interpretations, in particular as concerns whether the quantum-mechanical formalism in fact describes the behavior of quantum objects (including the quantum component of measuring instruments). My main point at the moment are qualifications, crucial but often disregarded, pertaining to Bohr's argument concerning the classical behavior of measuring instruments and its significance for quantum measurements. In short, the measuring apparatus in Bohr is not simply classical, as is nearly universally claimed, but both classical and and quantum. Rigorously considered, the situation has far reaching implications for both quantum epistemology and the specific character of the mathematical formalism of quantum mechanics (in particular, its dependence on complex rather than real mathematical entities), although the latter subject exceeds my, primarily epistemological, scope here. It is worth keeping in mind, however, that Bohr's claim that "in each experimental arrangement and measuring procedure we have only a free choice of this place within a region where the quantum-mechanical description of the process concerned is effectively equivalent with the classical description" exceeds the limits of his interpretation and appears to be uncircumventable in general. In sum, in Bohr's scheme the classical description applies only to physical effects manifest in certain parts of measuring instruments, which at the same time have an equally irreducible quantum component and (just as in von Neumann's scheme) interact quantum-mechanically with quantum objects. It is, as I have argued throughout, this interaction that is the efficacity of the classical effects in question.

It may appear that some interpretations of quantum mechanics allow us at least to directly refer to, if not quite to describe, quantum processes themselves and their key features, while bypassing measurement as an (irreducibly) constitutive component of quantum physics. Measurement is, now, seen as auxiliary, as in classical physics. Hence, such interpretations may be seen, and are often offered, as realist, and, accordingly, as epistemologically different from Bohr's. Thus, we would have at least competing and, to many, epistemologically more palatable alternatives to, if not counter-arguments against, Bohr's interpretation. I am, let me qualify, not referring to the hidden variables theories and related nonlocal theories (these are mathematically different from quantum mechanics) or the many-worlds interpretation of quantum mechanics. These theories do not retain the standard features of quantum mechanics under discussion (and Bohmian theories indeed employ a different formalism). I have in mind (in addition to some earlier versions, such Richard Feynman's) certain versions of the modal interpretation (specifically that of Bas van Frassen), of the consistent histories interpretation (specifically those of Robert Griffiths and Roland Omnés, which may be further linked to decoherence). These are, arguably, among the closest to Bohr's among such interpretation, except for Bohr's epistemology [7] [8] [16] [17]. (As I have indicated, von Neumann's and related interpretations, such as Mittelstaedt's minimal interpretation are far more grounded in the concept of measurement.) Both van Frassen and Omnés associate their interpretation with the Copenhagen interpretation; and Mermin's indicates further proximities between Bohr's and the histories interpretations of Griffith and Omnés ([18], p. 763).[16] I shall refrain from the ultimate judgement concerning such claims and offer instead the following set of observations, related more to what Bohr's or a Bohrian response would be as concerns the possibility of such claims in principle.

Bohr, I argue here, would deny that the quantum-mechanical formalism can unambiguously refer to quantum objects and processes in terms of space-time concepts in the way it can be done in classical physics or even in relativity, which already introduces significant complications in this respect ([6] 2, pp.40-41). As he said, "even in the indeterminacy relations, we are dealing with the formalism which defies unambiguous expression in words suited to describe classical physical [space-time] pictures" ([6] 2, p.40). Ultimately this formalism defies all "unambiguous use of space-time concepts," which are instead "confined to recording of observations which refer to marks on a photographic plate or similar practically irreversible amplification effects" ([6] 2, p.51). If one is concerned, as Bohr (or indeed Einstein) was, with the physical meaning of a theory as referring to the processes occurring in space and time, in Bohr's interpretation, what takes place in space and time is a particular configuration of phenomenal effects. Bohr saw the mathematical formalism of quantum mechanics as correlative to this situation, including as applied to the EPR experiment ([2], p.145).

This formalism is highly symbolic and nonvisualizable in its very nature, especially by virtue of using complex numbers, which could be related to observations, always recorded in real numbers, by means of artificial schemes, such as Born's "square moduli" rule for deriving probabilities from quantum amplitudes, projection postulate, and so forth. What would it mean to speak of the physical reality, of the physical meaning (in space and time) of such objects themselves, or of structures such as a linear superposition of vectors? Dirac, who may be seen as one of the originators of the approaches under discussion and the view that the quantum-mechanical formalism pertains to quantum object themselves, clearly denies that linear superposition can have any physical meaning. Accordingly, his epistemology may be much closer (which is not to say identical) to Bohr's than it may appear ([19], pp.10-14).[17] In other words, such mathematical objects cannot realistically represent or directly refer to space-time processes (even noncausal ones); they may not even constitute a mathematical, at least geometrical, phenomena in the sense of allowing us to conceive of them spatially.[18] Indeed, the practice (more or less standard since Dirac's and then von Neumann's versions of quantum-mechanical formalism) of using the Hilbert-space formalism and bra/ket-vector notation in explaining "what is actually going on" may be seen as a manifestation of this impossibility. This may also change our understanding of what it means to offer a description of a physical process, as (differently) Mermin and Haroche appear to suggest [18][43]. Such a change would, however, be consistent with Bohr's view insofar as one maintains that in speaking in terms of space-time behavior and classical-like physical variables or concepts we can only refer to the effects of the (quantum) interactions between (indescribable) quantum objects (and classically described) parts of measuring instruments.

Spin (whose ket-vector description in a two-dimensional Hilbert space is particularly elegant and simple) offers an especially illustrative case and a partial model for a general understanding of the situation (and it may have served Bohr as both in shaping his thinking on quantum epistemology). Spin has no classical counterpart and we can only access it through its effects upon measuring devices and speak of it only by way of loose metaphorical analogies, which inevitably break down at a certain point. Now, insofar as we can refer to, say, the position and momentum of a particle, the situation is in fact quite similar epistemologically. It is different insofar as the classically described parts of measuring instruments cannot be assigned spin, while they have of course momentum or position (which may also serve as a reference frame, as, say, in the double-slit experiment). (Hence, as I said, the analogy with spin is only partial.) More crucially, in Bohr's interpretation, the situation is analogous insofar as any "position" or "momentum" impact of the interactions between quantum object and measuring instruments upon the latter does not, in all rigor, measure a classical-like variable of "position" or "momentum" of the quantum object involved. (Once again, due to complementarity, we cannot measure both variables even as pertaining to measuring instruments, which are, in this sense, themselves subject to the uncertainty relations.) In this sense, it would not be quite accurate to speak, as Bohr does in his reply to EPR, of the momentum exchange between a particle and the apparatus, unless in provisional or symbolic terms ([2], p.146). The latter might have been the case in Bohr, who subsequently refined most such formulations in any event. In short, first, the standard quantum-mechanical formalism in any of its form does not offer a description of the behavior of quantum objects in space and time, but only a possible link to such a behavior (the nature of this link being subject to interpretation). This argument is not so surprising given the mathematical character of this formalism and (I say this with caution) may, accordingly, pertain to all interpretation of the standard quantum-mechanical formalism. Secondly, more radically, within Bohr's interpretation the quantum-mechanical formalism cannot refer or even relate in any way, however indirectly, to the independent behavior of quantum objects but only to their interactive impact upon the measuring instruments involved. Naturally, we can still use such references for the sake of convenience and economy of discourse, while, in all rigor, keeping their provisional (analogical, symbolic, metaphorical, and so forth) status in mind.

Accordingly, Bohr's phenomena may be more than merely a possible conception of physical, space-time, situations compatible with the quantum-mechanical formalism, using the observational data in question, and thus only a part of Bohr's interpretation. Instead, they may be more fundamentally correlative to the quantum-mechanical formalism, insofar as anything in space and time rigorously accountable by its means is possible, and thus exceeds the limits of Bohr's interpretation. By contrast, nothing in regard to quantum processes may be ascertained. That, however, would appear to imply (as I said, I refrain from a definitive claim here) the following for the interpretations in question at the moment, given that the mathematical formalism involved is that of standard quantum mechanics (certain nuances and rules of applications are introduced, especially in histories interpretations). Once their argument is related to anything that can actually occur in space-time, Bohr's phenomena and, hence, measurement may have to be reinstated as constitutive rather than merely auxiliary. For example, arguments, such Omnés's, that the macro-world is ultimately subject to quantum-mechanical, rather than classical, description may be retained, but may need to be adjusted insofar as we can consider the aspects of this description only at the level of effects [7][8].[19]

This might make classical physics indispensable, even though ultimately inadequate, and may make it impossible to apply quantum mechanics to objects, such as the universe as a whole, which we cannot, in principle, consider as objects under observation and measurement. But then it is far from self-evident that such concepts as object, or wholeness and the universe, can apply to "the universe" as a "whole," even leaving aside quantum-epistemological complications of applying all such terms, as here considered.

3. Quantum Entanglement and the EPR argument

Einstein, Podolsky and Rosen's 1935 article and Einstein's related arguments injected the question of nonlocality into the debate concerning quantum mechanics. EPR begin their article by proposing the following, apparently natural and minimal, criterion of physical reality: "If, without in any way disturbing a system, we can predict with certainty the value of a physical quantity, then there exists an element of physical reality corresponding to this physical quantity" ([2], p.138; emphasis added). It may appear that this criterion applies to quantum mechanics as well. Recall that, in view of uncertainty relations, it is only a joint simultaneous determination or prediction of two variables involved in the quantum-mechanical physical description, such as "position" and "momentum," that is impossible. Accordingly, a determination or prediction of the value of a single variable is always possible, with any degree of precision, including in the EPR experiment, which deals with the measurements of physical variables pertaining to two spatially separated quantum objects that have previously been in interaction. Hence, predictions concerning such variables now takes place without "disturbing" a quantum system under investigation by measurement, that is, without first performing a measurement upon it, which is how quantum-mechanical predictions are made in more standard cases. (Bohr speaks more cautiously of not "interfering" with this system, since, as we have seen, in his interpretation there is no classical-like or otherwise specifiable undisturbed configuration or properties that are then disturbed in the process.) This can indeed be done in quantum mechanics for (this point remains crucial) a single variable in certain cases, such as that considered by EPR. It is achieved by means of performing measurements on other systems (in the EPR argument, another particle), which have previously been in an interaction with the system (such as a particle) under investigation. (For the sake of economy, I speak for the moment of "particles," rather than, as would be more appropriate and as Bohr does, of "variables involved in the quantum-mechanical physical description" ([2], pp. 144, 145]). Indeed, as Bohr observes, in a certain sense, the EPR conditions apply to all quantum-mechanical predictions. In any quantum measurement, we can predict either the position or the momentum of a particle after a preceding measurement took place and on the basis of this measurement, and hence without interfering with the object and without assuming that we can define either quantity independently of measurement ([6] 2, p.57). In fact, the particle and those parts of the measuring instruments involved that interacted with it become entangled in the EPR way.[20] In the EPR situation, which involves two particles, we have a slightly more complicated, but not fundamentally different, case ([2], pp. 149-48; in this order). Hence, at the time of determination in question, there is no physical interaction either between the two particles or between any measuring apparatus and the second particle.

Bohr agrees that quantum mechanics allows for and enables such ("at-a-distance") predictions and by so doing rigorously accounts for quantum correlations and entanglement. (The entanglement/correlation part of the EPR-type argument, whether in its original terms or in terms of spin, can be adjusted so as to refer only to the outcomes of measurements rather than to quantum objects. Indeed, we recall that "spin" has no classical counterpart.) At the same time, however, it does not, in Bohr's interpretation, imply nonlocality, as some suggest.
In other words, entangled quantum objects or (Bohr, though, shuns this language) "states" exist, but this only means a) that particular forms of experimental detection (phenomena or effects), such as those considered in Ref. [43], are possible, and b) that the quantum-mechanical formalism allows us to predict such effects. Accordingly, nonlocality, which is entailed neither by a) nor by b), need not follow, unless of course it is independently derived from the formalism itself, which does not appear to be the case thus far.

It has been also argued that the latter issue is irrelevant for Bohr. It would be difficult to agree with either view. Bohr's concern with the relationships between quantum mechanics and relativity is manifest in his reply to EPR and most of his subsequent (and preceding) writings on the subject, including virtually all articles here cited. He states in a key (but usually ignored) note at the end of his reply, "the relativistic invariance of the uncertainty relations ensures the compatibility between the argument outlined in the present article [i.e. his reply to EPR] and all exigencies of relativity theory" ([2], pp.150 Note). Indeed, as the argument (mentioned as forthcoming in the same note) of "Discussion with Einstein" makes clear, under the impetus of

Einstein's criticism, the proper application of quantum-mechanical formalism and specifically of the uncertainty relations in fact depends on general relativity. This discussion between Einstein and Bohr took place well before the EPR article, in 1930, although, as Bohr observed, it contained all the key epistemological elements of the EPR argument ([6] 2, p.57).

For Bohr, relativity and hence, locality was a given, an "axiom" or "a law of nature," confirmed by experiment, just as the uncertainty relations were a given, "a law of nature," as a consequence of quantum theory, confirmed by experiment, or as (with qualifications given earlier) the classical description of measuring instruments were a given, corresponding to the fact that on the macro-level classical physics still holds. The latter view further reinforced by the "correspondence" argument, considered earlier, that the "cut" between classical and quantum-mechanical description can only be made in the region where both coincide. The first challenge answered by complementarity, as presented in the Como lecture, was an interpretation accounting for the peculiar character of the quantum-mechanical data and formalism, especially as manifest by the wave-particle complementarity and the uncertainty relations. (A number of important conceptual considerations and complementarities were introduced as well, such as that of coordination and causality, or of observation and definition of physical variables, which were supplanted by the more experimentally oriented "phenomena" in his later works.) The EPR and Einstein's later challenge was to develop an interpretation to prove that complementarity is, or may be refined to be, both complete, within its scope, and consistent with relativity. Bohr's ultimate interpretation allows him to see quantum theory and data as fully consistent with all three, and brings all three together, even if at the cost (again, to some exorbitant) of accepting Bohr's radical epistemology. The latter also deprives us of the possibility of knowing how quantum entanglement and correlations are ultimately possible in physical (space-time) terms. In this view (and with preceding analysis in mind), quantum mechanics describes only the effects of the ultimate constituents of matter upon their quantum interaction with measuring instruments but not each such effect itself. These effects are subject to a description in terms of classical physics, along with the behavior of those parts of the measuring instruments involved that constitute these effects (such as a change in momentum of a certain part of a given apparatus). At the same time, as I said, the (complementary) configuration of all effects involved is not describable by means of classical physics. It can only be described by means of quantum theory, thus allowing one to rigorously speak of quantum objects as the ultimate (micro) constituents of matter, subject (along with quantum theory itself) to the epistemology in question. Quantum behavior itself may, again, be macroscopic, while defined by the microscopic constituents of the objects involved.

Bohr argues that "the apparent contradiction [found by EPR] in fact discloses only the essential inadequacy of the customary viewpoint of natural philosophy for a rational account of physical phenomena with which we are concerned in quantum mechanics." Instead, in accordance with the preceding discussion, the irreducibility of this interaction "entails the necessity of the final renunciation of the classical ideal of causality and a radical revision of our attitude towards the problem of physical reality" ([2], pp. 145-46).[21] The last statement may be too strong, even though Bohr may be right after all. At the very least, however, this view makes it possible to develop a local interpretation of quantum mechanics, as well as, as a kind of philosophical bonus, a new form of epistemology.

Hence, as I stressed from the outset, quantum entanglement and the question of locality make "interpretation" even more significant in quantum mechanics. Short of a particular, very specific interpretation, it may not be possible to counter Einstein's arguments. Even if at an epistemological cost (acceptable to Bohr, exorbitant to Einstein), Bohr's interpretation ensures the compatibility of quantum mechanics with relativity, the desideratum equally defining his and Einstein's view. Is Bohr's interpretation the only such interpretation currently available? Not quite, I think, although, as I have suggested, it is at least useful, if not necessary, in helping other interpretations to make this compatibility more rigorous. Are more radically different interpretations of quantum theory possible, say, those that Einstein would accept as "more complete conceptions"? That is, is the epistemology in question inevitable given quantum theory or even already quantum data or only consistent with it? It is difficult to say, and it is the latter, quite enough in itself, that is my argument here. Bohr, too, especially in print, appears to argue primarily for the latter, although he did not appear to think (rightly) that there were other fully developed such interpretations at the time of his writing ([2], p.145). This view of Bohr's argument is not inconsistent with his claims, such as that of my epigraph here, that "in quantum mechanics we are not dealing with an arbitrary renunciation of a more detailed analysis of atomic phenomena, but with a recognition that such an analysis is, <u>in principle</u>, excluded." For, as is clear from Bohr's accompanying elaboration, this claim may be (and may need to be) read as applicable within Bohr's interpretation, seen by Bohr as enabling one to view quantum mechanics as a fully consistent and, within its proper scope, complete physical theory.

I would like to further illustrate this point by offering a reading of Bohr's passage on "the essential ambiguity" from his reply to EPR. Bohr writes (I shall explain my parsing below):

> **[B1]** From our point of view we now see that the above mentioned criterion of physical reality proposed by Einstein, Podolsky and Rosen contains an ambiguity as regards the meaning of the expression "without in any way disturbing a system."
> **[B2]** Of course there is in a case like that just considered no question of a mechanical disturbance of the system under investigation during the last critical stage of the measuring procedure.
> **[B3]** But even at this stage there is essentially the question of <u>an influence on the very conditions which define the possible types of predictions, regarding the future behavior of the system</u>.
> **[B4]** Since these conditions constitute an inherent element of the description of any phenomenon [of quantum physics] to which the term "physical reality" can be properly attached, we see that the argument of the mentioned authors does not justify their conclusion that quantum-mechanical description is essentially incomplete.
> **[B5]** On the contrary this description, as appears from the preceding discussion, may be characterized as a rational utilization of all possibilities of unambiguous interpretation of measurements, compatible with the finite and uncontrollable interaction between the object and the measuring instruments in the field of quantum theory.
> **[B6]** In fact it is only the mutual exclusion of any two experimental procedures, permitting the unambiguous definition of complementary physical quantities, which provides room for new physical laws, the coexistence of which might at first sight appear to be irreconcilable with the basic principles of science. It is just this entirely new situation as regards the description of physical phenomena, that the notion of <u>complementarity</u> aims at characterizing. ([2], p. 148; Bohr's emphasis)

This is arguably the most famous elaboration of Bohr's reply. It was seen as crucial by Bohr himself, who cited and commented on it in "Discussion with Einstein." This commentary in "Discussion with Einstein" tells us directly what is the nature of the ambiguity of the EPR criterion of reality and their argument. Bohr defines his argumentation in his reply as aiming at "bringing out the essential ambiguity involved in reference to physical attributes of objects when dealing with phenomena where no sharp separation can be made between the objects themselves and their interaction with measuring instruments" ([6] 2, p.61). The ambiguity of the EPR criteria is this ambiguity. This message is rarely heeded by most commentators on Bohr's argument, who (Bell, among them [4], pp. 189-90) try to understand this elaboration on the basis of experimental facts and general theoretical arguments of quantum mechanics rather than in Bohr's own terms, which is, I would argue, imperative.

It is also crucial how one parses the elaboration into units and groups these units, and I shall suggest a grouping different from that used in most readings. The invocation of "ambiguity" in **B[1]** and Bohr's claim that this ambiguity arises "as regards the meaning of the expression 'without in any way disturbing a system'" is sometimes taken (for example, by Stapp and possibly by Einstein) in conjunction with **B[3]**, that there is in fact, contrary to the EPR argument, some form of "disturbance" or "influence" with respect to the spatially separate system or quantum object—the second particle of the EPR pair ([10], p.162). It is difficult to argue that this is what Bohr has in mind either in **B[1]** or in **B[3]**. There does not appear to be anything to support such a view either in **B[3]** itself, or in the extended elaboration in question or in Bohr's preceding argument in his reply. The latter is important, as the opening phrase of **B[1]**—"from our point of view we now see"—indicates, in particular in stating that Bohr has a (local) interpretation within which the EPR criterion is ambiguous, but it, too, is often ignored by critics.[22] In order to properly understand the meaning of **B[1]**, one must consider other propositions in Bohr's reply and elsewhere on what can and cannot be unambiguously done in quantum mechanics. Once one does so, it becomes apparent that the "ambiguity" here need not be read as stating that, contrary to EPR, there is in fact a "disturbance" in making the EPR predictions, which, thus, invalidates their argument. Instead this ambiguity has to do with the conditions (including the necessary interpretation of the measurements involved) under which, in the EPR situation or in quantum mechanics in general, an unambiguous meaning can be assigned to terms and concepts, such as EPR's "elements of reality," that define the situation in question. These are the conditions of the interaction between quantum objects and measuring instruments, and ultimately of phenomena in Bohr's sense. Bohr does not have any "disturbance" in mind. As I said, he agrees with EPR that we cannot allow for any "disturbance," physical or other, of anything that is spatially separated from actual physical measurement.

To some degree, the reading of Bohr's elaboration under consideration and the key terms and concepts involved in it turns on how one reads Bohr's "But" in **B[3]**. One way of reading this "But" (and this is what is most often done) is to follow the sequence—call it **Sequence 1**—of **B[1]**, then **B[2]**, and then read **[B3]** as the negative of **B[2]**, and more or less to disregard **[B4]**, which is, I argue, in fact crucial to Bohr's argument and must be especially adhered to. What can and in my view should be done is, first of all, to take **B[4]** into account, and, then, to read Bohr's "But"—call it **Sequence 2**—as indicating the negation of **B[2]** not by **B[3]** but as indicating a negation of it by **B[3]** and **B[4]**, or perhaps still more accurately a conjunction of **B[2]**, with the combination of **B[3]** and **B[4]**. That is—this is the main point here—**B[3]** and **B[4]** must be considered together!

Thus, the customarily used **Sequence 1** is from **[B2]**: "Of course there is in a case like that just considered no question of a mechanical disturbance of the system under investigation during the last critical stage of the measuring procedure;" to **[B3]**: "But even at this stage there is essentially the question of an influence on the very conditions which define the possible types of predictions, regarding the future behavior of the system." **Sequence 2** is from **[B2]**: "Of course there is in a case like that just considered no question of a mechanical disturbance of the system under investigation during the last critical stage of the measuring procedure," thus, taking locality as an axiom, to **[B3]** and **B[4]**: "But even at this stage there is essentially the question of an influence on the very conditions which define the possible types of predictions, regarding the future behavior of the system. Since these conditions constitute an inherent element of the description of any phenomenon [of quantum physics] to which the term 'physical reality' can be properly attached ['unambiguously applied'], we see that the argument of the mentioned authors does not justify their conclusion that quantum-mechanical description is essentially incomplete."

If one takes the preceding discussion into account and proceeds via **Sequence 2**, it is clear that Bohr speaks of an ambiguity as concerns the expression "without in any way disturbing the system," not because there is a disturbance or influence of any kind, but because the whole reference to the system in question is "ambiguous" in Bohr's interpretation. This, then, is why Bohr says that the EPR criterion of physical reality contains "an [essential] ambiguity as regards the meaning of the expression 'without in any way disturbing the system'" ([2], p. 148). It concerns the question of an unambiguous definition of the system itself rather than indicates that the system is in fact disturbed or influenced and that the EPR application of their criteria in quantum mechanics is ambiguous for that reason. In fact, it is an obvious consequence of Bohr's interpretation that in any quantum-mechanical measurement the language of "disturbance" cannot apply in the sense that there are any undisturbed properties of quantum objects that are then disturbed by measurement ([6] 2, pp.63-64). Nor, however, does one need to conclude that, because we cannot speak of the properties of objects before experiments establishing them are performed (in Bohr's view we cannot do so even then), the two quantum systems in question in fact form an inseparable nonlocal whole (as they would in Bohm's theory). Consistently with the (Lorentz-invariant) formalism of quantum mechanics, these systems are seen as spatially separated or rather whatever can in fact be ascertained concerning them in space-time, that is, any phenomena (in Bohr's sense) associated with them, would be spatially separated.

It follows, then, that the "influence" in Bohr's statement does not refer to any influence, physical or not, upon the system associated with the second EPR particle (concerning which we make predictions but upon which we do not perform measurements). Instead it refers to the physical influence upon the (measuring) system associated with the first EPR particle (upon which we do perform measurements and with which we, hence, interfere) and on the conditions of such measurements as concerns the physical quantities involved. As a result, this influence of course—this is indeed Bohr's whole point—"influences" what kind of predictions we can or cannot make concerning the second particle. Hence it defines the conditions of the always irreducibly two mutually exclusive phenomena in the EPR-type experiments or (and they are, again, ultimately isomorphic) all quantum-mechanical predictions. If the apparatus is set up for measuring one complementary variable for the measuring system linked to the first particle, then within this setup the other variable cannot be unambiguously defined—we are absolutely precluded from doing so—for either particle of the EPR pair. Or more accurately, we cannot define it for the first system and hence cannot predict the corresponding variable for the second. For, "we have in each experimental arrangement suited for the study of proper quantum phenomena not merely to do with an ignorance of the value of certain physical quantities, but with the impossibility of defining these quantities in an unambiguous way" ([2], p. 149; emphasis added). Accordingly, "the procedure of measurement has an essential influence on the conditions on which the very definition of the physical quantities in question rests" ([2], p. 144;

emphasis added). This influence is thus that upon the conditions of an unambiguous definition of the quantities in question. These conditions physically pertain to the system associated with the first particle, where direct physical measurements are performed, even when the prediction and the possibility of definition themselves concern the second particle (which, or anything that can be associated with it, is not physically influenced). This is why, while Bohr indeed says that there is "an influence on the very conditions which define the possible types of prediction regarding the future behavior of the system," he never says that anything disturbs, interferes, or even influences the second system. The former need not, and in Bohr's argument does not entail, the latter. Nor, I would add, does he, as Stapp contends, see any other (nonphysical) disturbance or influence upon the second particle or, again, any system (at-a-distance) locally associated with it [10]. In view of the well-known theoretical and experimental findings, extending from those of Bell's theorems, we may consider as given that quantum entanglement is incompatible with local realism, a point accepted even by those who argue for quantum nonlocality.

We can now see that Bohr's "But" in the sentence introducing the influence in question need not be read as claiming that there is some influence, of whatever type, between two spatially separated systems, after Bohr's strong claim concerning the absence of any mechanical disturbance of the second particle, or whatever system is associated with it ([2], p. 148). Instead, if one takes into account Bohr's paragraph as a whole, this "But" indicates that Bohr questions the applicability of the EPR criterion of reality and argumentation, while remaining in agreement with EPR's locality requirement and pointing towards an interpretation of quantum mechanics compatible with it.

Accordingly, given Bohr's interpretation, it need not follow that quantum mechanics is incomplete by the EPR criteria (in this interpretation, inapplicable without ambiguity), which argument Einstein accepted to some degree. Nor does it follow that it is nonlocal, since, on the one hand, the formalism itself is consistent with relativity. On the other hand, all available arguments concerning nonlocality of (the standard) quantum mechanics, beginning with Einstein's, appear to make presuppositions which Bohr's interpretation does not make (which still appears to be the case), beginning and perhaps always amounting to making assumptions concerning the independent behavior of quantum objects. I would like to briefly comment in this context on counterfactual logic in quantum mechanics.

Bohr's interpretation would inhibit the applicability to quantum theory of EPR's and most standard (perhaps all) counterfactual arguments, that is, arguments based on what could, but actually did not, happen, if certain (unperformed) experiments were "in fact" (a word perhaps inapplicable under these conditions) performed. Thus, in examining an outcome of a given double-slit experiment we can only argue from within one among the possible complementary (mutually exclusive) situations, defined by the particular experimental arrangement at hand, say, when both slits open and no counters installed. Our examination, however, cannot rely on (counterfactual) considerations of the other complementary situation, defined by the alternative arrangement, here the one that would allow us to determine through which slit particles pass. We cannot draw conclusions concerning the situation at hand by considering what could, but actually did not, happen to the quantum objects in the alternative complementary situation. In the EPR situation the circumstances of complementarity manifest themselves in the fact that, in practice, two runs of any EPR type of experiment, and hence two actual physical situations are always required in order to confirm the EPR alternative predictions. Just as in (and equivalently to) the double-slit experiment, there is no single physical situation of which these two situations would be parts or aspects, or from which they could be derived, of which they would be effects, since, as we have seen, the efficacity of the complementary effects involved is as individual, unique each time as are the effects themselves. Hence, the EPR and Einstein's other nonlocality arguments cannot avoid counterfactual logic. Following Bell's theorem (which itself, however, does not prove the nonlocality of quantum mechanics, but only of certain, epistemologically realist hidden variables theories, which allow for counterfactual reasoning), the assumption that counterfactual logic applies in quantum mechanics appears to be correlative to deriving nonlocality from quantum theory or quantum-mechanical data. These assumptions, however, may be seen as in turn supplementary rather than structural in quantum physics, as Bohr perhaps knew, and as both Arthur Fine and more recently David Mermin emphasized [20] [18]. Indeed most of such arguments rely, directly or implicitly, on attributing physical variables or quantum-mechanical descriptions to quantum objects rather than to the outcome of quantum measurement registered in measuring instruments, even though counterfactual statements themselves may refer to the latter. Or, again, they rely on the presupposition that both complementary situations may, at least in principle, be derived from a single common situation, know or unknown (or even unknowable). I have just indicated, this presupposition is untenable in Bohr's interpretation and, it appears to follows (courtesy to Einstein), in any local interpretations of quantum mechanics. As Mermin observes, considered in terms of strictly statistical correlations, given a sufficient number of trials, quantum mechanics is local ([21], pp. 108-9). Einstein to some degree realized this point, although he considered this way out of the dilemma "too cheap": "The interpretation of the ψ-function as relating to an ensemble also eliminates the paradox that a measurement carried out in one part of space determines the kind of expectation for a measurement carried out later in another part of space (coupling parts of systems far apart in space)"([22], pp. 209). On the other hand, as Mermin shows, you cannot establish that quantum predictions, corresponding to these correlations, are strictly local at the level of individual events, which is of course what Einstein thought. However, as Mermin points out, the latter argument, just as Einstein's own, involves counterfactual reasoning ([23], pp. 171-74).

Here, again, given the possibility of Bohr's interpretation, we face a simple "either/or" logical choice—of giving up either locality/relativity, a well-confirmed experimental feature of the physical world, or certain arbitrary, even if commonsensical and in terms of classical physics natural, assumptions, such as the possibility of applying counterfactual reasoning. It follows, for example, that Einstein's alternative for quantum mechanics between "nonlocality and completeness" and "locality and incompleteness" depends on his use of counterfactual logic, and thus offers us the possibility of ascribing to quantum mechanics "locality and completeness," while suspending the applicability of counterfactual reasoning. Bohr's interpretation of course does not depend on the latter. An intriguing possibility would be to consider giving up the assumption that certain parts of measuring instruments and, by implication, the macro-world are in fact described in terms of classical physics. The question would require a separate discussion, although it may be pointed out that the applicability of classical physics at this level is difficult to contest, and quantum physics has relied on it in other respects as well throughout its history, beginning with the correspondence principle.[23] My main point at the moment is that Bohr's interpretation allows him to handle Einstein's argument, even if we adjust the latter argument to nonlocality, as Einstein did later, rather than only incompleteness. It is worth keeping in mind, however, that Bohr's reply to EPR addresses the EPR argument as such, and it would be inaccurate to disregard or diminish this fact, or the character of the EPR initial argument in considering and assessing Bohr's reply, as is sometimes done.

Obviously, there remains the fact that we make predictions concerning the future behavior of the second particle as a result of interfering, substantially (by physically, mechanically engaging with measuring devices) and unsubstantially (by making meaningful statements, predictions, etc.), with the local measuring system associated with the first particle. That, however, is quite different from and indeed incompatible with any reading of Bohr's argument as implying nonlocality. The system associated with the first particle enables us to unambiguously define either one or another complementary variable (never both simultaneously) associated, via a corresponding measurement or phenomenon, with the second particle. A measuring system of the same type may be introduced for the second particle and, then, (once a sufficient number of "identically" prepared experiments is performed) correlated with the one associated with the first particle, once this second system is being in turn interfered with, for example, in order to measure the value of a given variable for the second particle. Such correlations are involved in Bell's theorem and its refinements and the discussions surrounding these findings and their implications, more recently around Greenberger-Horne-Zeilinger and Lucien Hardy types of experiments (refinements of the EPR type of experiments). This, however, is not what Bohr refers to on this occasion, even if what he says is consistent with these "correlations," a point that, as Mermin suggests, Bohr perhaps missed, or missed making in this form ([18], p.765, n.31). Nor, of course, could Bohr have had in mind more subtle features in subsequent developments around the EPR experiment(s), but his argument can, I think, be easily adjusted to these cases as well. My main point here is that quantum entanglement and locality are both maintained, and can be experimentally and theoretically understood within the epistemology of "effects," as here discussed.

How, then, is this possible, especially given the apparent impossibility of applying Reichenbach's common cause principle, according to which such correlations should be explained in terms of the preceding common history of the events involved? ([24], p.159) The answer of Bohr's interpretation is that we do not know how this ultimately is or can be (physically) possible, indeed we cannot know, or even conceive of it, since any further analysis that would, in principle, allow us to do so is, in principle, excluded. It is the epistemology of knowable effects whose efficacity is ultimately inconceivable and, hence, the inaccessibility of "the quantum world" itself (to any knowledge or conceptualization that is or will ever be available to us, for example, as "quantum" or the "world," in any conceivable sense of either word) that allows one to rigorously maintain both the consistency and locality of quantum theory. But then, entanglement is only part of what we cannot conceive of in quantum theory, beginning with the double-slit experiment. How do electrons "know," individually or collectively, that both slits are open to arrange themselves in the interference-like pattern? Quantum theory appears to, and in Bohr's interpretation does, prohibit in principle what it does not deliver in practice. But it delivers plenty of what can be known, too, and different ways in which knowledge, including information in its technical sense, can be obtained and processed, for example, through quantum entanglement, which enables the whole field of quantum cryptography and computations. That it delivers the impossibility of knowledge or conception has a positive role to play, since this impossibility often enables knowledge and conceptions that would not be possible otherwise. This interpretation conforms to all standard requirements and desiderata of scientific research, and indeed enables us to advance where epistemologically classical accounts fall short. One can certainly regard most, indeed, it appears, so far all, of the advances in quantum theory in these terms, even though we often formulate them in more traditional terms. There is no special need to change this (and perhaps we can never quite do it), provided that we keep the provisional, rather than rigorous, nature of this classical-like terminology in mind.

4. Bohr's Epistemology and Decoherence

One may see an anticipation or a suggestion of the idea of decoherence in Bohr's concept of "irreversible amplification," correlative to and part of the conceptuality of the phenomena and individual quantum effects, as here considered. Indeed these effects are "irreversible implications effects," ineluctably left in a system and, thus, allowing us come to an agreement concerning the data. In this sense, quantum mechanics is objective and observer (but not an observational set-up) independent ([6] 2, pp. 73-74, 90-91; [6] 3, pp.6-7). As Bohr says, "it must be realized that ... all unambiguous use of space-time concepts in the description of atomic phenomena is confined to the recording of observations which refer to marks on photographic plates or to similar practically irreversible amplification effects like the building of a water drop around an ion in a cloud chamber" ([6] 2, p.51; emphasis added). What is now called "decoherence" is concerned with the physics of this process and more generally with the transition from the ("coherent" or linearly superpositioned) quantum level to the macro-level of the world that we experience. Bohr's appeals to "irreversibility" also indicate the potential thermodynamic aspects of decoherence, which have been part of the theoretical discussions concerning decoherence [25][26]. Most current decoherence theories exhibit marked affinities with Bohr's interpretation. Thus, the macro level, the level of phenomena in Bohr, is by and large described in terms of classical physics, keeping in mind the quantum aspects of the constitution of measuring instruments and their quantum interaction with quantum objects, as considered above.[24] However, the quantum level is presumed by most decoherence theories to be described by the formalism of quantum theory [25][26]. By contrast, as we have seen, in Bohr's interpretation the quantum-mechanical formalism could not be seen as unambiguously referring to the properties of quantum objects and processes in terms of spatial and temporal concepts (even noncausal ones), in the way it can be done in classical physics or even in relativity, which already introduces significant complications in this respect ([6] 2, p.41; [2], p. 150; [13] pp. 317-22).

Thus, in Bohr's view, while something like "decoherence" ("irreversible amplification") would take place, the formalism of quantum theory would describe the "effects," but, it follows, not the physical processes producing these effects. One might say, if one wants to apply to Bohr's view such terms, that for Bohr there is decoherence, but there is no (physical) "coherence" that decoheres. That is, one could speak of linearly superimposed ("coherent") states in referring to quantum processes themselves (that of a single "particle" or of an extended quantum system) only by convention or symbolically, without assuming an actual description of such processes in space and time. Such "states" will, of course, manifest themselves in particular phenomenal effects. In this view, there could be no physical "coherence" or anything that can be conceived of in terms of space-time processes corresponding to the coherence (say, again, a linear superposition of vectors in a Hilbert space) of the mathematical quantum-mechanical description, which would then "decohere" in the process of measurement.[25] The mathematical coherence of our description itself "arises" or applies due to the possibility and effects of measurement or preparation, which involves the same type of irreducible and ultimately uncontrollable interaction between quantum objects and measuring instruments, and in this structural sense is a measurement. (In this view, quantum mechanics is, again, a theory that allows one to predict the outcomes of future possible measurements on the basis of measurements already performed, and only this, rather than to say anything about the state of the system between measurements.) On Bohr's view, this interaction both allows for the particular mathematization of the overall situation in terms of quantum-mechanical formalism and epistemologically limits the meaning of the latter, which configuration becomes correlative to the inevitability of the uncertainty relations in quantum physics. Just as, and correlatively to, the application of specific probabilities to the quantum-mechanical data, the application of this mathematics cannot be explained in the way it would be or could be in classical physics. This irreducible inexplicability does not eliminate "decoherence," if we refer by the latter to the process of transition from the quantum world to the classical world, but it has an impact upon what kind of decoherence theory or its physical meaning one can speak of under these conditions.

Bohr, again, would not maintain that there is nothing more to say concerning the physics of transition from the quantum to the classical level, say, in the building of a water drop around an ion in a cloud chamber, or that he saw this problem as resolved. Even leaving aside Bohr's preoccupation with the correspondence principle, which, we recall, was indispensable to Heisenberg's invention of quantum mechanics and which clearly bears on this problem or rather this problem bears on it, Bohr was well aware that Heisenberg's paper on the uncertainty relations addressed this question. The latter (coupled with Heisenberg's matrix theory) may even have been a distant source and precursor of Bohr's concept of phenomenon itself [27]. He was also obviously well aware of the Ehrenfest theorem, the significance of which was so instructively commented on by Omnés in his recent work. In Bohr's view, the "gap" between the physical description of macroscopic physical phenomena and quantum objects and processes is insurmountable, whether the macro phenomena in question are, in general, those considered in classical physics or those in question in and described by quantum physics. The latter is, again, seen as describing only the effects of the interaction between quantum objects (which may be macroscopic in terms of the range of their effects) and measuring instruments.

Accordingly, while this view does not suspend the possibility or necessity of such theories, it entails that the "gap" in question would inevitably appear in any proper "decoherence" theory, a theory accounting for this transition. It would also appear in any theory accounting for the ultimate constitution of the macro-world (available to a classical description) out of the ultimate (quantum) micro-constituents of matter. The point may be put as follows. There can be, as it were, "macro" theories, say, accounting for the building emergence of a water drop around an ion in a cloud chamber, which is an interesting problem. Such theories may well be of a mixed, quantum and classical nature, including thermodynamics and chaos theory, which while different epistemologically, are both classical. There could be, however, no truly "micro" decoherence theory, which would or could account for this "gap," since such an account would entail an ultimate subdivision (and hence de facto elimination) of Bohr's phenomena, which is not possible. The phenomenal-effect structure (without an accessible underlying object-efficacity), such as that defining the building of a water drop around an ion in a cloud chamber, is both an actual physical phenomenon and a model of the quantum-mechanical situation. That is, nothing qualitatively different can be observed or be otherwise described by the mathematical formalism of quantum theory. This is in part why Bohr argues that the actual impossibility of performing some of the experiments he describes in his arguments is essentially irrelevant, since these "experiments" offer us such a structural model, realized in actual observations, such as those related to the Compton effect, one of his favorite examples. He clearly saw the

Heisenberg-microscope thought-experiment as exemplifying precisely this point—the impossibility of reducing the radical discontinuity in question between "quantum objects" and "processes" (even if these terms again apply) and observable macro (phenomenal) effects in any experiment or by means of any description. If there is "quantum discontinuity" (a term not especially favored by Bohr, in contrast to individuality or indivisibility), this would be the one. Accordingly, similarly to other quantum-theoretical phenomena (Bohr's sense becomes obligatory), as discussed earlier, new experimental and theoretical findings in that domain would have to be seen in terms of new effects. We can observe and can account for them in terms of particular quantum interactions of specific systems (and their quantum subsystems), even though, consistently with the scheme just described, we can, in all rigor, never describe the ultimate efficacity of these "effects." Quantum measurement is obviously itself such an effect, as, it follows, is any conceivable observation of "quantum objects," that is, of an effect with which we would associate such objects and from which we infer their existence, to begin with. In this area of investigation, too, we may be able to reach and in fact, from Wojciech Zurek's work (and indeed some earlier anticipations) on, have already reached quite far, experimentally and theoretically, and sometimes this nonclassical epistemology helps us [28].[26] Recent, often spectacular, experimental observations of quantum coherence would, I would argue, confirm this point, as they may be seen as presenting us with new effects of both coherence and decoherence (in this sense), and it appears to me that related claims concerning observations of coherent or linearly superposed states are, in all rigor, subject to this scheme [29] [30]. They may appear to some to exceed the constraints this scheme imposes [31]. This view does not seem to be justified, however, nor indeed is it expressed by the experimenters themselves just mentioned.

Could, by the same token, "decoherence" theories of that type be viable, effective, and significant, including for applications, say, in quantum computing? It would be difficult to doubt this, and I do not think that Bohr would. Such theories may, for example, be especially helpful in thinking how quantum computational devices can in principle be realized and actually built in practice. Indeed, similarly to certain interpretations of quantum mechanics, as considered earlier, some of the versions of decoherence may be argued to be consistent with or entail the view just outlined, even though their authors may not quite see it in this way.

In short, in Bohr's interpretation the phenomena or effects (in Bohr's sense) of decoherence would be subject to essentially the same considerations as the phenomena of quantum entanglement. The latter may of course itself be seen in terms of scale-extended coherence, while the observed effect may be seen (with the considerations just given in mind) in terms of "decoherence" or irreversible amplification effects, and both effects are often closely linked theoretically and experimentally [29] [30]. That is, either phenomenon must be seen as phenomena in Bohr's sense, rather referring to the level of quantum objects themselves, while everything just said concerning understanding and investigation, experimental or theoretical, would be retained. It also follows that Bohr's epistemology is in no way incompatible, quite the contrary, with the concept and theories or, more accurately (there are exceptions), most concepts and theories, of quantum computing, including those involving and even based on quantum entanglement. Perhaps it may ultimately help us to put it in place, for example by warning us to stay away from that which is incompatible with the laws of physics, such as and perhaps in particular nonlocality, but far from exclusively so.

5. Conclusion: The Bohr-Einstein Principle

I would like to conclude by offering the following proposition. Any consistent formalism and interpretation (including an interpretation of the standard mathematical formalism of quantum mechanics) providing an account of the quantum processes themselves, rather than of the effects of their interaction with measuring instruments upon the latter, is bound to be nonlocal. It may not even matter whether it would do so analogously to, or on the model of, classical physics or otherwise. Bohmian theories, which are theories of that type, are examples of this, while other theories just mentioned are in fact or in effect not, nor is, of course, Bohr's interpretation. I was unable to derive so general a result from Bell's theorem and its extensions, which, however, would seem to lead to this conclusion. In the case of the standard quantum-mechanical formalism, the nonlocality of quantum mechanics appears to follow from the EPR argument, once one assumes that the quantum-mechanical formalism describes the behavior of quantum objects themselves. So I must here state it as a conjecture or principle, which we may call the Bohr-Einstein principle. The principle both brings two views together as concerns locality and keeps them apart as concerns epistemology.

Acknowledgments. I would like to thank the NATO Scientific and Environmental Affairs Division and the organizers of the Workshop, and the participants for sharing their thoughts and knowledge. For productive discussions concerning the subject of this article I am grateful to Antonios Gonis, Peter Mittelstaedt, and Chris Fuchs, who also read and helped to improve the original draft. Finally, I owe a special debt of gratitude to David Mermin.

Notes

[1] Other terms, such as "separability," and different definitions of "locality" are sometimes used as well, but these variations and differences are not germane to and may be easily correlated with my argument. It may be pointed out, however, that "nonlocality" is sometimes invoked without implying a violation of the Lorentz invariance, and the readers do not always carefully follow these nuances, and the authors do not always adequately explain or qualify them. In this article, unless stated otherwise, nonlocality (and references to it elsewhere, as, for example, in Einstein, David Bohm, and Henry Stapp) is used strictly in this sense of the lack of the Lorentz invariance and incompatibility with relativity. Entanglement is used otherwise.

[2] Due to a printing error, the order of pages is reversed in this edition: page 149 should precede page 148. Throughout this article, reference to page numbers for Reference [2] are given in parentheses following [2].

[3] See especially the commentary by J. Bell ([4], pp. 155-56, 189-90) and H. Stapp ([10], pp.162). For a discussion of Bell and Einstein see Reference [32], pp. 160-63, 176-90. Bohr's views and writing encounter considerable resistance among the proponents of David Bohm's approach, who find their inspiration in Einstein's and Bell's critique, as, to give two recent prominent examples, in References [33] and [34]. To paraphrase Bohr on EPR's argument, the trends of the latter arguments do not seem to me adequately to meet the case Bohr's writings present to us. Bohm himself had a more balanced view and cogent sense of Bohr's argument and work, which he admired [35]. For arguably the final version of Bohm's theory itself see Reference [36].

[4] Throughout this article volume and page references to <u>The Philosophical Writings of Niels Bohr</u> (Reference [6]) are given in parentheses following [6].

[5] The best (comprehensive) summation of Bohr's ultimate argument is offered by him in his 1958 short essay "Quantum Physics and Philosophy: Causality and Complementarity" ([6] 3, pp. 1-7), which also contains some valuable nuances, to be considered below.

[6] An elegant exposition is given by Anthony J. Leggett [37].

[7] It is, incidentally, also easy to see that, if (at least given locality) we assume that complementary variables, while subject to uncertainty relations, nevertheless pertain to quantum objects, quantum-mechanical probabilities will not come out right ([38], p. 232]. As will be seen, however, Bohr's interpretation entails an even

more radical epistemology, whereby not even single properties can be ascribed to quantum objects themselves.

[8] Reference [11] offers an effective critique of Stapp's argument in [5].

[9] Among the more interesting discussions are References [7], [8], [39], [44], and [45].

[10] Cf. also Bohr's comments in Reference [6] 2, p.73.

[11] See Reference [46] for a further discussion by the present author.

[12] See an interesting remark by Einstein reported by Pauli in Reference [22], p.226.

[13] The apparent proximities and (perhaps less apparent) differences between this conception and Kant's phenomenal/noumenal conceptuality would require a separate analysis. Certainly, a number of key features of quantum mechanics and Bohr's interpretation, such as and in particular complementarity, give it further specificity not found in Kant, which, it may be argued, also lead to ultimately more radical epistemology as well. On more general relationships between Kant and quantum theory see Peter Mittelstaedt's article [47].

[14] See the final chapter, "The Measuring Process," of his *Mathematical Foundations of Quantum Mechanics*, [40] pp. 417-445, which refers to Bohr (p.420).

[15] See reference [39] for arguably the most refined version of this interpretations, seen as "the minimal interpretation" and for an elegant and illuminating discussion of its key features by Mittelstaedt, which, in my view, again, brings his interpretation closer to Bohr than he appears to see it.

[16] Mermin's own (Ithaca) interpretation, developed in Ref. [18], while also aiming to avoid the primary role of measurement, offers a more subtle case, since it does not attributes physical properties (the "correlata" of quantum correlations) to quantum objects, or indeed to anything. It sees only quantum correlations themselves (without correlata) as having objective reality that physics can speak of. The subject would require a separate discussion, however.

[17] It is not always easy to ascertain the degree of this proximity in the case of Dirac or others close to Bohr, such as Heisenberg and Pauli, since their epistemological views are not explicated to the same degree as Bohr's. By contrast the differences between Bohr's views and Einstein's (or those closer to Einstein, such as Schrödinger's or Bell's) are immediately manifest.

[18] These considerations are significant even in the case of spin, where we deal with finite-dimensional Hilbert spaces, but especially in the case of standard kinematical and dynamical variables and the infinite-dimensional Hilbert spaces.

[19] See, however, A. J. Leggett, Ref [37] and A. Garg's contribution to these proceedings, "Bypassing Decoherence: Macroscopic Coherence" [41]. These issues are the subject of continuing research. Cf., for example, as recent report in *Nature*, Ref. [42]

[20] Cf. Serge Haroche's article, "Experimental Aspects of Decoherence," in this volume [43].

[21] I omit Bohr's intermediate propositions concerning the interactions between quantum objects and measuring instruments. While they are important in explaining the reasons for Bohr's argument, they are fully consistent with the preceding analysis, presented above in terms of "Discussion with Einstein."

[22] Both Bell [4] and Stapp [10] are such examples. See Reference [32], pp. 176-90, for a discussion of Bell.

[23] See Reference [49] for a preliminary approach.

[24] In addition, there are also macroscopic quantum effects, meaning this now not so much in Bohr's sense just indicated (although the latter would apply and, as will be seen presently, is especially appropriate), but in terms of their size as macroscopic "coherent" quantum systems. Their behavior, however, is defined by their microscopic quantum constitution.

[25] A given state in a linear superposition is itself subject to this impossibility, as Dirac, the inventor of the concept, recognized ([19], pp. 10-14). For the same reason, the terminology of either the state vector (which concept has no physical meaning in Bohr's framework) or, especially, the collapse of the wave function is not favored by Bohr. At the mathematical level the terminology of coherence would, thus, be justified, while that of decoherence, although (terminologically) understandable and useful, would not have either mathematical or physical justification in Bohr's interpretation.

[26] See Ref. [50], p. 188 for an interesting earlier insight.

References

[1] A. Einstein, B. Podolsky, and N. Rosen "Can Quantum-mechanical Description of Physical Reality be Considered Complete?" In: J. A. Wheeler and W. H. Zurek, *Quantum Theory and Measurement*, Princeton University Press, Princeton, 1983, pp. 138–41.

[2] N. Bohr, "Can Quantum-mechanical Description of Physical Reality be Considered Complete?" In: J. A. Wheeler and W. H. Zurek, *Quantum Theory and Measurement*, Princeton University Press, Princeton, 1983, pp. 145-51.
[3] E. Schrödinger, "The Present Situation in Quantum Mechanics" In: J. A. Wheeler and W. H. Zurek, *Quantum Theory and Measurement*, Princeton University Press, Princeton, 1983, pp. 152-67.
[4] J. Bell, *Speakable and Unspeakable in Quantum Mechanics*, Cambridge University Press, Cambridge, 1987.
[5] H. P. Stapp, "Nonlocal Character of Quantum Theory," *American Journal of Physics* **65** (1997), pp. 300-304.
[6] N. Bohr, *The Philosophical Writings of Niels Bohr*, 3 vols., Ox Bow Woodbridge Press, Woodbridge, Conn., 1987.
[7] R. Omnès, *The Interpretation of Quantum Mechanics*, Princeton University Press, Princeton, 1994.
[8] R. Omnès, *Understanding Quantum Mechanics*, Princeton University Press, Princeton, 1999.
[9] P. A. Schilpp, *Albert Einstein: Philosopher-Scientist*, Tudor, New York, 1949.
[10] H. Stapp, "Quantum Nonlocality and the Description of Nature," In: J. T. Cushing and E. McMullin, eds., *Philosophical Consequences of Quantum Theory: Reflections on Bell's Theorem*, University of Notre Dame Press, Notre Dame, Ind., 1989, 154-174.
[11] N. D. Mermin, "Nonlocal Character of Quantum Theory?" *American Journal of Physics* **66** (1998), 920-924.
[12] N. Bohr, "On the Notion of Causality and Complementarity," *Dialectica* **2**, 312-19.
[13] N. Bohr, "The Causality Problem in Atomic Physics," *Niels Bohr: Collected Works*, Elsevier, Amsterdam, 1972-1996, vol. 7, 303-322.
[14] A. Einstein, *Journal of the Franklin Institute*, 221, **349** (1936).
[15] A. Einstein, "Reply to Criticism." In: P. A. Schilpp, *Albert Einstein: Philosopher-Scientist*, Tudor, New York, 1949.
[16] Bas C. Van Frassen, *Quantum Mechanics: An Empiricist View*, Clarendon, Oxford, 1991.
[17] R. B. Griffiths, "Consistent Histories and Quantum Reasoning," *Physical Review* A **54** (1996), 2759-2774.
[18] N. D. Mermin, "What Is Quantum Mechanics Trying To Tell Us?" *American Journal of Physics* **66** (9), 1998: 753-67.
[19] P. A. M. Dirac, *The Principles of Quantum Mechanics*, Clarendon, Oxford, 1995.
[20] A. Fine "Do Correlations Need to be Explained?" In: J. T. Cushing and E. McMullin, eds., *Philosophical Consequences of Quantum Theory: Reflections on Bell's Theorem*, Notre Dame University Press, Notre Dame, Ind., pp. 174-94.
[21] N. D. Mermin, "Can you help your team tonight by watching TV?" In: *Boojums All the Way Through*, Cambridge University Press, Cambridge, 1990, pp. 95-109.
[22] *The Einstein-Born Letters*, Walker, New York, 1971, p. 209.
[23] N. D. Mermin, "Spooky action at a distance: mysteries of quantum theory." In: *Boojums All the Way Through*, Cambridge University Press, Cambridge, 1990, pp. 110-177.
[24] H. Reichenbach, *The Direction of Time*, University of California Press, Los Angeles, 1956, p.159.
[25] W. H. Zurek, "Decoherence and Transition from Quantum to Classical," *Physics Today* **44** (1991), 36-44.
[26] W. H. Zurek, *Physics World* **January** (1997).
[27] W. Heisenberg, "The Physical Content of Quantum Kinematics and Mechanics." In: J. A. Wheeler and W. H. Zurek, *Quantum Theory and Measurement*, Princeton University Press, Princeton, 1983, pp. 62-86.
[28] W. H. Zurek, "Pointer basis of quantum apparatus: Into what mixture does the wave packet collapse," *Physics Review* D **24**, 1516-1525.
[29] S. Haroche, M. Brune, and J. M. Raimond, "Experiments with single atoms in cavity: Entanglement, Schrödinger's cats, and decoherence," *Philosophical Transactions of the Royal Society of London* **355** (1997), 2367-2380.
[30] C. J. Myatt, et. al. "Decoherence of quantum superpositions through coupling to engineered reservoirs," *Nature* **403** (2000), 269-273.

[31] M. Paty, "Are quantum systems physical objects with physical properties," *European Journal of Physics* **20** (1999), 373-388.
[32] A. Plotnitsky, *Complementarity: Anti-epistemology after Bohr and Derrida*, Duke University Press, Durham, NC., 1994.
[33] J. T. Cushing, *Quantum Mechanics, Historical Contingency and the Copenhagen Hegemony*, University of Chicago Press, Chicago, 1994.
[34] M. Beller, *Quantum Dialogue: The Making of a Revolution*, University of Chicago Press, Chicago, 1999.
[35] D. Bohm, *Wholeness and Implicate Order*, Routledge, London, 1995.
[36] D. Bohm and B. J. Hiley, *The Undivided Universe: An Ontological Interpretation of Quantum Mechanics*, Routledge, London, 1993.
[37] A. J. Leggett, "Experimental Approaches to the Quantum Measurement Paradox," *Foundations of Physics*, **18**, 9 (1988), 940-41.
[38] P. Mittelstaedt, "Language and Reality in Quantum Physics." In P. Lahti and P. Mittelstaedt, ed. *Symposium on the Foundations of Modern Physics 1987*, World Scientific, Singapour, 1987
[39] P. Mittelstaedt, *The Interpretation of Quantum Mechanics and the Measuring Process*, Cambridge University Press, Cambridge, 1997.
[40] John von Neumann, *Mathematical Foundations of Quantum Mechanics*, trans. Robert T. Beyer. Princeton University Press, Princeton, 1983.
[41] A. Garg, "Bypassing Decoherence: Macroscopic Coherence," in this volume.
[42] J.R. Friedman, et. al. "Quantum Superposition of Distinct Macroscopic States," *Nature* **406** (2000), 43-46.
[43] S. Haroche, "Experimental Aspects of Decoherence," in this volume.
[44] C. Rovelli, "Relational Quantum Mechanics," *International Journal of Theoretical Physics*, **35** (1996), 1637-78.
[45] L. Smolin, *The Life of the Cosmos*, Oxford University Press, New York, 1997.
[46] A. Plotnitsky, "Techno-Atoms: The Ultimate Constituents of Matter and the Technological Constitution of Phenomena in Quantum Physics." *Tekhnema: Journal of Philosophy and Technology* 5 (1999), 36-95.
[47] P. Mittelstaedt, "Kant and the Quantum Theory." In P. Parrini, ed. *Kant and Contemporary Epistemology*, Kluwer, Dordrecht, 1994.
[48] Stephen Hawking and Roger Penrose, *The Nature of Space and Time*, Princeton University Press, (Princeton, NJ, 1996, pp. 59-60.
[49] A. Plotnitsky, "Landscapes of Sibylline Strangeness: Complementarity, Quantum Measurement and Classical Physics." In: G.C. Cornelis, J. P. Van Bendegem, and D. Aerts, eds., *Metadebates*, Kluwer, Dordrecht, 1999.
[50] K. Gottfried, *Quantum Mechanics*, Vol. 1. Benjamin: New York, 1966.

Decoherence and its Implications in Quantum Computation and Information Transfer
T. Gonis and P.E.A. Turchi (Eds.)
IOS Press, 2001

Quantum Foundations in the Light of Quantum Information

Christopher A. Fuchs
Computing Science Research Center
Bell Labs, Lucent Technologies
Room 2C-420, 600–700 Mountain Ave.
Murray Hill, New Jersey 07974, USA

Abstract

In this paper, I try to cause some good-natured trouble. The issue at stake is when will we ever stop burdening the taxpayer with conferences and workshops devoted—explicitly or implicitly—to the quantum foundations? The suspicion is expressed that no end will be in sight until a means is found to reduce quantum theory to two or three statements of crisp *physical* (rather than abstract, axiomatic) significance. In this regard, no tool appears to be better calibrated for a direct assault than quantum information theory. Far from being a strained application of the latest fad to a deep-seated problem, this method holds promise precisely because a large part (but not all) of the structure of quantum theory has always concerned information. It is just that the physics community has somehow forgotten this.

1 Imprimatur

> **im·pri·ma·tur** (ĭm′pre-mä¹ter, -mâ¹ter) **1.** Official approval or license to print or publish, especially under conditions of censorship.
>
> — *American Heritage Dictionary*

The title of the NATO Advanced Research Workshop that gave birth to this volume was "Decoherence and its Implications in Quantum Computation and Information Transfer." It was a wonderful meeting—the kind most of us lick our lips for year after year, with little hope of ever tasting. It combined the best of science with the exotic solitude of an island far, far away. One could not help but have a creative thought shaken loose with each afternoon's gusty wind. Indeed, it was a meeting that will make NATO proud. But, as any attendee can tell you, the most popular pastime—in spite of those windy beaches and dark tans—was an activity just half-devoted to the conference title. The life of the party was all the talks and conversations on "Decoherence and its Implications in Quantum *Foundations*."

In this article, I am going to make an admission at the outset. Despite the industry of work it has spawned for so many conscientious colleagues, and the absolute importance its understanding holds for implementing quantum information technologies, *I simply cannot see that decoherence bears in any way on the quantum foundation problem.* In saying this, I am well aware of the jeopardy I bring upon myself. For I am but one person—one, in fact, who makes no bones about how limited his knowledge is—while my decoherence colleagues tell me that they are many. The press tells me that they are the "new orthodoxy." Try as I might, though, I just do not get it. Something about their program does not click in my head.

The root of this problem could be many things, of course—not least of which might be my supreme thickheadedness. There is, however, one thing I know for sure: My want of understanding does not come from a lack of *trying* to understand. To the extent that I am a relatively reasonable person—and presumably as conscientious as the enthusiasts of the "new orthodoxy"—it seems to me this is a datum that should not go unnoticed. Why might it be that an honest outsider is having so much trouble coming to grips with something the indoctrinated profess to see with great clarity? I cannot stop myself from thinking, "Where there is smoke, there is smoke."

But let me put that aside: What I wish to get at with this self-sanctioned imprimatur is not a detailed criticism of the decoherence-based quantum-foundations programs. I say all these things instead to provide a particular instantiation (and a little background) for what I deem to be a significantly larger problem in the quantum-foundations efforts to date. My reluctance, or more accurately, my inability, to toe the line for any of the rival quantum political parties—be they the Bohmians [1], the Consistent Historians [2], the Einselectionists [3], the Spontaneous Collapsicans [4], or the Everettistas [5]—springs from a distrust captured vividly by the image of a political convention. The relevant point is, what are their platforms?

Throughout the 2000 presidential campaign in America, the Green Party accused the Democrats and the Republicans of having no difference *whatsoever* in their platforms. The Democrats and Republicans were appalled. Likewise, to what I suspect will be a jaw-dropping shock to the quantum-party leaders, I declare that I can see little to no difference in *any* of their beliefs. They all look equally pale to me. For though everyone seems to want a little reality—i.e., something in the theory that they can point to and say, "There, that term is what is real in the universe even when there are no physicists about"—none are willing to dig very deep for it.

What I mean by this deliberately provocative statement is that in spite of the differences in what the various parties are willing to label[1] "real" in quantum theory,[2] they nonetheless

[1]Or *add* to the theory, as the case may be.

[2]Very briefly, a *cartoon* of the positions might be as follows. For the Bohmians, "reality" is captured by supplementing the state vector with an actual trajectory in coordinate space. For the Everettistas, it is the universal wave function and the universe's Hamiltonian. (Depending upon the faction, though, these two entities are sometimes supplemented with the terms in various Schmidt decompositions of the universal state vector with respect to a preconceived tensor-product structure.) For the Spontaneous Collapsicans it is again the state vector—though now for the individual system—but Hamiltonian dynamics is supplemented

all proceed from essentially the same *abstract* starting point. It is nothing other than the standard textbook accounts of the *axioms* of quantum theory.[3]

The Platform for Most Quantum Foundations Ventures:
The Axioms (plain and simple)

1. For every system, there is a complex Hilbert space $\mathcal{H}$.
2. States of the system correspond to projection operators onto $\mathcal{H}$.
3. Those things that are observable *somehow* correspond to the eigenprojectors of Hermitian operators.
4. Isolated systems evolve according to the Schrödinger equation.

⋮

"But what nonsense is this," you must be asking. "Where else could they start!?!" The main issue is this, and no one has said it more clearly than Carlo Rovelli [6]. Where present-day quantum-foundation studies have stagnated in the stream of history is not so unlike where the physics of length contraction and time dilation stood before Einstein's 1905 paper on special relativity.

The Lorentz transformations have the name that they do, rather than, say, the Einstein transformations, for a simple historical reason: Lorentz had published some of them as early as 1895.[4] Indeed one could say that most of the empirical predictions of special relativity were in place well before Einstein came onto the scene. But that was of little consolation to the pre-Einsteinian physics community striving so hard to make sense of electromagnetic phenomena and the luminiferous ether. Precisely because the *only* justification for the Lorentz transformations appeared to be their *empirical adequacy*, they remained a mystery to be conquered. More particularly, this was a mystery that heaping further *ad hoc* (mathematical) structure onto could not possibly solve.

What was being begged for in the years between 1895 and 1905 was an understanding of the *origin* of that abstract, mathematical structure—some simple, crisp *physical* statements with respect to which the necessity of the mathematics would be indisputable. Einstein supplied that and became one of the greatest physicists of all time. He reduced the mysterious structure of the Lorentz transformations to two simple statements that could be written in any common language:

with an objective collapse mechanism. For the Consistent Historians "reality" is captured with respect to an initial quantum state and a Hamiltonian by the addition of a set of preferred positive-operator valued measures (POVMs)—they call them consistent sets of histories—along with a truth-value assignment within each of those sets. For the Einselectionists, I leave it as an exercise to the reader.

[3]To be fair, they do, each in their own way, contribute minor modifications to the *meanings* of a few *words* in the axioms. But that is essentially where the effort stops.

[4]Though, FitzGerald had considered length contraction as early as 1889.

1) the speed of light in empty space is independent of the speed of its source,
2) physics should appear the same in all inertial reference frames.

The deep significance of this for the quantum problem should stand out and speak overpoweringly to anyone who admits the simplicity of these principles.

Einstein's move effectively stopped all further debate on the origins of the Lorentz transformations. Outside of the time of the Nazi regime in Germany, I suspect there have been less than a handful of conferences devoted to "interpreting" them. More importantly, with the supreme simplicity of Einstein's principles, physics became ready for "the next step." Is it possible to imagine that any mind—even Einstein's—could have made the leap to general relativity directly from the original, abstract structure of the Lorentz transformations? A structure that was only empirically adequate? I would say no. Indeed, one might question what wonders we will find by pursuing the same strategy of simplification for the quantum foundations.

Symbolically, where we are:	Where we need to be:
$x' = \dfrac{x - vt}{\sqrt{1 - v^2/c^2}}$	Speed of light is constant.
$t' = \dfrac{t - vx/c^2}{\sqrt{1 - v^2/c^2}}$	Physics is the same in all inertial frames.

The task is not to make sense of the quantum axioms by heaping more structure, more definitions, more science-fiction imagery on top of them, but to throw them away wholesale and start afresh. We should be relentless in asking ourselves: From what deep *physical* principles might we *derive* this exquisite mathematical structure? Those principles should be crisp; they should be compelling. They should stir the soul. When I was in junior high school, I sat down with Martin Gardner's book *Relativity for the Million* and came away with an understanding of the subject that sustains me even today: The concepts were strange to my everyday world, but they were clear enough that I could get a grasp of them knowing little more mathematics than arithmetic. One should expect nothing less for a proper foundation to the quantum. Until we can explain the essence of the theory to a junior high-school or high-school student—*the essence, not the mathematics!*—and have them walk away with a deep, lasting memory, I well believe we will have not understood a thing about quantum foundations.

But I am not fooling myself. I know that anyone with a vested interest in any of the existing quantum interpretations will be quick to point out every hole, every nonnecessity in the sermon I just gave. Indeed, I can feel the upcoming wrath of their email as I write this sentence. I have no retort. Only a calm confidence that if progress is not made in this direction, 100 years from now the political pollster will still have a niche at the latest quantum foundations meeting [7].

So, throw the existing axioms of quantum mechanics away and start afresh! But how to proceed? I myself see no alternative but to contemplate deep and hard the tasks, the techniques, and the implications of quantum information theory. The reason is simple, and I think inescapable. Quantum mechanics has always been about information. It is just that the physics community has somehow forgotten this.

Quantum Mechanics: *The Axioms and Our Imperative!*	
1. States correspond to density operators ρ over a Hilbert space $\mathcal{H}$.	*Give an information theoretic reason if possible!*
2. Measurements correspond to positive operator-valued measures (POVMs) $\{E_b\}$ on $\mathcal{H}$.	*Give an information theoretic reason if possible!*
3. $\mathcal{H}$ is a complex vector space, not a real vector space, not a quaternionic module.	*Give an information theoretic reason if possible!*
4. Systems combine according to the tensor product of their separate vector spaces, $\mathcal{H}_{\text{AB}} = \mathcal{H}_{\text{A}} \otimes \mathcal{H}_{\text{B}}$.	*Give an information theoretic reason if possible!*
5. Between measurements, states evolve according to trace-preserving completely positive linear maps.	*Give an information theoretic reason if possible!*
6. By way of measurement, states evolve (up to normalization) via outcome-dependent completely positive linear maps.	*Give an information theoretic reason if possible!*
7. Probabilities for the outcomes of a measurement obey the Born rule for POVMs $\text{tr}(\rho E_b)$.	*Give an information theoretic reason if possible!*

This table is my plea to the community. Our foremost task should be to go to each and every axiom of quantum theory and give it an information theoretic justification if we can. Only when we are finished picking off all the terms (or combinations of terms) that can be interpreted as information—subjective information—will we be in a position to make real progress. The raw distillate that is left behind, miniscule though it may be, will be our first glimpse of what quantum mechanics is trying to tell us about nature itself.

2 Introduction

This paper is about taking that plea to heart, though it contributes only a small amount to the labor it asks. Just as in the founding of quantum mechanics, this is not something that will spring forth from a lone mind contemplating the universe from the cloisters of a medieval college. It is a task for a community with diverse but productive points of view. The quantum information community is nothing if not that. "Philosophy is too important to be left to the philosophers," John Archibald Wheeler once said. Likewise, I am apt to say the same for the quantum foundations.

The structure of the remainder of the paper is as follows. In Section 3 "**Why Information?**," I reiterate the cleanest argument I know of that the quantum state is solely an expression of information—the information one has about a quantum system. It has no objective reality in and of itself. The argument is then refined by considering the phenomenon of quantum teleportation [8].

In Section 4 "**Information About What?**," I tackle that very question [9] head on. The answer is, "nothing more than the potential consequences of our experimental interventions into nature." Once freed from the notion that quantum measurement ought to be about revealing traces of some preexisting property [10] (or beable [11]), one finds no particular reason to take the standard account of measurement (in terms of complete sets of orthogonal projection operators) as a basic notion. Indeed quantum information theory, with its emphasis on the utility of generalized measurements or positive operator-valued measures (POVMs) [12], suggests one should take those entities as the basic notion instead. The productivity of this point of view is demonstrated by the beautifully simple Gleason-like derivation of the quantum probability rule recently found by Paul Busch [14] and, independently, by Joseph Renes and collaborators [15]. Contrary to Gleason's original theorem [16], this theorem works just as well for two-dimensional Hilbert spaces, and even for Hilbert spaces over the field of rational numbers.

In Section 5 "**Whither Bayes Rule?**," I ask why one should expect the rule for updating quantum state assignments upon the completion of a measurement to take the form it actually does. Along the way, I give a simple derivation that one's information always increases on average for *any* quantum mechanical measurement that does not itself discard information. (Despite the appearance otherwise, this is not a tautology!) More importantly, the proof technique used for showing the theorem indicates an extremely strong analogy between quantum collapse and Bayes' rule in classical probability theory: Up to an overall unitary "readjustment" of one's knowledge (that takes into account details of the measurement interaction as well as one's initial state of knowledge), quantum collapse is *precisely* Bayesian conditionalization. This in turn gives even more impetus for the assumptions behind the Gleason-like theorem of the previous section.

In Section 6 "**Wither Entanglement?**," I ask whether entanglement is all it is cracked-up to be as far as quantum foundations are concerned. In particular, I give a simple derivation of the tensor-product rule for combining Hilbert spaces of individual systems into a larger composite system. To no surprise, once again this comes from Gleason-like considerations for local measurements in the presence of classical communication—the very bread and butter of much of quantum information theory.

In Section 7 **"Unknown Quantum States?,"** I tackle the conundrum posed by these very words. Despite the phrase's ubiquitous use in the quantum information literature, what can an *unknown* state possibly be? A quantum state—from my point of view, explicitly someone's information—must always be known by someone, if it exists at all. On the other hand, for many an application in quantum information, it would be quite a contrivance to imagine there is always someone in the background with extra knowledge of the system being measured or manipulated. The solution, at least in the case of quantum state tomography [13], is found through a quantum mechanical version of de Finetti's classic theorem on "unknown probabilities." This reports work from Refs. [17] and [18]. Maybe one of the most interesting things about the theorem is that it fails for Hilbert spaces over the field of real numbers, suggesting that perhaps the whole discipline of quantum information might not be well defined in that imaginary world.

Finally, in Section 8 **"The Oyster and the Quantum,"** I flirt with the most tantalizing question of all: Why the quantum? There are no answers here, but I do not discount that there will be one within 20 years. In this regard no platform seems firmer for the leap than the very existence of quantum cryptography and quantum computing. The world is sensitive to our touch. It has a kind of "Zing!"[5] that makes it fly off in ways that were not imaginable classically. The whole structure of quantum mechanics—*it is speculated*—may be nothing more than the optimal method of reasoning and processing information in the light of such a fundamental (wonderful) sensitivity.

3 Why Information?

It may be, as one French physicist put it, "the fog from the north," but the Copenhagen interpretation remains the best interpretation of the quantum that we have.

— *John Archibald Wheeler*
New York Times, 12 December 2000

Einstein was the master of clear thought. I have already expressed my reasons for thinking this in the arena of electromagnetic phenomena. Likewise, I would say he possessed the same great penetrating power when it came to analyzing the quantum. For even there, he was immaculately clear and concise in his expression. In particular, he was the first person to say in absolutely unambiguous terms why the quantum state should be viewed as information (or, to say the same thing, as a representation of one's knowledge).

His argument was simply that a quantum-state assignment for a system can be forced to go one way or the other by interacting with a part of the world that should have no causal connection with the system of interest. The paradigm here is of course the one well

[5]Dash, verve, vigor, vim, zip, pep, punch, pizzazz!

known through the Einstein, Podolsky, Rosen paper [19], but simpler versions of the train of thought had a long pre-history with Einstein [20].

The best was in essence this. Take two spatially separated systems A and B prepared in some entangled quantum state $|\psi^{AB}\rangle$. By performing the measurement of one or another of two observables on system A alone, one can *immediately* write down a new state for system B. Either the state will be drawn from one set of states $\{|\phi_i^B\rangle\}$ or another $\{|\eta_i^B\rangle\}$, depending upon which observable is measured.[6] The key point is that it does not matter how distant the two systems are from each other, what sort of medium they might be immersed in, or any of the other fine details of the world. Einstein concluded that whatever these things called quantum states *be*, they cannot be "real states of affairs" for system B alone. For, whatever the real, objective state of affairs at B is, it should not depend upon the measurements one can make on a causally unconnected system A.

Thus one must take it seriously that the new state (either a $|\phi_i^B\rangle$ or a $|\eta_i^B\rangle$) represents partial knowledge about system B. In making a measurement on A, one learns something about B, but that is where the story ends. The state change cannot be construed to be something more physical than that. More particularly, the state itself cannot be viewed as more than a reflection of the knowledge gained through the measurement. Expressed in the language of Einstein, the quantum state cannot be a "complete" description of the quantum system.

Here is the way Einstein put it to Michele Besso in a 1952 letter [22]:

> What relation is there between the "state" ("quantum state") described by a function ψ and a real deterministic situation (that we call the "real state")? Does the quantum state characterize completely (1) or only incompletely (2) a real state? ...
>
> I reject (1) because it obliges us to admit that there is a rigid connection between parts of the system separated from each other in space in an arbitrary way (instantaneous action at a distance, which doesn't diminish when the distance increases). Here is the demonstration: ...
>
> If one considers the method of the present quantum theory as being in principle definitive, that amounts to renouncing a complete description of real states. One could justify this renunciation if one assumes that there is no law for real states—i.e., that their description would be useless. Otherwise said, that would mean: laws don't apply to things, but only to what observation teaches us about them. (The laws that relate to the temporal succession of this partial knowledge are however entirely deterministic.)
>
> Now, I can't accept that. I think that the statistical character of the present theory is simply conditioned by the choice of an incomplete description.

There are two issues in this letter that are worth disentangling. 1) Rejecting the rigid connection of all nature[7]—that is to say, admitting that the very notion of a *system* (as

[6]Generally there need be hardly any relation between the two sets of states: only that when the states are weighted by their probabilities, they mix together to form the initial density operator for system B alone. For a precise statement of this freedom, see Ref. [21].

[7]The rigid connection of all nature, on the other hand, is exactly what the Bohmians and Everettistas *do* embrace, even glorify. So, I suspect these words will fall on deaf ears with them. But similarly would they fall on deaf ears with the believer who says that God wills each and every event in the universe and no further explanation is needed. No point of view should be dismissed out of hand: the overriding issue is simply which view will lead to the most progress, which view has the potential to close the debate, which view will give the most new phenomena for the physicist to have fun with?

conceptually separate from the totality of existence) has any meaning at all—one is led to the conclusion that a quantum state cannot be a complete specification of a system. It must be information, at least in part. This point should be placed in contrast to the other well-known facet of Einstein's thought: namely, 2) an unwillingness to accept such an "incompleteness" as a necessary trait of the physical world.

It is quite important to recognize that the first issue does not entail the second. Einstein had that firmly in mind, but he wanted more. His reason for going the further step was, I think, well justified *at the time* [23]:

> There exists [...] a simple psychological reason for the fact that this most nearly obvious interpretation is being shunned. For if the statistical quantum theory does not pretend to describe the individual system (and its development in time) completely, it appears unavoidable to look elsewhere for a complete description of the individual system; in doing so it would be clear from the very beginning that the elements of such a description are not contained within the conceptual scheme of the statistical quantum theory. With this one would admit that, in principle, this scheme could not serve as the basis of theoretical physics.

But the world has seen much in the mean time. The last seventeen years have given confirmation after confirmation that the Bell inequality (and several variations of it) are indeed violated by the physical world. The Kochen-Specker no-go theorems have been meticulously clarified to the point where simple textbook pictures can be drawn of them [24]. Incompleteness, it seems, is here to stay: The theory prescribes that no matter how much we know about a quantum system—even when we have *maximal* information about it—there will always be a statistical residue. There will always be questions that we can ask of a system for which we cannot predict the outcomes. In quantum theory, maximal information is simply not complete information [25]. But neither can it be completed. As Wolfgang Pauli once wrote to Markus Fierz [26], "The well-known 'incompleteness' of quantum mechanics (Einstein) is certainly an existent fact somehow-somewhere, but certainly cannot be removed by reverting to classical field physics." Nor, I would add, will the mystery of that "existent fact" be removed by attempting to give the quantum state anything resembling an ontological status.

The complete disconnectedness of the quantum-state change rule from anything to do with spacetime considerations is telling us something deep: The quantum state is information. Subjective, incomplete information. Put in the right mindset, this is *not* so intolerable. It is a statement about our world. There is something about the world that keeps us from ever getting more information than can be captured through the formal structure of quantum mechanics. Einstein had wanted us to look further—to find out how the incomplete information could be completed—but perhaps the real question is, "Why can it *not* be completed?"

Indeed I think this is one of the deepest questions we can ask and still hope to answer. But first things first. The more immediate question for anyone who has come this far—and one that deserves to be answered forthright—is what is this information symbolized by a $|\psi\rangle$ actually about? I have hinted that I would not dare say that it is about some kind of hidden variable (as the Bohmian might) or even about our place within the universal wavefunction (as the Everettista might).

Perhaps the best way to build up to an answer is to be true to the title of this paper: quantum foundations in the light of quantum information. Let us forage the phenomena of quantum information to see if we might first refine Einstein's argument. One need look no further than to the phenomenon of quantum teleportation [8]. Not only can a quantum-state assignment for a system be forced to go one way or the other by interacting with another part of the world of no causal significance, but, for the cost of two bits, one can make that quantum state assignment anything one wants it to be.

Such an experiment starts out with Alice and Bob sharing a maximally entangled pair of qubits in the state $|\psi^{AB}\rangle = \sqrt{\frac{1}{2}}\,(|0\rangle|0\rangle + |1\rangle|1\rangle)$. Bob then goes to any place in the universe he wishes. Alice in her laboratory prepares another qubit with any state $|\psi\rangle$ that she ultimately wants to impart onto Bob's system. She performs a Bell-basis measurement on the two qubits in her possession. In the same vein as Einstein's thought experiment, Bob's system immediately takes on the character of one of the states $|\psi\rangle$, $\sigma_x|\psi\rangle$, $\sigma_y|\psi\rangle$, or $\sigma_z|\psi\rangle$. But that is only insofar as Alice is concerned.[8] Since there is no (reasonable) causal connection between Alice and Bob, it must be that these states represent the possibilities for Alice's new knowledge of Bob's system.

If now Alice broadcasts the result of her measurement to the world, Bob may complete the teleportation protocol by performing one of the four Pauli rotations (I, σ_x, σ_y, σ_z) on his system, conditioning it on the information he receives. The result, as far as Alice is concerned, is that Bob's system finally resides predictably in the state $|\psi\rangle$.[9]

How can Alice convince herself that such is the case? Well, if Bob is willing to reveal his location, she just need walk to his site and perform the **YES-NO** measurement: $|\psi\rangle\langle\psi|$ vs. $I - |\psi\rangle\langle\psi|$. The outcome will be a **YES** with probability one if all has gone well in carrying out the protocol. Thus, for the cost of a measurement on a causally disconnected system and two bits worth of causal action on the system of actual interest—i.e., one of the four Pauli rotations—Alice can sharpen her predictability to complete certainty for *any* **YES-NO** observable she wishes.

Roger Penrose argues in his book *The Emperor's New Mind* [27] that when a system "has" a state $|\psi\rangle$ there ought to be some property in the system (in and of itself) that corresponds to its "$|\psi\rangle$'ness." For how else could the system be prepared to reveal a **YES** in the case that Alice actually checks it? Asking this rhetorical question with a sufficient amount of command is enough to make many a would-be informationist weak in knees. But there is a crucial oversight implicit in its confidence, and we have already caught it in action. If Alice fails to reveal her information to anyone else in the world, there is no one else who can predict the qubit's ultimate revelation with certainty. More importantly, there is nothing in quantum mechanics that gives the qubit the power to stand up and say **YES** all by itself: If Alice does not take the time to walk over to it and interact with it, there is no revelation. There is only the confidence in Alice's mind that, *should* she interact with it, she *could* predict the consequence of that interaction.

---

[8] As far as Bob is concerned, nothing whatsoever changes about the system in his possession: It started in the completely mixed state $\rho = \frac{1}{2}I$ and remains that way.

[9] As far as Bob is concerned, nothing whatsoever changes about the system in his possession: It started in the completely mixed state $\rho = \frac{1}{2}I$ and remains that way.

4 Information About What?

> [S]urely, the existence of [the] world is the primary experimental fact of all, without which there would be no point to physics or any other science; and for which we all receive new evidence every waking minute of our lives. This direct evidence of our senses is vastly more cogent than are any of the deviously indirect experiments that are cited as evidence for the Copenhagen interpretation.
>
> — *E. T. Jaynes, 1986*

> The criticism of the Copenhagen interpretation of the quantum theory rests quite generally on the anxiety that, with this interpretation, the concept of "objective reality" which forms the basis of classical physics might be driven out of physics. ... [T]his anxiety is groundless ... At this point we realize the simple fact that natural science is not Nature itself but a part of the relation between Man and Nature, and therefore is dependent on Man.
>
> — *Werner Heisenberg, 1955*

There are great rewards in being a new parent. Not least of all is the opportunity to have a close-up look at a mind in formation. I have been watching my two year old daughter learn things at a fantastic rate, and though there have been untold numbers of lessons for her, there have also been a sprinkling for me. For instance, I am just starting to see her come to grips with the idea that there is a world independent of her desires. What strikes me is the contrast between this and the concomitant gain in confidence I see grow in her everyday that there are aspects of existence she actually *can* control. The two go hand in hand. She pushes on the world, and sometimes it gives in a way that she has learned to predict, and sometimes it pushes back in a way she has not foreseen (and may never be able to). If she could manipulate the world to the complete desires of her will, I am quite sure, there would be little difference between wake and dream.

But the main point is that she learns from her forays into the world. In my more cynical moments, I find myself thinking, "How can she think that she's learned anything at all? She has no theory of measurement. She leaves measurement completely undefined. How can she have any true stake to knowledge?"

Hideo Mabuchi once told me, "The quantum measurement problem refers to a set of people." And though that is a bit harsh, maybe it also contains a bit of the truth. With the physics community making use of theories that tend to last between 100 and 300 years, we are apt to forget that scientific views of the world are built from the top down, not from the bottom up. The experiment is the basis of all that we know to be firm. But an experiment is an active intervention into the course of nature on the part of the experimenter; it is not contemplation of nature from afar [28]. We set up this or that experiment to see how nature reacts. It is the conjunction of myriads of such interventions and their consequences that we record into our data books.[10]

[10]But I must stress that I am not so positivistic as to think that physics should somehow be grounded on a primitive notion of "sense impression" as the philosophers of the Vienna Circle did. The interventions and their consequences that an experimenter records, have no option but to be thoroughly theory-laden. It is just that, in a sense, they are by necessity at least one theory behind. No one got closer to the salient point than Heisenberg (in a quote he attributed to Einstein many years after the fact) [29]:

> It is quite wrong to try founding a theory on observable magnitudes alone. In reality the very opposite happens. It is the theory which decides what we can observe. You must appreciate

We tell ourselves that we have learned something new when we can distill from the data a compact description of all that was seen and—even more tellingly—when we can dream up further experiments to corroborate that description. This is the minimal requirement of science. If, however, from such a description we can *further* distill a model of a free-standing "reality" independent of our interventions, then so much the better. I have no bone to pick with reality. It is the most solid thing we can hope for from a theory.[11] Classical physics is the ultimate example in that regard. It gives us a compact description, but it can give much more if we want it to.

The important thing to realize, however, is that there is no logical necessity that such a worldview always be obtainable. If the world is such that we can never identify a reality—a *free-standing* reality—independent of our experimental interventions, then we must be prepared for that too. That is where quantum theory in its most minimal and conceptually simplest dispensation seems to stand [30]. It is a theory whose terms refer predominately to our interface with the world. It is a theory that cannot go the extra step that classical physics did without "writing songs I can't believe, with words that tear and strain to rhyme" [31]. It is a theory not about observables, not about beables, but about "dingables." We tap the bell with our gentle touch and listen for its beautiful ring.

So what are the ways we can intervene on the world? What are the ways we can push it and wait for its unpredictable reaction? The usual textbook story is that those things that are measurable correspond to Hermitian operators. Or perhaps to say it in more modern language, to each observable there corresponds a set of orthogonal projection operators $\{\Pi_i\}$ over a complex Hilbert space $\mathcal{H}_d$ that form a complete resolution of the identity,

$$\sum_i \Pi_i = I \ . \tag{1}$$

The index i labels the potential outcomes of the measurement (or *intervention*, to slip back into the language promoted above). When an observer possesses a state of knowledge ρ—captured most generally by a mixed-state density operator—quantum mechanics dictates that he can expect the various outcomes with a probability

$$P(i) = \mathrm{tr}(\rho\Pi_i) \ . \tag{2}$$

that observation is a very complicated process. The phenomenon under observation produces certain events in our measuring apparatus. As a result, further processes take place in the apparatus, which eventually and by complicated paths produce sense impressions and help us to fix the effects in our consciousness. Along this whole path—from the phenomenon to its fixation in our consciousness—we must be able to tell how nature functions, must know the natural laws at least in practical terms, before we can claim to have observed anything at all. Only theory, that is, knowledge of natural laws, enables us to deduce the underlying phenomena from our sense impressions. When we claim that we can observe something new, we ought really to be saying that, although we are about to formulate new natural laws that do not agree with the old ones, we nevertheless assume that the existing laws—covering the whole path from the phenomenon to our consciousness—function in such a way that we can rely upon them and hence speak of "observation."

[11] Woody Allen said it best: "I hate reality, but, you know, where else can you get a good steak dinner?" Spending my childhood in Texas, where beef is the staple of most meals, the reader should realize that it has been no easy journey for me to come to my present view of quantum mechanics!

The best justification for this probability rule comes by way of Andrew Gleason's amazing 1957 theorem [16]. For, it states that the standard rule is the *only* rule that satisfies a very simple kind of noncontextuality for measurement outcomes [33]. In particular, if one contemplates measuring two distinct observables $\{\Pi_i\}$ and $\{\tilde{\Pi}_i\}$ which happen to share a single projector Π_k, then the probability of outcome k is independent of which observable it is associated with. More formally, the statement is this. Let $\mathcal{P}_d$ be the set of projectors associated with a (real or complex) Hilbert space $\mathcal{H}_d$ for $d \geq 3$, and let $f : \mathcal{P}_d \longrightarrow [0,1]$ be such that

$$\sum_i f(\Pi_i) = 1 \tag{3}$$

whenever a set of projectors $\{\Pi_i\}$ forms an observable. The theorem concludes that there exists a density operator ρ such that

$$f(\Pi) = \mathrm{tr}(\rho\Pi) \ . \tag{4}$$

In fact, in a single blow, Gleason's theorem derives not only the probability rule, but also the state-space structure for quantum mechanical states (i.e., that it corresponds to the convex set of density operators).

In itself this is no small feat, but the thing that makes the theorem an "amazing" theorem is the shear difficulty required to prove it [32]. Note that no restrictions have been placed upon the function f beyond the ones mentioned above. There is no assumption that it need be differentiable, nor that it even need be continuous. All of that, and linearity too, comes from the structure of the observables—i.e., that they are complete sets of orthogonal projectors onto a linear vector space.

Nonetheless, one should ask: Does this theorem really give *the physicist* a clearer vision of where the probability rule comes from? Astounding feats of mathematics are one thing; insight into physics is another. The two are often at opposite ends of the spectrum. As fortunes turn however, a unifying strand can indeed be drawn by viewing quantum foundations in the light of quantum information.

The place to start is to drop the fixation that the basic set of observables in quantum mechanics are complete sets of orthogonal projectors. In quantum information theory it has been found to be extremely convenient to expand the notion of measurement to also include general positive operator-valued measures (POVMs) [24, 34]. In other words, in place of the usual textbook notion of measurement, *any* set $\{E_b\}$ of positive-semidefinite operators on $\mathcal{H}_d$ that forms a resolution of the identity, i.e., that satisfies

$$\langle\psi|E_b|\psi\rangle \geq 0 \,, \quad \text{for all } |\psi\rangle \in \mathcal{H}_d \tag{5}$$

and

$$\sum_b E_b = I \,, \tag{6}$$

counts as a measurement. The outcomes of the measurement are identified with the indices b, and the probabilities of the outcomes are computed according to a generalized Born rule,

$$P(b) = \mathrm{tr}(\rho E_b) \ . \tag{7}$$

The set $\{E_b\}$ is called a POVM, and the operators E_b are called POVM elements. (In the nonstandard language promoted earlier, the set $\{E_b\}$ signifies an intervention into nature, while the individual E_b represent the potential consequences of that intervention.) Unlike standard measurements, there is no limitation on the number of values the index b can take. Moreover, the E_b may be of any rank, and there is no requirement that they be mutually orthogonal.

The way this expansion of the notion of measurement is usually justified is that any POVM can be represented formally as a standard measurement on an ancillary system that has interacted in the past with the system of actual interest. Indeed, suppose the system and ancilla are initially described by the density operators ρ_S and ρ_A respectively. The conjunction of the two systems is then described by the initial quantum state

$$\rho_{SA} = \rho_S \otimes \rho_A \, . \tag{8}$$

An interaction between the systems via some unitary time evolution leads to a new state

$$\rho_{SA} \longrightarrow U\rho_{SA}U^\dagger \, . \tag{9}$$

Now, imagine a standard measurement on the ancilla. It is described on the total Hilbert space via a set of orthogonal projection operators $\{I \otimes \Pi_b\}$. An outcome b will be found, by the standard Born rule, with probability

$$P(b) = \mathrm{tr}\Big(U(\rho_S \otimes \rho_A)U^\dagger(I \otimes \Pi_b)\Big) \, . \tag{10}$$

The number of outcomes in this seemingly indirect notion of measurement is limited only by the dimensionality of the ancilla's Hilbert space—in principle, there can be arbitrarily many.

As advertised, it turns out that the probability formula above can be expressed in terms of operators on the system's Hilbert space alone: This is the origin of the POVM. If we let $|s_\alpha\rangle$ and $|a_c\rangle$ be an orthonormal basis for the system and ancilla respectively, then $|s_\alpha\rangle|a_c\rangle$ will be a basis for the composite system. Using the cyclic property of the trace in Eq. (10), we get

$$\begin{aligned} P(b) &= \sum_{\alpha c} \langle s_\alpha|\langle a_c|\Big((\rho_S \otimes \rho_A)U^\dagger(I \otimes \Pi_b)U\Big)|s_\alpha\rangle|a_c\rangle \\ &= \sum_{\alpha} \langle s_\alpha|\,\rho_S\left(\sum_c \langle a_c|\Big((I \otimes \rho_A)U^\dagger(I \otimes \Pi_b)U\Big)|a_c\rangle\right)|s_\alpha\rangle \, . \end{aligned} \tag{11}$$

Letting tr_A and tr_S denote partial traces over the system and ancilla, respectively, it follows that

$$P(b) = \mathrm{tr}_S(\rho_S E_b) \, , \tag{12}$$

where

$$E_b = \mathrm{tr}_A\Big((I \otimes \rho_A)U(I \otimes \Pi_b)U^\dagger\Big) \tag{13}$$

is an operator acting on the Hilbert space of the original system. This proves half of what is needed, but it is also straightforward to go in the reverse direction—i.e., to show that for any POVM $\{E_b\}$, one can pick an ancilla and find operators ρ_A, U, and Π_b such that Eq. (13) is true.

Putting this all together, there is a sense in which standard measurements capture everything that can be said about quantum measurement theory [34]. As became clear above, a way to think about this is that by learning something about the ancillary system through a standard measurement, one in turn learns something about the system of real interest. Indirect though it may seem, this can be a powerful technique, sometimes revealing information that could not have been revealed otherwise [35]. A very simple example is where a sender has only a single qubit available for the sending one of three potential messages. She therefore has a need to encode the message in one of three preparations of the system, even though the system is a two-state system. To recover as much information as possible, the receiver might (just intuitively) like to perform a measurement with three distinct outcomes. If, however, he were limited to a standard quantum measurement, he would only be able to obtain two outcomes. This—perhaps surprisingly—generally degrades his opportunities for recovery.

What I would like to bring up is whether this standard way of justifying the POVM is the most productive point of view one can take. Might any of the mysteries of quantum mechanics be alleviated by taking the POVM as a basic notion of measurement? Does the POVM's utility portend a larger role for it in the foundations of quantum mechanics?

Standard Measurements	*Generalized Measurements*
$\{\Pi_i\}$	$\{E_b\}$
$\langle\psi\|\Pi_i\|\psi\rangle \geq 0\,,\ \forall\|\psi\rangle$	$\langle\psi\|E_b\|\psi\rangle \geq 0\,,\ \forall\|\psi\rangle$
$\sum_i \Pi_i = I$	$\sum_b E_b = I$
$P(i) = \mathrm{tr}(\rho\Pi_i)$	$P(b) = \mathrm{tr}(\rho E_b)$
$\Pi_i\Pi_j = \delta_{ij}\,\Pi_i$	———

I try to make the point dramatic in my lectures by exhibiting a transparency of the table above. On the left-hand side there is a list of various properties for the standard notion of a quantum measurement. On the right-hand side, there is an almost identical list of properties for the POVMs. The only difference between the two columns is that the right-hand one is *missing* the orthonormality condition required of a standard measurement. The question I ask the audience is this: Does the addition of that one extra assumption really make the process of measurement any less mysterious? Indeed, I imagine myself teaching quantum mechanics for the first time and taking a vote with the best audience of all, the students.[12] "Which set of postulates for quantum measurement would you prefer?" I am quite sure they

[12]I am making the safe bet that they will be lucky enough to not yet be conditioned by years of squabbles in quantum foundations.

would respond with a blank stare. But that is the point! It would make no difference to them, and it should make no difference to us. The only issue worth debating is which notion of measurement will allow us to see more deeply into quantum mechanics.

Therefore let us pose the question that Gleason did, but with POVMs. In other words, let us suppose that the sum total of ways an experimenter can intervene on a quantum system corresponds to the full set of POVMs on its Hilbert space $\mathcal{H}_d$. It is the task of the theory to give him probabilities for the various consequences of his interventions. Concerning those probabilities, let us (in analogy to Gleason) assume only that whatever the probability for a given consequence E_c is, it does not depend upon whether E_c is associated with the POVM $\{E_b\}$ or, instead, any other one $\{\tilde{E}_b\}$. This means we can assume there exists a function

$$f : \mathcal{E}_d \longrightarrow [0,1] \,, \tag{14}$$

where

$$\mathcal{E}_d = \Big\{ E : 0 \leq \langle\psi|E|\psi\rangle \leq 1 \,,\ \forall\,|\psi\rangle \in \mathcal{H}_d \Big\} \,, \tag{15}$$

such that whenever $\{E_b\}$ forms a POVM,

$$\sum_b f(E_b) = 1 \,. \tag{16}$$

(In general, we will call any function satisfying $f(E) \geq 0$ and $\sum_b f(E_b) = \text{const}$ a *frame function*, in analogy to Gleason's nonnegative frame functions.)

It will come as no surprise, of course, that a Gleason-like theorem must hold for the function in Eq. (14). Namely, it can be shown that there must exist a density operator ρ for which

$$f(E) = \mathrm{tr}(\rho E) \,. \tag{17}$$

This was recently shown by Paul Busch [14] and, independently, by Joseph Renes and collaborators [15]. What *is* surprising however is the utter simplicity of the proof. Let us exhibit the whole thing right here and now.

> First, consider the case where $\mathcal{H}_d$ and the operators on it are defined *only* over the field of (complex) rational numbers. It is no problem to see that f is "linear" with respect to positive combinations of operators that never go outside $\mathcal{E}_d$. For consider a three-element POVM $\{E_1, E_2, E_3\}$. By assumption $f(E_1) + f(E_2) + f(E_3) = 1$. However, we can also group the first two elements in this POVM to obtain a new POVM, and must therefore have $f(E_1 + E_2) + f(E_3) = 1$. In other words, the function f must be additive with respect to a fine-graining operation:
>
> $$f(E_1 + E_2) = f(E_1) + f(E_2) \,. \tag{18}$$
>
> Similarly for any two integers m and n,
>
> $$f(E) = m\, f\left(\frac{1}{m}E\right) = n\, f\left(\frac{1}{n}E\right) \tag{19}$$
>
> Suppose $\frac{n}{m} \leq 1$. Then if we write $E = nG$, this statement becomes:
>
> $$f\left(\frac{n}{m}G\right) = \frac{n}{m} f(G) \,. \tag{20}$$

Thus we immediately have a kind of limited linearity on $\mathcal{E}_d$.

One *might* imagine using this property to cap off the theorem in the following way. Clearly the full d^2-dimensional vector space $\mathcal{O}_d$ of Hermitian operators on $\mathcal{H}_d$ is spanned by the set $\mathcal{E}_d$ since that set contains, among other things, all the projection operators. Thus, we can write any operator $E \in \mathcal{E}_d$ as a linear combination

$$E = \sum_{i=1}^{d^2} \alpha_i E_i \tag{21}$$

for some fixed operator-basis $\{E_i\}_{i=1}^{d^2}$. "Linearity" of f would then give

$$f(E) = \sum_{i=1}^{d^2} \alpha_i f(E_i)\,. \tag{22}$$

So, if we define ρ by solving the d^2 linear equations

$$\mathrm{tr}(\rho E_i) = f(E_i)\,, \tag{23}$$

we would have

$$f(E) = \sum_i \alpha_i \mathrm{tr}(\rho E_i) = \mathrm{tr}\left(\rho \sum_i \alpha_i E_i\right) = \mathrm{tr}(\rho E) \tag{24}$$

and essentially be done. (Positivity and normalization of f would require ρ to be an actual density operator.) But the *problem* is that in expansion (21) there is no guarantee that the coefficients α_i can be chosen so that $\alpha_i E_i \in \mathcal{E}_d$.

What remains to be shown is that f can be extended uniquely to a function that is truly linear on $\mathcal{O}_d$. This too is rather simple. First, take any positive semi-definite operator E. We can always find a positive rational number g such that $E = gG$ and $G \in \mathcal{E}_d$. Therefore, we can simply define $f(E) \equiv gf(G)$. To see that this definition is unique, suppose there are two such operators G_1 and G_2 (with corresponding numbers g_1 and g_2) such that $E = g_1 G_1 = g_2 G_2$. Further suppose $g_2 \geq g_1$. Then $G_2 = \frac{g_1}{g_2} G_1$ and, by the homogeneity of the original unextended definition of f, we obtain $g_2 f(G_2) = g_1 f(G_1)$. Furthermore this extension retains the additivity of the original function. For suppose that neither E nor G, though positive semi-definite, are necessarily in $\mathcal{E}_d$. We can find a positive rational number $c \geq 1$ such that $\frac{1}{c}(E+G)$, $\frac{1}{c}E$, and $\frac{1}{c}G$ are all in $\mathcal{E}_d$. Then, by the rules we have already obtained,

$$f(E+G) = cf\left(\frac{1}{c}(E+G)\right) = cf\left(\frac{1}{c}E\right) + cf\left(\frac{1}{c}G\right) = f(E) + f(G). \tag{25}$$

Let us now further extend f's domain to the full space $\mathcal{O}_d$. This can be done by noting that any operator H can be written as the difference $H = E - G$ of two positive semi-definite operators. Therefore define $f(H) \equiv f(E) - f(G)$, from which it also follows that $f(-G) = -f(G)$. To see that this definition is unique suppose there are four operators E_1, E_2, G_1, and G_2, such that $H = E_1 - G_1 = E_2 - G_2$. It follows that $E_1 + G_2 = E_2 + G_1$. Applying f (as extended in the previous paragraph) to this equation, we obtain $f(E_1) + f(G_2) = f(E_2) + f(G_1)$ so that $f(E_1) - f(G_1) = f(E_2) - f(G_2)$. Finally, with this new extension, full linearity can be checked immediately. This completes the proof as far as the (complex) rational number field is concerned: Because f extends uniquely to a linear functional on $\mathcal{O}_d$, we can indeed go through the steps of Eqs. (21) through (24) without worry.

There are two things that are significant about this much of the proof. First, in contrast to Gleason's original theorem, there is nothing to bar the same logic from working when $d = 2$. This is quite nice because much of the community has gotten in the habit of thinking that there is nothing particularly "quantum mechanical" about a single qubit. Indeed, because orthogonal projectors on $\mathcal{H}_2$ can be mapped onto antipodes of the Bloch sphere, it is known that the measurement-outcome statistics for any standard measurement can be mocked-up through a noncontextual hidden-variable theory. What this result shows, is that that simply is not the case when one considers the full set of POVMs as one's potential measurements.

The other important thing is that the theorem works for Hilbert spaces over the rational number field: one does not need to invoke the full power of the continuum. This contrasts with the surprising result of Meyer [36] that the standard Gleason theorem fails in such a setting. The present theorem hints at a kind of resiliency to the structure of quantum mechanics that falls through the mesh of the standard Gleason result: The probability rule for POVMs does not actually depend so much upon the detailed workings of the number field.

The final step of the proof, indeed, is to show that nothing goes awry when we go the extra step of reinstating the continuum.

> In other words, we need to show that the function f (now defined on the set $\mathcal{E}_d$ complex operators) is a continuous function. This comes about in simple way from f's additivity. Suppose for two positive semi-definite operators E and G that $E \leq G$ (i.e., $G - E$ is positive semi-definite). Then trivially there exists a positive semi-definite operator H such that $E + H = G$ and through which the additivity of f gives $f(E) \leq f(G)$. Let c be an irrational number, and let a_n be an increasing sequence and b_n a decreasing sequence of rational numbers that both converge to c. It follows for any positive semi-definite operator E, that
>
> $$f(a_n E) \leq f(cE) \leq f(b_n E)\ , \tag{26}$$
>
> which implies
>
> $$a_n f(E) \leq f(cE) \leq b_n f(E)\ . \tag{27}$$
>
> Since $\lim a_n f(E)$ and $\lim b_n f(E)$ are identical, by the "pinching theorem" of elementary calculus, they must equal $f(cE)$. This establishes that we can consistently define
>
> $$f(cE) = cf(E)\ . \tag{28}$$
>
> Reworking the extensions of f in the last inset (but with this enlarged notion of homogeneity), one completes the proof in a straightforward manner.

Of course we are not getting something from nothing. The reason the present derivation is so easy in contrast to the standard proof is that *mathematically* the assumption of POVMs as the basic notion of measurement is significantly stronger than the usual assumption. *Physically*, though, I would say it is just the opposite. Why add extra restrictions to the notion of measurement when they only make the route from basic assumption to practical usage more circuitous than it need be?

Still, no assumption should be left unanalyzed if it stands a chance of bearing fruit. Indeed, one can ask what is so very compelling about the noncontextuality property (of probability assignments) that both Gleason's original theorem and the present version make use of. Given the picture of measurement as a kind of invasive intervention into the world, one might expect the very opposite. One is left wondering why measurement probabilities do not depend upon the whole context of the measurement interaction. Why is $p(b)$ not of the form $f(b, \{E_c\})$? Is there any precedent for the usual assumption?

5 Whither Bayes' Rule?

And so you see I have come to doubt
All that I once held as true
I stand alone without beliefs
The only truth I know is you.
— *Paul Simon*, Kathy's Song

Quantum states are states of knowledge, not states of nature. That statement is the cornerstone of this paper. Thus, in searching to make sense of the remainder of quantum mechanics, one strategy ought to be to seek guidance [37] from the most developed avenue of "knowledge theory" to date—Bayesian probability theory [38, 39, 40]. Indeed, the very aim of Bayesian theory is to develop reliable methods of reasoning and making decisions in the light of incomplete knowledge. To what extent does that structure mesh with the seemingly independent structure of quantum mechanics? To what extent are there analogies; to what extent distinctions?

This section is about turning a distinction into an analogy. The core of the matter is the manner in which states of knowledge are updated in the two theories. At first sight, they appear to be quite different in character. To see this, let us first explore how quantum mechanical states change when information is gathered.

In older accounts of quantum mechanics, one often encounters the "collapse postulate" as a basic statement of the theory. One hears things like, "Axiom 5: Upon the completion of a measurement of a Hermitian observable H, the system is left in an eigenstate of H." In quantum information, however, it has become clear that it is useful to broaden the notion of measurement, and with it, the analysis of how a state can change in the process. The foremost reason for this is that the collapse postulate is simply not true in general: Depending upon the exact nature of the measurement interaction, there may be any of a large set of possibilities for the final state of a system.

The broadest (consistent) notion of state change arose in the theory of "effects and operations" [34]. The statement is this. Suppose one's initial state for a quantum system is a density operator ρ, and a POVM $\{E_b\}$ is measured on that system. Then, according to this formalism, the state after the measurement can be *any* state ρ_b of the form

$$\rho_b = \frac{1}{\mathrm{tr}(\rho E_b)} \sum_i A_{bi} \rho A_{bi}^\dagger \,, \tag{29}$$

where

$$\sum_i A_{bi}^\dagger A_{bi} = E_b \ . \tag{30}$$

Note the immense generality of this formula. There is no constraint on the number of indices i in the A_{bi} and these operators need not even be Hermitian.

The usual justification for this kind of generality—just as in the case of the commonplace justification for the POVM formalism—comes about by imagining that the measurement arises in an indirect fashion rather than as a direct attack. In other words, the primary system is pictured to interact with an ancilla first, and only then subjected to a "real" measurement on the ancilla alone. The trick is that one posits a kind of projection postulate on the primary system due to this process. This assumption has a much safer feel than the raw projection postulate since, after the interaction, no measurement on the ancilla should cause a physical perturbation to the primary system.

More formally, we can start out by following Eqs. (8) and (9), but in place of Eq. (10) we must make an assumption on how the system's state changes. For this one invokes a kind of "projection-postulate-at-a-distance."[13] Namely, one takes

$$\rho_b = \frac{1}{P(b)} \, \mathrm{tr}_{\mathrm{A}}\Big((I \otimes \Pi_b) U (\rho_{\mathrm{S}} \otimes \rho_{\mathrm{A}}) U^\dagger (I \otimes \Pi_b)\Big) \ . \tag{31}$$

The reason for invoking the partial trace is to make sure any hint of a state change for the ancilla remains unaddressed.

To see how expression (31) makes connection to Eq. (29), denote the eigenvalues and eigenvectors of ρ_{A} by λ_α and $|a_\alpha\rangle$ respectively. Then $\rho_{\mathrm{S}} \otimes \rho_{\mathrm{A}}$ can be written as

$$\rho_{\mathrm{S}} \otimes \rho_{\mathrm{A}} = \sum_\alpha \sqrt{\lambda_\alpha}\, |a_\alpha\rangle \rho_{\mathrm{S}} \langle a_\alpha| \sqrt{\lambda_\alpha} \ , \tag{32}$$

and, expanding Eq. (31), we have

$$\rho_b = \frac{1}{P(b)} \sum_\beta \langle a_\beta | (I \otimes \Pi_b) U^\dagger (\rho_{\mathrm{S}} \otimes \rho_{\mathrm{A}}) U (I \otimes \Pi_b) | a_\beta \rangle \tag{33}$$

$$= \frac{1}{P(b)} \sum_{\alpha\beta} \left(\sqrt{\lambda_\alpha}\, \langle a_\beta | (I \otimes \Pi_b) U^\dagger | a_\alpha \rangle \right) \rho_{\mathrm{S}} \left(\langle a_\alpha | U (I \otimes \Pi_b) | a_\beta \rangle \sqrt{\lambda_\alpha} \right) \ . \tag{34}$$

A representation of the form in Eq. (29) can be made by taking

$$A_{b\alpha\beta} = \sqrt{\lambda_\alpha}\, \langle a_\alpha | U (I \otimes \Pi_b) | a_\beta \rangle \tag{35}$$

and lumping the two indices α and β into the single index i. Indeed, one can easily check that Eq. (30) holds.[14] This completes what we had set out to show. However, just as with the case of the POVM $\{E_b\}$, one can always find a way to reverse engineer the derivation:

[13] David Mermin has also recently emphasized this point in Ref. [41].

[14] As an aside, it should be clear from the construction in Eq. (35) that there are many equally good representations of ρ_b. For a precise statement of the latitude of this freedom, see Ref. [42].

Given a set of A_{bi}, one can always find a U, a ρ_A, and set of Π_b such that Eq. (31) becomes true.

Of course the old collapse postulate is contained within the extended formalism as a special case: There, one just takes both sets $\{E_b\}$ and $\{A_{bi} = E_b\}$ to be sets of orthogonal projectors. Let us take a moment to think about this special case in isolation. What is distinctive about it is that it captures in the extreme a common folklore associated with the measurement process. For it tends to convey the image that measurement is a kind of gut wrenching violence: In one moment the state is a $\rho = |\psi\rangle\langle\psi|$, while in the very next it is a $\Pi_i = |i\rangle\langle i|$. Moreover, such a wild transition need depend upon no details of $|\psi\rangle$ and $|i\rangle$; in particular the two states may even be almost orthogonal to each other. In density-operator language, there is no sense in which Π_i is contained in ρ: the two states are simply in distinct places of the operator space.

Contrast this with the description of information gathering that arises in Bayesian probability theory. There, an initial state of knowledge is captured by a probability distribution $p(h)$ for some hypothesis H. The way gathering a piece of data d is taken into account in assigning one's new state of knowledge is through Bayes' conditionalization rule. That is to say, one expands $p(h)$ in terms of the relevant joint probability distribution and picks off the appropriate term:

$$p(h) = \sum_d p(h,d) \tag{36}$$

$$= \sum_d p(d)p(h|d) \tag{37}$$

$$\downarrow$$

$$p(h) \quad \xrightarrow{d} \quad p(h|d)\ , \tag{38}$$

where $p(h|d)$ satisfies the tautology

$$p(h|d) = \frac{p(h,d)}{p(d)}\ . \tag{39}$$

How gentle this looks in comparison to quantum collapse! When one gathers new information, one simply refines one's old knowledge in the most literal of senses. It is not as if the new state is incommensurable with the old. *It was always there*; it was just initially averaged in with various other potential states of knowledge.

Why does quantum collapse not look more like Bayes' rule? Is quantum collapse really a more violent kind of change, or might it be an artifact of a problematic representation? By this stage, it should come as no surprise to the reader that dropping the ancilla from our image of generalized measurements will be the first step to progress. Taking the transition from ρ to ρ_b in Eqs. (29) and (30) as the basic statement of what quantum measurement *is* is a good starting point.

To accentuate a similarity between Eq. (29) and Bayes' rule, let us first contemplate cases of it where the index i takes on a single value. Then, we can conveniently drop that index and write

$$\rho_b = \frac{1}{P(b)} A_b \rho A_b^\dagger\ , \tag{40}$$

where

$$E_b = A_b^\dagger A_b \,. \tag{41}$$

In a loose way, one can say that measurements of this sort are the most efficient they can be for a given POVM $\{E_b\}$: For, a measurement interaction with an explicit i-dependence may be viewed as "more truly" a measurement of a finer-grained POVM that just happens to throw away some of the information it gained. Let us make this point more precise.

Notice that Bayes' rule has the property that one's uncertainty about a hypothesis can be expected to decrease upon the acquisition of data. This can be made rigorous, for instance, by gauging uncertainty with the Shannon entropy function [43],

$$S(H) = -\sum_h p(h) \log p(h) \,. \tag{42}$$

This number is bounded between 0 and the logarithm of the number of hypotheses in H, and there are several reasons to think of it as a good measure of uncertainty. Perhaps the most important of these is that it quantifies the number of **YES-NO** questions one can expect to ask per instance of H, if one's only means to ascertain the outcome is from a colleague who knows the actual result [44]. Under this quantification, the lower the Shannon entropy, the more predictable a measurement's outcomes.

Because the function $f(x) = -x \log x$ is concave on the interval $[0,1]$, it follows that,

$$\begin{aligned} S(H) &= -\sum_h \left(\sum_d p(d)p(h|d)\right) \log \left(\sum_d p(d)p(h|d)\right) && (43)\\ &\geq -\sum_d p(d) \sum_h p(h|d) \log p(h|d) \,. && (44)\\ &= \sum_d p(d) S(H|d) && (45)\\ &\equiv S(H|D) \,. && (46) \end{aligned}$$

Indeed we hope to find a similar statement for how the result of efficient quantum measurements decrease uncertainty or impredictability. But, what can be meant by a decrease of uncertainty through quantum measurement? I have argued strenuously that the information gain in a measurement cannot be information about a preexisting reality. The way out of the impasse is simple: The uncertainty that decreases in quantum measurement is the uncertainty one expects for the results of any potential future measurements.

There are at least two ways of quantifying this that are worthy of note. The first has to do with the von Neumann entropy of a density operator ρ:

$$S(\rho) = -\mathrm{tr}\, \rho \log \rho = -\sum_{k=1}^{d} \lambda_k \log \lambda_k \,, \tag{47}$$

where the λ_k signify the eigenvalues of ρ. (We use the convention that $\lambda \log \lambda = 0$ whenever $\lambda = 0$ so that $S(\rho)$ is always well defined.)

The intuitive meaning of the von Neumann entropy can be found by first thinking about the Shannon entropy. Consider any standard measurement $\mathcal{P}$ consisting of d one-dimensional orthogonal projectors Π_i. The Shannon entropy for the outcomes of this measurement is given by

$$H(\mathcal{P}) = -\sum_{i=1}^{d} (\mathrm{tr}\rho\Pi_i) \log (\mathrm{tr}\rho\Pi_i) \ . \tag{48}$$

A natural question to ask is: With respect to a given density operator ρ, which measurement $\mathcal{P}$ will give the most predictability over its outcomes? As it turns out, the answer is any $\mathcal{P}$ that forms a set of eigenprojectors for ρ [45]. When this obtains, the Shannon entropy of the measurement outcomes reduces to simply the von Neumann entropy of the density operator. The von Neumann entropy, then, signifies the amount of impredictability one achieves by way of a standard measurement in a best case scenario. Indeed, true to one's intuition, one has the most knowledge by this account when ρ is a pure state—for then $S(\rho) = 0$. Alternatively, one has the least knowledge when ρ is proportional to the identity operator—for then any measurement $\mathcal{P}$ will have outcomes that are all equally likely.

The best case scenario for predictability, however, is a limited case, and not very indicative of the density operator as a whole. Since the density operator contains, in principle, all that can be said about every possible measurement, it seems a shame to throw away the vast part of that information in our considerations.

This leads to a second method for quantifying uncertainty in the quantum setting. For this, we again rely on the Shannon information as our basic notion of impredictability. The difference is we evaluate it with respect to a "typical" measurement rather than the best possible one. But typical with respect to what? The notion of typical is only defined with respect to a given *measure* on the the set of measurements.

Regardless, there is a fairly canonical answer. There is a unique measure $d\Omega_\Pi$ on the space of one-dimensional projectors that is invariant with respect to all unitary operations. That in turn induces a canonical measure $d\Omega_\mathcal{P}$ on the space of von Neumann measurements $\mathcal{P}$ [46]. Using this measure leads to the following quantity

$$\overline{S}(\rho) = \int H(\Pi)\, d\Omega_\mathcal{P} \tag{49}$$

$$= -d \int (\mathrm{tr}\rho\Pi) \log (\mathrm{tr}\rho\Pi)\, d\Omega_\Pi \ , \tag{50}$$

which is intimately connected to the so-called quantum "subentropy" of Ref. [47]. This mean entropy can be evaluated explicitly in terms of the eigenvalues of ρ and takes on the expression

$$\overline{S}(\rho) = \frac{1}{\ln 2}\left(\frac{1}{2} + \frac{1}{3} + \cdots + \frac{1}{d}\right) + Q(\rho) \tag{51}$$

where the subentropy $Q(\rho)$ is defined by

$$Q(\rho) = -\sum_{k=1}^{d} \left(\prod_{i\neq k} \frac{\lambda_k}{\lambda_k - \lambda_i}\right) \lambda_k \log \lambda_k \ . \tag{52}$$

In the case where ρ has degenerate eigenvalues, $\lambda_l = \lambda_m$ for $l \neq m$, one need only reset them to $\lambda_l + \epsilon$ and $\lambda_m - \epsilon$ and consider the limit as $\epsilon \to 0$. The limit is convergent and hence

$Q(\rho)$ is finite for all ρ. With this, one can also see that for a pure state ρ, $Q(\rho)$ vanishes. Furthermore, since $\overline{S}(\rho)$ is bounded above by $\log d$, we know that

$$0 \leq Q(\rho) \leq \log d - \frac{1}{\ln 2}\left(\frac{1}{2} + \cdots + \frac{1}{d}\right) \leq \frac{1-\gamma}{\ln 2}\,, \tag{53}$$

where γ is Euler's constant. This means that for any ρ, $Q(\rho)$ never exceeds approximately 0.60995 bits.

The interpretation of this result is the following. Even when one has maximal information about a quantum system—i.e., one has a pure state for it—one can predict almost nothing about the outcome of a typical measurement [25]. In the limit of large d, the outcome entropy for a typical measurement is just a little over a half bit away from its maximal value. Having a mixed state for a system, reduces one's predictability even further, but indeed not by that much: The small deviation is captured by the function in Eq. (52), which becomes a quantification of uncertainty in its own right.

The way to get at a quantum statement of Eq. (46) is to make use of the fact that $S(\rho)$ and $Q(\rho)$ are both concave in the variable ρ [48]. That is, for either function, we have

$$F(t\rho_0 + (1-t)\rho_1) \geq tF(\rho_0) + (1-t)F(\rho_1)\,, \tag{54}$$

for any density operators ρ_0 and ρ_1 and any real number $t \in [0,1]$. However, the result does not arise in the trivial fashion it did for classical case of Eq. (46). This is because generally, as already emphasized,

$$\rho \neq \sum_b P(b)\rho_b \tag{55}$$

for ρ_b defined as in Eq. (40). One must be slightly more roundabout.

The key is in noticing that

$$\begin{aligned} \rho &= \rho^{1/2} I \rho^{1/2} && (56)\\ &= \sum_b \rho^{1/2} E_b \rho^{1/2} && (57)\\ &= \sum_b P(b)\tilde{\rho}_b && (58) \end{aligned}$$

where

$$\tilde{\rho}_b = \frac{1}{P(b)}\rho^{1/2}E_b\rho^{1/2} = \frac{1}{P(b)}\rho^{1/2}A_b^\dagger A_b\rho^{1/2}\,. \tag{59}$$

What is special about this decomposition of ρ is that for each b, ρ_b and $\tilde{\rho}_b$ have the same eigenvalues. This follows since $X^\dagger X$ and $XX^\dagger$ have the same eigenvalues, for any operator X. In the present case, setting $X = A_b\rho^{1/2}$ does the trick. Using the fact that both $S(\rho)$ and $Q(\rho)$ depend only upon the eigenvalues of ρ we obtain:

$$\begin{aligned} S(\rho) &\geq \sum_b P(b)S(\rho_b) && (60)\\ Q(\rho) &\geq \sum_b P(b)Q(\rho_b)\,, && (61) \end{aligned}$$

as we had been hoping for. Thus, in performing an efficient quantum measurement of a POVM $\{E_b\}$, an observer can expect to be left with less uncertainty than he started with.[15]

In this sense, quantum "collapse" does indeed have some of the flavor of Bayes' rule. But we can expect more, and the derivation above hints at just the right ingredient: ρ_b and $\tilde{\rho}_b$ have the same eigenvalues! To see the impact of this, let us once again explore the content of Eqs. (40) and (41). A common way to describe their meaning is to use the operator polar-decomposition theorem [50] to rewrite Eq. (40) in the form

$$\rho_b = \frac{1}{P(b)} U_b E_b^{1/2} \rho E_b^{1/2} U_b^\dagger \,, \tag{62}$$

where U_b is a unitary operator. Since—subject only to the constraint of efficiency—the operators A_b are not determined any further than Eq. (41), U_b can be *any* unitary operator whatsoever. Thus, a customary way of thinking of the state-change process is to break it up into two conceptual pieces. First there is a "raw collapse":

$$\rho \longrightarrow \sigma_b = \frac{1}{P(b)} E_b^{1/2} \rho E_b^{1/2} \,. \tag{63}$$

Then, subject to the details of the measurement interaction and the particular outcome b, one imagines the measuring device enforcing a further kind of "feedback" on the measured system:

$$\sigma_b \longrightarrow \rho_b = U_b \sigma_b U_b^\dagger \,. \tag{64}$$

But this break down of transition is a purely conceptual game.

Since the U_b are arbitrary to begin with, we might as well break down the state-change process into the following (nonstandard) conceptual components. First one imagines an observer refining his initial state of knowledge and simply plucking out a term corresponding to the "data" collected:

$$\rho \;=\; \sum_b P(b)\tilde{\rho}_b \tag{65}$$

$$\downarrow$$

$$\rho \quad \stackrel{b}{\longrightarrow} \quad \tilde{\rho}_b \,. \tag{66}$$

Finally, there may be a further "mental readjustment" of the observer's knowledge, taking into account details both of the measurement interaction and the observer's initial state of knowledge. This is enacted via some (formal) unitary operation V_b:

$$\tilde{\rho}_b \longrightarrow \rho_b = V_b \tilde{\rho}_b V_b^\dagger \,. \tag{67}$$

Putting the two processes together, one has the same result as the usual picture.

The resemblance between the process in Eq. (66) and the classical Bayes' rule of Eq. (38) is unmistakable.[16] By this way of viewing things, quantum collapse is indeed not such a

[15]By differing methods, a strengthening of this result in terms of a majorization property can be found in Refs. [48] and [49].

[16]Earlier allusions to a resemblance between quantum collapse and Bayes' rule can be found in Ref. [51].

violent state of affairs after all. Quantum measurement is nothing more, and nothing less, than a refinement and a readjustment of one's state of knowledge. More general state changes of the form Eq. (29) come about similarly, but with a further step of coarse-graining (i.e., throwing away information that was in principle accessible).

Let us look at two limiting cases of efficient measurements. In the first, we imagine an observer whose initial state of knowledge $\rho = |\psi\rangle\langle\psi|$ is a maximal state of knowledge. By this account, no measurement whatsoever can *refine* that state of knowledge. This follows because, no matter what $\{E_b\}$ is,

$$\rho^{1/2} E_b \rho^{1/2} = P(b)|\psi\rangle\langle\psi| \,. \tag{68}$$

The only state change that can come about from such a measurement must be purely of the "mental readjustment" sort: We learn nothing new; we just change what we can predict as a consequence of experimental intervention. In particular, when the POVM is an orthogonal set of projectors $\{\Pi_i = |i\rangle\langle i|\}$ and the state-change mechanism is the von Neumann collapse postulate, this simply corresponds to a readjustment according to the unitary operators

$$U_i = |i\rangle\langle\psi| \,. \tag{69}$$

At the opposite end of things, we can contemplate measurements that have no causal connection at all to the system being measured. This could come about, for instance, by interacting with one side of an entangled pair of systems and using the consequence of that intervention to update one's knowledge about the other side. In such a case, one can show that the state change is purely of the refinement variety (with no further mental readjustment). For instance, consider a pure state $|\psi^{AB}\rangle$ whose Schmidt decomposition takes the form

$$|\psi^{AB}\rangle = \sum_i \sqrt{\lambda_i}|a_i\rangle|b_i\rangle \,. \tag{70}$$

An efficient measurement on the A side of this leads to a state update of the form

$$|\psi^{AB}\rangle\langle\psi^{AB}| \longrightarrow (A_b \otimes I)|\psi^{AB}\rangle\langle\psi^{AB}|(A_b^\dagger \otimes I) \,. \tag{71}$$

Tracing out the A side, then gives

$$\begin{aligned}
\mathrm{tr}_A\Big(A_b \otimes I|\psi^{AB}\rangle\langle\psi^{AB}|A_b^\dagger \otimes I\Big) &= \sum_{ijk} \sqrt{\lambda_j}\sqrt{\lambda_k}\langle a_i|A_b \otimes I|a_j\rangle|b_j\rangle\langle a_k|\langle b_k|A_b^\dagger \otimes I|a_i\rangle && (72)\\
&= \sum_{ijk} \sqrt{\lambda_j}\sqrt{\lambda_k}\langle a_k|A_b^\dagger|a_i\rangle\langle a_i|A_b|a_j\rangle|b_j\rangle\langle b_k| && (73)\\
&= \sum_{jk} \sqrt{\lambda_j}\sqrt{\lambda_k}\langle a_k|A_b^\dagger A_b|a_j\rangle|b_j\rangle\langle b_k| && (74)\\
&= \sum_{jk} \sqrt{\lambda_j}\sqrt{\lambda_k}\langle b_k|UA_b^\dagger A_b U^\dagger|b_j\rangle|b_j\rangle\langle b_k| && (75)\\
&= \sum_{jk} \sqrt{\lambda_j}\sqrt{\lambda_k}\langle b_j|\Big(UA_b^\dagger A_b U^\dagger\Big)^{\mathrm{T}}|b_k\rangle|b_j\rangle\langle b_k| && (76)\\
&= \rho^{1/2}\Big(UA_b^\dagger A_b U^\dagger\Big)^{\mathrm{T}}\rho^{1/2} && (77)
\end{aligned}$$

where ρ is the initial quantum state on the B side, U is the unitary operator connecting the $|a_i\rangle$ basis to the $|b_i\rangle$ basis, and ^{T} represents taking a transpose with respect to the $|b_i\rangle$ basis. Since the operators

$$F_b = \left(UA_b^\dagger A_b U^\dagger\right)^{\mathrm{T}} \tag{78}$$

go together to form a POVM, we indeed have the claimed result.

In summary, the lesson here is that it turns out to be rather easy to think of quantum collapse as a noncommutative variant of Bayes' rule. In fact it is just in this that one starts to get a feel for (at least) a partial reason for Gleason's noncontextuality assumption. In the setting of classical Bayes' conditionalization we have just that: The probability of the transition $p(h) \longrightarrow p(h|d)$ is governed solely by the local probability $p(d)$. The transition does not care about how we have partitioned the rest of the potential transitions. That is, it does not care whether d is embedded in a two outcome set $\{d, \neg d\}$ or whether it is embedded in a three outcome set, $\{d, e, \neg(d \vee e)\}$. Similarly with the quantum case. The probability for a transition from ρ to ρ_0 cares not whether our refinement is of the form

$$\rho = P(0)\rho_0 + \sum_{b=1}^{17} P(b)\rho_b \qquad \text{or of the form} \qquad \rho = P(0)\rho_0 + P(18)\rho_{18}\,, \tag{79}$$

as long as

$$P(18)\rho_{18} = \sum_{b=1}^{17} P(b)\rho_b \tag{80}$$

What could have been a simpler generalization of Bayes' rule?

Indeed, leaning on that, we can restate the discussion of the "measurement problem" at the beginning of Section 4 in slightly more technical terms. Go back to the classical setting of Eqs. (36) through (39) where an agent has a probability distribution $p(h,d)$ over two sets of hypotheses. Marginalizing over the possibilities for d, one obtains the agent's initial state of knowledge $p(h)$ for the hypothesis h. If he gathers an explicit piece of data d, he should use Bayes' rule to update his knowledge about h to $p(h|d)$.

The question is this: Is the transition

$$p(h) \longrightarrow p(h|d) \tag{81}$$

a mystery we should contend with? If someone asked for a *physical description* of that transition, would we be able to give an explanation? After all, one value for h is true and always remains true: there is no transition in it. One value for d is true and always remains true: there is no transition in it. The only transition is in the *knowledge* $p(h)$. To put the issue into perspective for the quantum measurement problem, let us ask: Should we not have a detailed theory of how the brain works before we can trust in the validity of Bayes' rule?

The answer is, "Of course not!" Bayes' rule, and with it all of probability theory, is an intellectual construct that stands beyond the details of physics. George Boole called probability theory a *law of thought* [52]. Its calculus specifies the optimal way an agent should reason and make decisions when faced with incomplete information. In this way, probability theory is but a generalization of Aristotelian logic[17]—a construct very few would

[17] In addition to Ref. [39], many further materials concerning this point of view can be downloaded from the *Probability Theory As Extended Logic* web site maintained by G. L. Bretthorst, `http://bayes.wustl.edu/`.

accept as being tied to the explicit details of the physical world.[18]

The formal similarities between Bayes' rule and quantum collapse may be telling us how to finally cut the Gordian knot of the measurement problem. Namely, it may be telling us that it is simply not a problem at all! Indeed, drawing on the analogies between the two theories, one is left with a spark of insight: perhaps the better part of quantum mechanics is simply "law of thought" [53]. Perhaps the structure of the theory denotes the optimal way to reason and make decisions in light of *some* fundamental situation, waiting to be ferreted out in a more satisfactory fashion.

This much we know: That "fundamental situation"—whatever it is—must be an ingredient Bayesian probability theory does not have. There must be something to drive a wedge between the two theories. Probability theory alone is too general of a structure. Narrowing it will require input from the world about us.

6 Whither Entanglement?

> When two systems, of which we know the states by their respective representatives, enter into temporary physical interaction due to known forces between them, and when after a time of mutual influence the systems separate again, then they can no longer be described in the same way as before, viz. by endowing each of them with a representative of its own. I would not call that *one* but rather *the* characteristic trait of quantum mechanics, the one that enforces its entire departure from classical lines of thought. By the interaction the two representatives (or ψ-functions) have become entangled.
>
> — *Erwin Schrödinger*, 1935

Quantum entanglement certainly gets a load of airplay these days. By most accounts it is the main ingredient in quantum information theory and quantum computing [54], and it is the main mystery of the quantum foundations [55]. But what is it? Where does it come from?

The predominant purpose it has served in this paper has been as a kind of background. For it, more than any other ingredient in quantum mechanics, has clinched the issue of "information about what?" in the author's mind: That information cannot be about a preexisting reality (a hidden variable) unless we are willing to renege on our very reason for rejecting the quantum state's objective reality in the first place. What I am alluding to here is the conjunction of the Einstein argument reported in Section 3 and the phenomena of the Bell inequality violations by quantum mechanics. Putting those points together gave us that the information symbolized by a $|\psi\rangle$ must be information about the potential consequences of our interventions into the world.

[18] We have, after all, used simple Aristotelian logic in making deductions from *all* our physical theories to date: from Aristotle's physics to quantum mechanics to general relativity and even string theory.

But, now I would like to turn the tables and ask whether the structure of our potential interventions—the POVMs—can tell us something about the origin of entanglement. Could it be that the concept of entanglement is just a minor addition to the much deeper point that measurements correspond to refinements of density operators (i.e., the substance of the two preceding sections)?

The technical translation of this question is, why do we combine systems according to the tensor product rule? There are certainly innumerable ways to combine two Hilbert spaces $\mathcal{H}_A$ and $\mathcal{H}_B$ to obtain a third $\mathcal{H}_{AB}$. We could take the tensor sum of the two spaces $\mathcal{H}_{AB} = \mathcal{H}_A \oplus \mathcal{H}_B$. We could take their Grassmann product $\mathcal{H}_{AB} = \mathcal{H}_A \wedge \mathcal{H}_B$ [56]. We could take scads of other things. But instead we take their tensor product,

$$\mathcal{H}_{AB} = \mathcal{H}_A \otimes \mathcal{H}_B \; . \tag{82}$$

Why?

Could it arise from the selfsame considerations we have already made our mainstay—from a noncontextuality property for measurement-outcome probabilities? The answer is yes, and the theorem I am about demonstrate owes much in inspiration to Ref. [57].

Here is the scenario. Suppose we have two quantum systems, and we can make a measurement on each.[19] On the first, we can measure any POVM on the $d_{\rm A}$-dimensional Hilbert space $\mathcal{H}_A$; on the second, we can measure any POVM on the $d_{\rm B}$-dimensional Hilbert space $\mathcal{H}_B$. Moreover, suppose we may condition the the second measurement on the nature and the outcome of the first, and vice versa. That is to say—walking from A to B—we could first measure $\{E_i\}$ on A, and then, depending on the outcome i, measure $\{F_j^i\}$ on B. Similarly—walking from B to A—we could first measure $\{F_j\}$ on B, and then, depending on the outcome j, measure $\{E_i^j\}$ on A. So that we have valid POVMs, we must have

$$\sum_i E_i = I \qquad \text{and} \qquad \sum_j F_j^i = I \quad \forall i \; , \tag{83}$$

and

$$\sum_i E_i^j = I \quad \forall j \qquad \text{and} \qquad \sum_j F_j = I \; , \tag{84}$$

for these sets of operators. Let us denote by S_{ij} an *ordered pair* of operators, either of the form (E_i, F_j^i) or of the form (E_i^j, F_j), as appearing above. Let us call a set of such operators $\{S_{ij}\}$ a *locally-measurable POVM tree.*

Suppose now that—just as with the POVM-version of Gleason's theorem in Section 4—the joint probability $P(i,j)$ for the outcomes of such a measurement should not depend upon which tree S_{ij} is embedded in: This is essentially the same assumption we made there, but now applied to local measurements on the separate systems. In other words, let us suppose that there exists a function

$$f : \mathcal{E}_{d_{\rm A}} \times \mathcal{E}_{d_{\rm B}} \longrightarrow [0,1] \tag{85}$$

such that

$$\sum_{ij} f(S_{ij}) = 1 \tag{86}$$

[19]This, one might think, is the very essence of having *two* systems rather than one—that we can probe them independently.

whenever the S_{ij} satisfy either Eq. (83) or Eq. (84).

Note in particular that Eq. (85) makes no use of the tensor product: The domain of f is the *Cartesian product* of the two sets $\mathcal{E}_{d_A}$ and $\mathcal{E}_{d_B}$. The notion of a *local* measurement on the separate systems is enforced by the requirement that the ordered pairs S_{ij} satisfy the side conditions of Eqs. (83) and (84). This, of course, is not the most general kind of local measurement one can imagine—more sophisticated measurements could involve multiple ping-pongings between A and B as in Ref. [58]—but the present restricted class is already sufficient for fixing the probability rule for local measurements.

The *theorem* is this: If f satisfies Eqs. (85) and (86) for all locally-measurable POVM trees, then there exists a density operator $\tilde{\rho}$ on $\mathcal{H}_A \otimes \mathcal{H}_B$ such that

$$f(E,F) = \mathrm{tr}\Big(\tilde{\rho}(E \otimes F)\Big) . \tag{87}$$

If $\mathcal{H}_A$ and $\mathcal{H}_B$ are defined over the field of complex numbers, then $\tilde{\rho}$ is unique. Uniqueness does not hold, however, if the underlying field is the real numbers.

The proof of this statement is almost a trivial extension of the proof in Section 4. One again starts by showing additivity, but this time in the two variables E and F separately. For instance, for a fixed $E \in \mathcal{E}_{d_A}$, define

$$g_E(F) = f(E,F) , \tag{88}$$

and consider two locally-measurable POVM trees

$$\{(I-E,F_i),(E,G_\alpha)\} \qquad \text{and} \qquad \{(I-E,F_i),(E,H_\beta)\} , \tag{89}$$

where $\{F_i\}$, $\{G_\alpha\}$, and $\{H_\beta\}$ are arbitrary POVMs on $\mathcal{H}_B$. Then Eq. (86) requires that

$$\sum_i g_{I\text{-}E}(F_i) + \sum_\alpha g_E(G_\alpha) = 1 \tag{90}$$

and

$$\sum_i g_{I\text{-}E}(F_i) + \sum_\beta g_E(H_\beta) = 1 . \tag{91}$$

From this it follows that,

$$\sum_\alpha g_E(G_\alpha) = \sum_\beta g_E(H_\beta) = \text{const.} \tag{92}$$

That is to say, $g_E(F)$ is a frame function in the sense of Section 4. Consequently, we know that we can use the same methods as there to uniquely extend $g_E(F)$ to a linear functional on the complete set of Hermitian operators on $\mathcal{H}_B$. Similarly, for fixed $F \in \mathcal{E}_{d_B}$, we can define

$$h_F(E) = f(E,F) , \tag{93}$$

and prove that this function too can be uniquely extended to a linear functional on the Hermitian operators on $\mathcal{H}_A$.

The linear extensions of $g_E(F)$ and $h_F(E)$ can be put together in a simple way to give a full bilinear extension to the function $f(E,F)$. Namely, for any two Hermitian operators

A and B on $\mathcal{H}_A$ and $\mathcal{H}_B$, respectively, let $A = \alpha_1 E_1 - \alpha_2 E_2$ and $B = \beta_1 F_1 - \beta_2 F_2$ be decompositions such that $\alpha_1, \alpha_2, \beta_1, \beta_2 \geq 0$, $E_1, E_2 \in \mathcal{E}_{d_A}$, and $F_1, F_2 \in \mathcal{E}_{d_B}$. Then define

$$f(A,B) \equiv \alpha_1 g_{E_1}(B) - \alpha_2 g_{E_2}(B) \,. \tag{94}$$

To see that this definition is unique, take any other decomposition $A = \tilde{\alpha}_1 \tilde{E}_1 - \tilde{\alpha}_2 \tilde{E}_2$. Then we have

$$\begin{aligned}
f(A,B) &= \tilde{\alpha}_1 g_{\tilde{E}_1}(B) - \tilde{\alpha}_2 g_{\tilde{E}_2}(B) & (95)\\
&= \tilde{\alpha}_1 f(\tilde{E}_1, B) - \tilde{\alpha}_2 f(\tilde{E}_2, B) & (96)\\
&= \beta_1\big(\tilde{\alpha}_1 f(\tilde{E}_1, F_1) - \tilde{\alpha}_2 f(\tilde{E}_1, F_1)\big) - \beta_2\big(\tilde{\alpha}_1 f(\tilde{E}_1, F_2) - \tilde{\alpha}_2 f(\tilde{E}_2, F_2)\big) & (97)\\
&= \beta_1 h_{F_1}(A) - \beta_2 h_{F_2}(A) & (98)\\
&= \beta_1\big(\alpha_1 f(E_1, F_1) - \alpha_2 f(E_1, F_1)\big) - \beta_2\big(\alpha_1 f(E_1, F_2) - \alpha_2 f(E_2, F_2)\big) & (99)\\
&= \alpha_1 f(E_1, B) - \alpha_2 f(E_2, B) & (100)\\
&= \alpha_1 g_{E_1}(B) - \alpha_2 g_{E_2}(B) \,, & (101)
\end{aligned}$$

which is as desired.

With bilinearity for the function f established, we have essentially the full story [56, 59]. For, let $\{E_i\}$, $i = 1, \ldots, d_A^2$, be a complete basis for the Hermitian operators on $\mathcal{H}_A$ and let $\{F_j\}$, $j = 1, \ldots, d_B^2$, be a complete basis for the Hermitian operators on $\mathcal{H}_B$. If $E = \sum_i \alpha_i E_i$ and $F = \sum_j \beta_j F_j$, then

$$f(E,F) = \sum_{ij} \alpha_i \beta_j f(E_i, F_j) \,. \tag{102}$$

Define $\tilde{\rho}$ to be a linear operator on $\mathcal{H}_A \otimes \mathcal{H}_B$ satisfying the $(d_A d_B)^2$ linear equations

$$\mathrm{tr}\big(\tilde{\rho}(E_i \otimes F_j)\big) = f(E_i, F_j) \,. \tag{103}$$

Such an operator always exists. Consequently we have,

$$\begin{aligned}
f(E,F) &= \sum_{ij} \alpha_i \beta_j \mathrm{tr}\big(\tilde{\rho}(E_i \otimes F_j)\big) & (104)\\
&= \mathrm{tr}\big(\tilde{\rho}(E \otimes F)\big) \,. & (105)
\end{aligned}$$

Enforcing positivity and normalization for the function f proves the main point of the theorem.

For complex Hilbert spaces $\mathcal{H}_A$ and $\mathcal{H}_B$, the uniqueness of $\tilde{\rho}$ comes about because the set $\{E_i \otimes F_j\}$ forms a complete basis for the Hermitian operators on $\mathcal{H}_A \otimes \mathcal{H}_B$. For real Hilbert spaces, however, the analog of the Hermitian operators are the symmetric operators. The dimensionality of the space of symmetric operators on a real Hilbert space $\mathcal{H}_d$ is $\frac{1}{2}d(d+1)$, rather than the d^2 it is for the complex case. This means that in the steps above only

$$\frac{1}{4} d_A d_B (d_A + 1)(d_B + 1) \tag{106}$$

equations will appear in Eq. (103), whereas

$$\frac{1}{2} d_A d_B (d_A d_B + 1) \tag{107}$$

are needed to uniquely specify a $\tilde{\rho}$. For instance take $d_A = d_B = 2$. Then Eq. (106) gives nine equations, while Eq. (107) requires ten.

Let us emphasize the striking feature of this way of deriving the tensor product rule for combining separate quantum systems: It is built on the very concept of local measurement. There is nothing "spooky" or "nonlocal" about it; there is nothing in it resembling "passion at a distance" [61]. Indeed, one need not even consider probability assignments for the outcomes of measurements of the "nonlocality without entanglement" variety [58] in order to uniquely fix the probability rule. That is—to give an example on $\mathcal{H}_3 \otimes \mathcal{H}_3$—one need not consider standard measurements like $\{E_b = |\psi_b\rangle\langle\psi_b|\}$, where

$$|\psi_1\rangle = |1\rangle|1\rangle \tag{108}$$

$$|\psi_2\rangle = |0\rangle|0+1\rangle \qquad |\psi_6\rangle = |1+2\rangle|0\rangle \tag{109}$$

$$|\psi_3\rangle = |0\rangle|0-1\rangle \qquad |\psi_7\rangle = |1-2\rangle|0\rangle \tag{110}$$

$$|\psi_4\rangle = |2\rangle|1+2\rangle \qquad |\psi_8\rangle = |0+1\rangle|2\rangle \tag{111}$$

$$|\psi_5\rangle = |2\rangle|1-2\rangle \qquad |\psi_9\rangle = |0-1\rangle|2\rangle \tag{112}$$

with $|0\rangle$, $|1\rangle$, and $|2\rangle$ forming an orthonormal basis on $\mathcal{H}_3$, and $|0+1\rangle = \frac{1}{\sqrt{2}}(|0\rangle + |1\rangle)$, etc. This is a measurement that takes neither the form of Eq. (83) nor (84). It stands out instead in that, even though all its POVM elements are tensor product operators—i.e., they have no quantum entanglement—it still *cannot* be measured by local means, even with the elaborate ping-ponging strategies mentioned earlier.

Thus, the tensor product rule, and with it quantum entanglement, seems to be more a statement of locality than anything else. It, like the probability rule, is more a product of the structure of the observables—that they are POVMs—combined with noncontextuality. In searching for the secret ingredient to drive a wedge between general Bayesian probability theory and quantum mechanics, it seems that the direction *not* to look is toward quantum entanglement. Perhaps the trick instead is to dig deeper into the Bayesian toolbox.

7 Unknown Quantum States?

> My thesis, paradoxically, and a little provocatively, but nonetheless genuinely, is simply this:
>
> QUANTUM STATES DO NOT EXIST.
>
> The abandonment of superstitious beliefs about the existence of Phlogiston, the Cosmic Ether, Absolute Space and Time, ..., or Fairies and Witches, was an essential step along the road to scientific thinking. The quantum state, too, if regarded as something endowed with some kind of objective existence, is no less a misleading conception, an illusory attempt to exteriorize or materialize the information we possess.
>
> — *the ghost of Bruno de Finetti*

The hint of a more fruitful direction can be found in tackling one of the most commonly used phrases in quantum information theory head-on. It is the *unknown quantum state.*

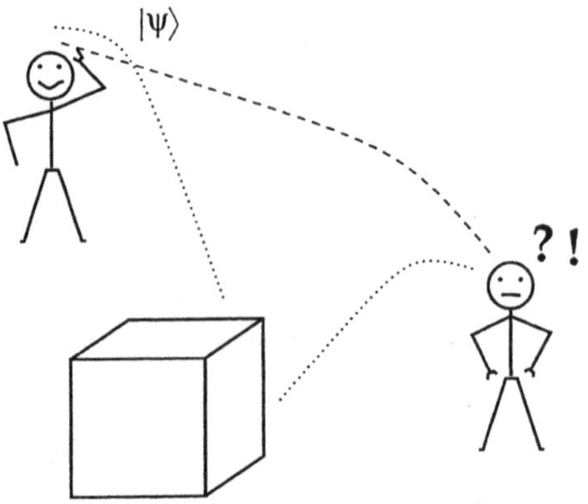

Figure 1: What can the term "unknown state" mean if quantum states are taken exclusively to be states of knowledge rather than states of nature? When we say that a system has an unknown state, must we always imagine a further observer whose state of knowledge is symbolized by some $|\psi\rangle$, and it is the identity of the symbol that we are ignorant of?

There is hardly a paper in quantum information that does not make use of it. Unknown quantum states are teleported [8], protected with quantum error correcting codes [62], and used to check for quantum eavesdropping [63]. The list of uses grows each day. But what can the term possibly mean? In an information-based interpretation of quantum mechanics, it is an overt oxymoron: If quantum states, by their very definition, are states of knowledge and not states of nature, then the state is *known* by someone—at the very least, by the person who wrote it down.

Thus, if a phenomenon ostensibly invokes the concept of an unknown state in its formulation, that unknown state had better be shorthand for a more basic situation (even if that basic situation still awaits a complete analysis). This means that for any phenomenon using the idea of an unknown quantum state in its description, we should demand that either

1. The owner of the unknown state—a further decision-making agent or observer—be explicitly identified. (In this case, the unknown state is merely a stand-in for the unknown *state of knowledge* of an essential player who went unrecognized in the original formulation.) Or,

2. If there is clearly no further agent or observer on the scene, then a way must be found to reexpress the phenomenon with the term "unknown state" completely banished from its formulation. (In this case, the end-product of the effort will be a single quantum state used for describing the phenomenon—namely, the state that actually captures the describer's state of knowledge throughout.)

This Section reports the work of Ref. [18], where such a project is carried out for the experimental practice of *quantum-state tomography* [13]. The usual description of tomography

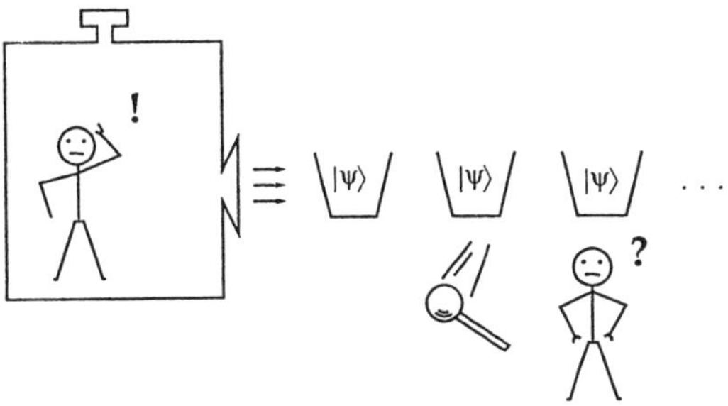

Figure 2: To make sense of quantum tomography, must we go to the extreme of imagining a "man in the box" who has a better description of the systems than we do? How contrived our usage would be if that were so!

is this. A device of some sort, say a nonlinear optical medium driven by a laser, repeatedly prepares many instances of a quantum system, say many temporally distinct modes of the electromagnetic field, in a fixed quantum state ρ, pure or mixed [64]. An experimentalist who wishes to characterize the operation of the device or to calibrate it for future use may be able to perform measurements on the systems it prepares even if he cannot get at the device itself. This can be useful if the experimenter has some prior knowledge of the device's operation that can be translated into a probability distribution over states. Then learning about the state will also be learning about the device. Most importantly, though, this description of tomography assumes that the precise state ρ is unknown. The goal of the experimenter is to perform enough measurements, and enough kinds of measurements (on a large enough sample), to estimate the identity of ρ.

This is clearly an example where there is no further player on whom to pin the unknown state as a state of knowledge. Any attempt to find such a missing player would be entirely artificial: Where would the player be placed? On the inside of the device the tomographer is trying to characterize?[20] The only available course is the second strategy above—to banish the idea of the unknown state from the formulation of tomography.

To do this, we once again take our cue from Bayesian probability theory [38, 39, 40]. As emphasized previously, in Bayesian theory probabilities—just like quantum states—are not objective states of nature, but rather measures of credible belief, reflecting one's state

[20]Placing the player here would be about as respectable as George Berkeley's famous patch to his philosophical system of idealism. The difficulty is captured engagingly by a limerick of Ronald Knox and its anonymous reply:

> There was a young man who said, "God : Must think it exceedingly odd : If he finds that this tree : Continues to be : When there's no one about in the Quad." REPLY: "Dear Sir: Your astonishment's odd. : I am always about in the Quad. : And that's why the tree : Will continue to be, : Since observed by Yours faithfully, God."

of knowledge. In light of this, it comes as no surprise that one of the most overarching Bayesian themes is to identify the conditions under which a set of decision-making agents can come to a common belief or probability assignment for a random variable even though their initial beliefs may differ [40]. Following that theme is the key to understanding the essence of quantum-state tomography.

Indeed, classical Bayesian theory encounters almost precisely the same problem as our unknown quantum state through the widespread use of the phrase "unknown probability" in its domain. This is an oxymoron every bit as egregious as unknown state.

The procedure analogous to quantum-state tomography in Bayesian theory is the estimation of an unknown probability from the results of repeated trials on "identically prepared systems." The way to eliminate unknown probabilities from this situation was introduced by Bruno de Finetti in the early 1930s [66]. His method was simply to focus on the equivalence of the repeated trials—namely, that what is really of importance is that the systems are indistinguishable as far as probabilistic predictions are concerned. Because of this, any probability assignment $p(x_1, x_2, \ldots, x_N)$ for multiple trials should be symmetric under permutation of the systems. As innocent as this conceptual shift may sound, de Finetti was able to use it to powerful effect. For, with his *representation theorem*, he showed that any multi-trial probability assignment that is permutation-symmetric for an arbitrarily large number of trials—de Finetti called such multi-trial probabilities *exchangeable*—is equivalent to a probability for the "unknown probabilities."

Let us outline this in a little more detail. In an objectivist description of N "identically prepared systems," the individual trials are described by discrete random variables $x_n \in \{1, 2, \ldots, k\}$, $n = 1, \ldots, N$, and the probability in the multi-trial hypothesis space is given by an independent identically distributed distribution

$$p(x_1, x_2, \ldots, x_N) = p_{x_1} p_{x_2} \cdots p_{x_N} = p_1^{n_1} p_2^{n_2} \cdots p_k^{n_k} . \tag{113}$$

The numbers p_j describe the objective, "true" probability that the result of a single experiment will be j ($j = 1, \ldots, k$). The variable n_j, on the other hand, describes the number of times outcome j is listed in the vector $(x_1, x_2, \ldots, x_N)$. But this description—for the objectivist—only describes the situation from a kind of "God's eye" point of view. To the experimentalist, the "true" probabilities $p_1, \ldots, p_k$ will very often be *unknown* at the outset. Thus, his burden is to estimate the unknown probabilities by a statistical analysis of the experiment's outcomes.

In the Bayesian approach, however, it does not make sense to talk about estimating a true probability. Instead, a Bayesian assigns a prior probability distribution $p(x_1, x_2, \ldots, x_N)$ on the multi-trial hypothesis space and uses Bayes' theorem to update the distribution in the light of his measurement results. The content of de Finetti's theorem is this. Assuming *only* that

$$p(x_{\pi(1)}, x_{\pi(2)}, \ldots, x_{\pi(N)}) = p(x_1, x_2, \ldots, x_N) \tag{114}$$

for any permutation π of the set $\{1, \ldots, N\}$, and that for any integer $M > 0$, there is a distribution $p_{N+M}(x_1, x_2, \ldots, x_{N+M})$ with the same permutation property such that

$$p(x_1, x_2, \ldots, x_N) = \sum_{x_{N+1}, \ldots, x_{N+M}} p_{N+M}(x_1, \ldots, x_N, x_{N+1}, \ldots, x_{N+M}) , \tag{115}$$

then $p(x_1, x_2, \ldots, x_N)$ can be written uniquely in the form

$$p(x_1, x_2, \ldots, x_N) = \int_{\mathcal{S}_k} P(\vec{p})\, p_{x_1} p_{x_2} \cdots p_{x_N}\, d\vec{p} \tag{116}$$

$$= \int_{\mathcal{S}_k} P(\vec{p})\, p_1^{n_1} p_2^{n_2} \cdots p_k^{n_k}\, d\vec{p}\,, \tag{117}$$

where $\vec{p} = (p_1, p_2, \ldots, p_k)$, and the integral is taken over the simplex of such distributions

$$\mathcal{S}_k = \left\{ \vec{p} : \;\; p_j \geq 0 \text{ for all } j \text{ and } \sum_{j=1}^{k} p_j = 1 \right\}. \tag{118}$$

Furthermore, the function $P(\vec{p}) \geq 0$ is required to be a probability density function on the simplex:

$$\int_{\mathcal{S}_k} P(\vec{p})\, d\vec{p} = 1\,, \tag{119}$$

With this representation theorem, the unsatisfactory concept of an unknown probability vanishes from the description in favor of the fundamental idea of assigning an exchangeable probability distribution to multiple trials.

This cue in hand, it is easy to see how to reword the description of quantum-state tomography to meet our goals. What is relevant is simply a judgment on the part of the experimenter—notice the essential subjective character of this "judgment"—that there is no distinction between the systems the device is preparing. In operational terms, this is the judgment that *all the systems are and will be the same as far as observational predictions are concerned.* At first glance this statement might seem to be contentless, but the important point is this: To make this statement, one need never use the notion of an unknown state—a completely operational description is good enough. Putting it into technical terms, the statement is that if the experimenter judges a collection of N of the device's outputs to have an overall quantum state $\rho^{(N)}$, he will also judge any permutation of those outputs to have the same quantum state $\rho^{(N)}$. Moreover, he will do this no matter how large the number N is. This, complemented only by the consistency condition that for any N the state $\rho^{(N)}$ be derivable from $\rho^{(N+1)}$, makes for the complete story.

The words "quantum state" appear in this formulation, just as in the original formulation of tomography, but there is no longer any mention of *unknown* quantum states. The state $\rho^{(N)}$ is known by the experimenter (if no one else), for it represents his state of knowledge. More importantly, the experimenter is in a position to make an unambiguous statement about the structure of the whole sequence of states $\rho^{(N)}$: Each of the states $\rho^{(N)}$ has a kind of permutation invariance over its factors. The content of the *quantum de Finetti representation theorem* [17, 18] is that a sequence of states $\rho^{(N)}$ can have these properties, which are said to make it an *exchangeable* sequence, if and only if each term in it can also be written in the form

$$\rho^{(N)} = \int_{\mathcal{D}_d} P(\rho)\, \rho^{\otimes N}\, d\rho\,, \tag{120}$$

where

$$\rho^{\otimes N} = \underbrace{\rho \otimes \rho \otimes \cdots \otimes \rho}_{\substack{N\text{-fold tensor}\\ \text{product}}} \tag{121}$$

Here $P(\rho) \geq 0$ is a fixed probability distribution over the density operator space $\mathcal{D}_d$, and

$$\int_{\mathcal{D}_d} P(\rho)\, d\rho = 1 \,, \tag{122}$$

where $d\rho$ is a suitable measure.

The interpretive import of this theorem is paramount. For it alone gives a mandate to the term unknown state in the usual description of tomography. It says that the experimenter can act *as if* his state of knowledge $\rho^{(N)}$ comes about because he knows there is a "man in the box," hidden from view, repeatedly preparing the same state ρ. He does not know which such state, and the best he can say about the unknown state is captured in the probability distribution $P(\rho)$.

The quantum de Finetti theorem furthermore makes a connection to the overarching theme of Bayesianism stressed above. It guarantees for two independent observers—as long as they have a rather minimal agreement in their initial beliefs—that the outcomes of a sufficiently informative set of measurements will force a convergence in their state assignments for the remaining systems [67]. This "minimal" agreement is characterized by a judgment on the part of both parties that the sequence of systems is exchangeable, as described above, and a promise that the observers are not absolutely inflexible in their opinions. Quantitatively, the latter means that though $P(\rho)$ may be arbitrarily close to zero, it can never vanish.

This coming to agreement works because an exchangeable density operator sequence can be updated to reflect information gathered from measurements by a another quantum version of Bayes's rule for updating probabilities [67]. Specifically, if measurements on K systems yield results D_K, then the state of additional systems is constructed as in Eq. (120), but using an updated probability on density operators given by

$$P(\rho|D_K) = \frac{P(D_K|\rho)P(\rho)}{P(D_K)} \,. \tag{123}$$

Here $P(D_K|\rho)$ is the probability to obtain the measurement results D_K, given the state $\rho^{\otimes K}$ for the K measured systems, and $P(D_K) = \int P(D_K|\rho)\, P(\rho)\, d\rho$ is the unconditional probability for the measurement results. For a sufficiently informative set of measurements, as K becomes large, the updated probability $P(\rho|D_K)$ becomes highly peaked on a particular state ρ_{D_K} dictated by the measurement results, regardless of the prior probability $P(\rho)$, as long as $P(\rho)$ is nonzero in a neighborhood of ρ_{D_K}. Suppose the two observers have different initial beliefs, encapsulated in different priors $P_i(\rho)$, $i = 1, 2$. The measurement results force them to a common state of knowledge in which any number N of additional systems are assigned the product state $\rho_{D_K}^{\otimes N}$, i.e.,

$$\int P_i(\rho|D_K)\, \rho^{\otimes N}\, d\rho \quad \longrightarrow \quad \rho_{D_K}^{\otimes N} \,, \tag{124}$$

independent of i, for K sufficiently large.

This shifts the perspective on the purpose of quantum-state tomography: It is not about uncovering some "unknown state of nature," but rather about the various observers' coming to agreement over future probabilistic predictions. In this connection, it is interesting to note that the quantum de Finetti theorem and the conclusions just drawn from it work only within the framework of complex vector-space quantum mechanics. For quantum mechanics based on real Hilbert spaces, the connection between exchangeable density operators and unknown quantum states does not hold.

A simple counterexample is the following. Consider the N-system state

$$\rho^{(N)} = \frac{1}{2}\rho_+^{\otimes N} + \frac{1}{2}\rho_-^{\otimes N} , \tag{125}$$

where

$$\rho_+ = \frac{1}{2}(I + \sigma_2) \qquad \text{and} \qquad \rho_- = \frac{1}{2}(I - \sigma_2) \tag{126}$$

and σ_1, σ_2, and σ_3 are the Pauli matrices. In complex-Hilbert-space quantum mechanics, Eq. (125) is clearly a valid density operator: It corresponds to an equally weighted mixture of N spin-up particles and N spin-down particles in the y-direction. The state $\rho^{(N)}$ is thus exchangeable, and the decomposition in Eq. (125) is unique according to the quantum de Finetti theorem.

But now consider $\rho^{(N)}$ as an operator in real-Hilbert-space quantum mechanics. Despite its ostensible use of the imaginary number i, it remains a valid quantum state. This is because, upon expanding the right-hand side of Eq. (125), all the terms with an odd number of σ_2's cancel away. Yet, even though it is an exchangeable density operator, it cannot be written in de Finetti form Eq. (120) using only real symmetric operators. This follows because $i\sigma_2$ cannot be written as a linear combination of I, σ_1, and σ_3, while a real-Hilbert-space de Finetti expansion as in Eq. (120) can *only* contain those three operators. Hence the de Finetti theorem does not hold in real-Hilbert-space quantum mechanics.

In classical probability theory, exchangeability characterizes those situations where the only data relevant for updating a probability distribution are frequency data, i.e., the numbers n_j in Eq. (117). The quantum de Finetti representation shows that the same is true in quantum mechanics: Frequency data (with respect to a sufficiently robust measurement)[21] are sufficient for updating an exchangeable state to the point where nothing more can be learned from sequential measurements. That is, one obtains a convergence of the form Eq. (124), so that ultimately any further measurements on the individual systems will be statistically independent. That there is no quantum de Finetti theorem in real Hilbert space means that there are fundamental differences between real and complex Hilbert spaces with respect to learning from measurement results.

Finally, in summary, let us hang on the point of learning for just a little longer. The quantum de Finetti theorem shows that the essence of quantum-state tomography is not in revealing an "element of reality" but in deriving that various agents (who agree some minimal amount) can come to agreement in their ultimate quantum-state assignments. This is not the same thing as the stronger statement that "reality does not exist." It is simply that one need not go to the extreme of taking the "unknown quantum state" as being objectively real to make sense of the experimental practice of tomography.

One is left with the feeling—an almost salty feeling[22]—that perhaps this is the whole point to quantum mechanics. That is: Perhaps the missing ingredient for narrowing the

[21]Technically, this means any POVM $\{E_b\}$ whose elements span the space of Hermitian operators. See Ref. [18] for details.

[22]Working under the presumption that no interpretation of quantum mechanics is worth its salt unless it raises as many technical questions as it answers philosophical ones.

structure of Bayesian probability down to the structure of quantum mechanics has been in front of us all along. It finds no better expression than in the taking account of the limitations the physical world poses to our ability to come to agreement.

8 The Oyster and the Quantum

> The significance of this development is to give us insight into the logical possibility of a new and wider pattern of thought. This takes into account the observer, including the apparatus used by him, differently from the way it was done in classical physics ... In the new pattern of thought we do not assume any longer the *detached observer*, ... but an observer who by his indeterminable effects creates a new situation, theoretically described as a new state of the observed system. ...
>
> Nevertheless, there remains still in the new kind of theory an *objective reality*, inasmuch as these theories deny any possibility for the observer to influence the results of a measurement, once the experimental arrangement is chosen. Particular qualities of an individual observer do not enter the conceptual framework of the theory.
>
> — *Wolfgang Pauli*, 1954

A grain of sand falls into the shell of an oyster and the result is a pearl. The oyster's sensitivity to the touch is the source of one of our most beautiful gems. In the 75 years that have passed since the founding of quantum mechanics, only the last 10 have turned to a view and an attitude that may finally reveal the essence of the theory. The quantum world is sensitive to the touch, and that may well be one of the best things about it. Quantum information science—with its three prongs of quantum information theory, quantum cryptography, and quantum computing—leads the way in telling us how to quantify that sentence. Quantum algorithms can be exponentially faster that classical algorithms. Secret keys can be encoded into physical systems in such a way as to reveal whether information has been gathered about them. The list of triumphs keeps growing.

The key to so much of this has been simply in a change of attitude. This can be seen by going back to almost any older textbook on quantum mechanics: Nine times out of ten, the Heisenberg uncertainty relation is presented in a way that conveys the feeling that we have been short-changed by the physical world.

> "Look at classical physics, how nice it is: We can measure a particle's position and momentum with as much accuracy as we would like. How limiting quantum theory is instead. We are stuck with
>
> $$\Delta x \Delta p \geq \frac{1}{2}\hbar ,$$
>
> and there is just nothing we can do about it. The task of physics is just to sober up to this state of affairs and make the best of it."

How this contrasts with the point of departure of quantum information science! There the task is not to ask what limits quantum mechanics places upon us, but what novel, productive things we can do in the quantum world that we could not have done otherwise. In what ways is the quantum world fantastically better than the classical one?

If one is looking for something "real" in quantum theory, what more direct attack could one take than to look to its technologies? People may argue about the objective reality of the wave function ad infinitum, but few would argue about the existence of quantum cryptography as a solid prediction of the theory. Why not take that or a similar effect as the grounding for what quantum mechanics is trying to tell us about nature?

Let us try to give this imprecise set of thoughts some shape by trying to mold quantum cryptography into the vision built in the previous sections. For quantum key distribution it is essential to be able to prepare a physical system in one or another quantum state drawn from some fixed *nonorthogonal* set [63, 68]. These nonorthogonal states are used to encode a potentially secret cryptographic key to be shared between the sender and receiver. The information an eavesdropper seeks is about which quantum state was actually prepared in each individual transmission. What is novel here is that the encoding of the proposed key into nonorthogonal states forces the information-gathering process to induce a disturbance to the overall *set* of states. That is, the presence of an active eavesdropper transforms the initial pure states into a set of mixed states or, at the very least, into a set of pure states with larger overlaps than before. This action ultimately boils down to a loss of predictability for the sender over the outcomes of the receiver's measurements and, so, is directly detectable by the receiver (who reveals some of those outcomes for the sender's inspection). More importantly, there is a direct connection between the statistical information gained by an eavesdropper and the consequent disturbance she must induce to the quantum states in the process. As the information gathered goes up, the necessary disturbance also goes up in a precisely formalizable way [69].

Note the two ingredients that appear in this scenario. First, the information gathering or measurement is grounded with respect to one observer (in this case, the eavesdropper), while the disturbance is grounded with respect to another (here, the sender). In particular, the disturbance is a disturbance to the sender's previous information—this is measured by her diminished ability to predict the outcomes of certain measurements the legitimate receiver might perform. No hint of any variable intrinsic to the system is made use of in this formulation of the idea of "measurement causing disturbance."

The second ingredient is that one must consider at least two possible nonorthogonal preparations in order for the formulation to have any meaning. This is because the information gathering is not about some classically-defined observable—i.e., about some unknown hidden variable or reality intrinsic to the system—but is instead about which of the unknown states the sender actually prepared. The lesson is this: Forget about the unknown preparation, and the random outcome of the quantum measurement is information about nothing. It is simply "quantum noise" with no connection to any preexisting variable.

How crucial is this second ingredient—that is, that there be at least two nonorthogonal states within the set under consideration? We can address its necessity by making a slight shift in the account above: One might say that the eavesdropper's goal is not so much to uncover the identity of the unknown quantum state, but to sharpen her predictability over the receiver's measurement outcomes. In fact, she would like to do this at the same time as disturbing the sender's predictions as little as possible. Changing the language still further to the terminology of Section 4, the eavesdropper's actions serve to sharpen her information about the potential consequences of the receiver's further interventions upon the system. (Again, she would like to do this while minimally diminishing the sender's previous information about those same consequences.) In the cryptographic context, a byproduct of this effort is that the eavesdropper ultimately comes to a more sound prediction of the secret key. From the present point of view, however, the importance of this change of language is that it leads to an almost Bayesian perspective on the information–disturbance problem.

As already emphasized, within Bayesian probability the most significant theme is to identify the conditions under which a set of decision-making agents can come to a common probability assignment for some random variable in spite of the fact that their initial probabilities differ [40]. One might similarly view the process of quantum eavesdropping. The sender and the eavesdropper start off initially with differing quantum state assignments for a single physical system. In this case it so happens that the sender can make sharper predictions than the eavesdropper about the outcomes of the receiver's measurements. The eavesdropper, not satisfied with this situation, performs a measurement on the system in an attempt to sharpen those predictions. In particular, there is an attempt to come into something of an agreement with the sender but without revealing the outcomes of her measurements or, indeed, her very presence.

It is at this point that a distinct *property* of the quantum world makes itself known. The eavesdropper's attempt to surreptitiously come into alignment with the sender's predictability is always shunted away from its goal. This shunting of various observer's predictability is the subtle manner in which the quantum world is sensitive to our experimental interventions.

And maybe this is our crucial hint! The wedge that drives a distinction between Bayesian probability theory in general and quantum mechanics in particular is perhaps nothing more than this "Zing!" of a quantum system that is manifested when an agent interacts with it. It is this wild sensitivity to the touch that keeps our knowledge and beliefs from ever coming into too great of an alignment. The most our knowledge about the potential consequences of our interventions on a system can come into alignment is captured by the mathematical structure of a pure quantum state $|\psi\rangle$. Take all possible information-disturbance curves for a quantum system, tie them into a bundle, and *that* is the long-awaited property, the input we have been looking for from nature.

Or, at least, that is the speculation. Look at that bundle long and hard and we might just find that it stays together without the help of our tie.

9 Acknowledgments

I thank Carl Caves, Greg Comer, and Rüdiger Schack for inspiration and years of conversation, David Mermin for suggestions on the manuscript, and Sam Braunstein for trying to turn my e's upside down.

References

[1] J. T. Cushing, A. Fine, and S. Goldstein, editors, *Bohmian Mechanics and Quantum Theory: An Appraisal*, (Kluwer, Dordrecht, 1996).

[2] R. B. Griffiths and R. Omnès, "Consistent Histories and Quantum Measurements," Phys. Today **52**(8), 26–31 (1999).

[3] W. H. Zurek, "Decoherence, Einselection and the Existential Interpretation (The Rough Guide)," Phil. Trans. R. Soc. Lond. A **356**, 1793–1821 (1998).

[4] G. C. Ghirardi and P. Pearle, "Dynamical Reduction Theories: Changing Quantum Theory so the Statevector Represents Reality," *PSA 1990: Proceedings of the Biennial Meeting of the Philosophy of Science Association, Vol. I*, edited by A. Fine, M. Forbes, and L. Wessels (Philosophy of Science Association, East Lansing, MI, 1990), pp. 19–34.

[5] D. Deutsch, *The Fabric of Reality: The Science of Parallel Universes—and its Implications*, (Allen Lane, New York, 1997).

[6] C. Rovelli, "Relational Quantum Mechanics," Int. J. Theor. Phys. **35**, 1637–1678 (1996).

[7] M. Tegmark, "The Interpretation of Quantum Mechanics: Many Worlds or Many Words?," Fort. der Phys. **46**, 855–862 (1998).

[8] C. H. Bennett G. Brassard, C. Crépeau, R. Jozsa, A. Peres, and W. K. Wootters, "Teleporting an Unknown Quantum State via Dual Classical and Einstein-Podolsky-Rosen Channels," Phys. Rev. Lett. **70**, 1895 (1993).

[9] J. Bub, "Quantum Mechanics as a Principle Theory," Stud. Hist. Phil. Mod. Phys. **31**, 74–95 (2000).

[10] J. Bub, *Interpreting the Quantum World*, (Cambridge University Press, Cambridge, 1997).

[11] J. S. Bell, *Speakable and Unspeakable in Quantum Mechanics: Collected Papers on Quantum Philosophy* (Cambridge U. Press, Cambridge, 1987).

[12] M. A. Nielsen and I. L. Chuang, *Quantum Computation and Quantum Information*, (Cambridge University Press, Cambridge, 2000).

[13] K. Vogel and H. Risken, "Determination of Quasiprobability Distributions in Terms of Probability Distributions for the Rotated Quadrature Phase," Phys. Rev. A **40**, 2847 (1989); U. Leonhardt, "Quantum-State Tomography and Discrete Wigner Function," Phys. Rev. Lett. **74**, 4101 (1995).

[14] P. Busch, "Quantum States as Effect Valuations: Giving Operational Content to von Neumann's No-Hidden-Variables Theorem," `quant-ph/9909073`.

[15] C. M. Caves, C. A. Fuchs, K. Manne, and J. Renes, "Gleason's Theorem with POVMs," unpublished, 2000.

[16] A. M. Gleason, "Measures on the Closed Subspaces of a Hilbert Space," J. Math. Mech. **6**, 885–894 (1957).

[17] R. L. Hudson and G. R. Moody, "Locally Normal Symmetric States and an Analogue of de Finetti's Theorem," Z. Wahrschein. verw. Geb. **33**, 343 (1976).

[18] C. M. Caves, C. A. Fuchs and R. Schack, "Unknown Quantum States: The Quantum de Finetti Representation," `quant-ph/0104088`.

[19] A. Einstein, B. Podolsky, and N. Rosen, "Can Quantum-Mechanical Description of Physical Reality Be Considered Complete?," Phys. Rev. **47**, 777–780 (1935).

[20] A. Fine, *The Shaky Game: Einstein Realism and the Quantum Theory*, (University of Chicago Press, Chicago, 1986); M. Jammer, "The EPR Problem in Its Historical Development," in *Symposium on the Foundations of Modern Physics: 50 years of the Einstein-Podolsky-Rosen Gedankenexperiment*, edited by P. Lahti and P. Mittelstaedt (World Scientific, Singapore, 1985), pp. 129–149.

[21] L. P. Hughston, R. Jozsa, and W. K. Wootters, "A Complete Classification of Quantum Ensembles Having a Given Density Matrix," Phys. Lett. A **183**, 14–18 (1993).

[22] J. Bernstein, *Quantum Profiles*, (Princeton University Press, Princeton, NJ, 1991).

[23] A. Einstein, "Remarks Concerning the Essays Brought Together in this Co-Operative Volume," in *Albert Einstein: Philosopher-Scientist*, third edition, edited by P. A. Schilpp (Open Court, La Salle, IL, 1970), pp. 665–688.

[24] A. Peres, *Quantum Theory: Concepts and Methods*, (Kluwer, Dordrecht, 1993).

[25] C. M. Caves and C. A. Fuchs, "Quantum Information: How Much Information in a State Vector?," in *The Dilemma of Einstein, Podolsky and Rosen – 60 Years Later*, edited by A. Mann and M. Revzen, Ann. Israel Phys. Soc. **12**, 226–257 (1996).

[26] W. Pauli, letter to M. Fierz dated 10 August 1954, reprinted and translated in K. V. Laurikainen, *Beyond the Atom: The Philosophical Thought of Wolfgang Pauli*, (Springer-Verlag, Berlin, 1988), p. 226.

[27] R. Penrose, *The Emperor's New Mind: Concerning Computers, Minds, and the Laws of Physics*, (Oxford University Press, Oxford, 1989), pp. 268–269.

[28] M. Jammer, "The Experiment in Classical and in Quantum Physics," in *Proceedings of the International Symposium, Foundations of Quantum Mechanics in the Light of New Technology*, edited by S. Kamefuchi, H. Ezawa, Y. Murayama, M. Namiki, S. Nomura, Y. Ohnuki, and T. Yajima (Physical Society of Japan, Tokyo, 1984), pp. 265–276.

[29] W. Heisenberg, *Physics and Beyond: Encounters and Conversations*, translated by A. J. Pomerans (Harper & Row, New York, 1971), pp. 63–64.

[30] C. A. Fuchs and A. Peres, "Quantum Theory Needs No 'Interpretation'," Phys. Today **53**(3), 70 (2000); "Quantum Theory – Interpretation, Formulation, Inspiration: Fuchs and Peres Reply," **53**(9), 14 (2000).

[31] P. Simon, "Kathy's Song," (Columbia Records, New York, 1965).

[32] R. Cooke, M. Keane, and W. Moran, "An Elementary Proof of Gleason's Theorem," Math. Proc. Camb. Phil. Soc. **98**, 117–128 (1981); I. Pitowsky, "Infinite and Finite Gleason's Theorems and the Logic of Indeterminacy," J. Math. Phys. **39**, 218–228 (1998).

[33] H. Barnum, C. M. Caves, J. Finkelstein, C. A. Fuchs, and R. Schack, "Quantum Probability from Decision Theory?," Proc. Roy. Soc. Lond. A **456**, 1175–1182 (2000).

[34] K. Kraus, *States, Effects, and Operations. Fundamental Notions of Quantum Theory* (Springer, Berlin, 1983), Lecture Notes in Physics Vol. 190.

[35] A. S. Holevo, "Information-Theoretical Aspects of Quantum Measurement," Prob. Info. Trans. **9**, 110–118 (1973).

[36] D. A. Meyer, "Finite Precision Measurement Nullifies the Kochen-Specker Theorem," Phys. Rev. Lett. **83**, 3751–3754 (1999); A. Kent, "Non-Contextual Hidden Variables and Physical Measurements," Phys. Rev. Lett. **83** 3755–3757 (1999).

[37] C. M. Caves, C. A. Fuchs, and R. Schack, "Bayesian Probability in Quantum Mechanics," in preparation (2001).

[38] H. E. Kyburg, Jr. and H. E. Smokler, eds., *Studies in Subjective Probability*, Second Edition (Robert E. Krieger Publishing, Huntington, NY, 1980).

[39] E. T. Jaynes, *Probability Theory: The Logic of Science*, available for download on the World Wide Web at `http://bayes.wustl.edu/`. This huge book was unfortunately never finished due to Prof. Jaynes' death in 1998. Nevertheless, it stands as perhaps the most complete and persuasive document in the field.

[40] J. M. Bernardo and A. F. M. Smith, *Bayesian Theory* (Wiley, Chichester, 1994).

[41] N. D. Mermin, "From Classical State-Swapping to Quantum Teleportation," `quant-ph/0105117`.

[42] B. Schumacher, "Sending Quantum Entanglement Through Noisy Channels," Phys. Rev. A **54**, 2614–2628 (1996).

[43] T. M. Cover and J. A. Thomas, *Elements of Information Theory*, (John Wiley & Sons, New York, 1991).

[44] R. B. Ash, *Information Theory*, (Dover, New York, 1965).

[45] A. Wehrl, "General Properties of Entropy," Rev. Mod. Phys. **50**, 221–259 (1978).

[46] W. K. Wootters, "Random Quantum States," Found. Phys. **20**, 1365–1378 (1990); K. R. W. Jones, "Riemann-Liouville Fractional Integration and Reduced Distributions on Hyperspheres," J. Phys. A **24**, 1237–1244 (1991).

[47] R. Jozsa, D. Robb, and W. K. Wootters, "Lower Bound for Accessible Information in Quantum Mechanics," Phys. Rev. A **49**, 668–677 (1994).

[48] C. A. Fuchs and K. Jacobs, "Information Tradeoff Relations for Finite-Strength Quantum Measurements," in Phys. Rev. A **63**, 062305-1–062305-15 (2001).

[49] M. A. Nielsen, "Characterizing Mixing and Measurement in Quantum Mechanics," Phys. Rev. A **63**, 022114-1–022114-11 (2001).

[50] R. Schatten, *Norm Ideals of Completely Continuous Operators*, (Springer-Verlag, Berlin, 1960).

[51] J. Bub, "Von Neumann's Projection Postulate as a Probability Conditionalization Rule in Quantum Mechanics," J. Phil. Logic **6**, 381–390 (1977); S. L. Braunstein and C. M. Caves, "Wringing Out Better Bell Inequalities," Ann. Phys. **202**, 22–56 (1990).

[52] G. Boole, *An Investigation of the Laws of Thought*, (Dover, New York, 1958).

[53] C. A. Fuchs, *Notes on a Paulian Idea: Foundational, Historical, Anecdotal and Forward-Looking Thoughts on the Quantum*, `quant-ph/0105039`.

[54] R. Jozsa, "Entanglement and Quantum Computation," in *The Geometric Universe: Science, Geometry, and the Work of Roger Penrose*, edited by S. A. Huggett, L. J. Mason, K. P. Tod, S. T. Tsou, and N. M. J. Woodhouse, (Oxford U. Press, Oxford, 1998), p. 369.

[55] N. D. Mermin, *Boojums All the Way Through: Communicating Science in a Prosaic Age*, (Cambridge U. Press, Cambridge, 1990).

[56] R. Bhatia, *Matrix Analysis*, (Springer-Verlag, New York, 1997).

[57] N. R. Wallach, "An Unentangled Gleason's Theorem," `quant-ph/0002058`.

[58] C. H. Bennett, D. P. DiVincenzo, C. A. Fuchs, T. Mor, E. Rains, P. W. Shor, J. A. Smolin, and W. K. Wootters, "Quantum Nonlocality without Entanglement," Phys. Rev. A **59**, 1070–1091 (1999).

[59] S. MacLane and G. Birkhoff, *Algebra*, (Macmillan, New York, 1967).

[60] H. Araki, "On a Characterization of the State Space of Quantum Mechanics," Comm. Math. Phys. **75**, 1–24 (1980); W. K. Wootters, "Local Accessibility of Quantum Information," in *Complexity, Entropy and the Physics of Information*, edited by W. H. Zurek (Addison-Wesley, Redwood City, CA, 1990), p. 29–46.

[61] R. S. Cohen, M. Horne, and J. Stachel, editors, *Potentiality, Entanglement and Passion-at-a-Distance*, (Kluwer, Dordrecht, 1997).

[62] P. W. Shor, "Scheme for Reducing Decoherence in Quantum Computer Memory," Phys. Rev. A **52**, R2493 (1995); A. M. Steane, "Error Correcting Codes in Quantum Theory," Phys. Rev. Lett. **77**, 793 (1996).

[63] C. H. Bennett and G. Brassard, "Quantum Cryptography: Public Key Distribution and Coin Tossing," in *Proc. IEEE International Conference on Computers, Systems and Signal Processing* (IEEE Press, New York, 1984), p. 175, IEEE, 1984; C. H. Bennett, "Quantum Cryptography Using Any Two Nonorthogonal States," Phys. Rev. Lett. **68**, 3121 (1992).

[64] S. J. van Enk and C. A. Fuchs, "The Quantum State of a Laser Field," `quant-ph/0104036`.

[65] E. C. G. Stueckelberg, "Quantum Theory in Real Hilbert Space," Helv. Phys. Acta **33**, 727 (1960); S. L. Adler, *Quaternionic Quantum Mechanics and Quantum Fields* (Oxford U. Press, New York, 1995).

[66] B. de Finetti, *Theory of Probability* (Wiley, New York, 1990); For a collection de Finetti's original papers and their translations into English, see P. Monari and D. Cocchi, editors, *Probabilità e Induzione—Induction and Probability* (Biblioteca di Statistica, CLUEB, Bologna, 1993).

[67] R. Schack, T. A. Brun, and C. M. Caves, "Quantum Bayes Rule," `quant-ph/000113`.

[68] C. H. Bennett, G. Brassard, and N. D. Mermin, "Quantum Cryptography without Bell's Theorem," Phys. Rev. Lett. **68**, 557–559 (1992); C. A. Fuchs, "Information Gain vs. State Disturbance in Quantum Theory," Fort. der Phys. **46**, 535–565 (1998).

[69] C. A. Fuchs and A. Peres, "Quantum State Disturbance vs. Information Gain: Uncertainty Relations for Quantum Information," Phys. Rev. A **53**, 2038–2045 (1996); C. A. Fuchs, N. Gisin, R. B. Griffiths, C.-S. Niu, and A. Peres, "Optimal Eavesdropping in Quantum Cryptography. I. Information Bound and Optimal Strategy," Phys. Rev. A **56**, 1163–1172 (1997); D. Bruss, "Optimal Eavesdropping in Quantum Cryptography with Six States," Phys. Rev. Lett. **81**, 2598–2601 (1998).

Decoherence and its Implications in Quantum Computation and Information Transfer
T. Gonis and P.E.A. Turchi (Eds.)
IOS Press, 2001

Formal Aspects of the Particle Concept

Meinard Kuhlmann
Studiengang Philosophie, University of Bremen
mkuhlmann@uni-bremen.de

Abstract

The scope and relation of two contributions on the (non-)localizability of relativistic particles by Malament and Redhead will be explored. The assumptions, the employed theoretical tools (mathematical and physical) and the interpretations of their results will be analysed and compared. Two things will be shown. First, both results are equivalent despite of their different appearance. Second, the intended interpretations of the respective authors are not conclusive since there are various alternatives which cannot be immediately excluded. The broader conceptual concern is the question which ontological concepts of basic entities are compatible with the structure of quantum field theory (QFT).

1 Introduction

The wider context of the present investigation is the search of a consistent understanding of the nature of the most basic building blocks to which QFT refers. In short, the general topic is the search of an ontology for QFT. The issue of the following analysis are theorems and reflections about one prominent possibility for an ontology of QFT, viz. the particle interpretation. More specifically, I will explore two recent mathematical argumentations with respect to their mutual relation and the legitimacy of their conclusions. I will show, first, that both results are equivalent from an interpretive point of view and, second, that the conclusions drawn by the respective authors regarding the tenability of a particle interpretation of QFT are not inevitable.

There are a number of features which are discussed as indispensible ingredients of the particle concept. Probably the most obvious characteristic of particles is their discreteness. Particles are countable entities. Just as obvious as the fact that discreteness is a necessary condition for being a particle is the fact that it cannot be a sufficient condition. Modes of a standing wave or dollars on a bank account are countable without being particles. What is missing for them to be particles is individuality. The specific feature is sometimes called primitive thisness, i. e. something has to be identifiable as this one and not that one where its thisness is taken to be a brute fact which cannot be analyzed any further. The last example brings me to another characteristic of particles, namely their having a finite number of degrees of freedom in contrast to fields which have an infinite number. The last feature of particles which I wish to mention and which is most relevant for the following investigation is the localizability of particles. While some premodern atomistic theories conceived of the fundamental "atoms" as structureless point particles we think of elementary particles as entities which are extended. Nevertheless, again in contrast to fields, particles are generally taken to have a finite extension. It is often argued that a physical entity which is not restricted to a finite region is, by definition, a field.[1] I will come back to this argument in the concluding discussion.

Provided we accept the above-mentioned features as necessary conditions we have enough to start with despite of the lack of a proper definition for the concept of a particle. Clear-cut necessary conditions allow us to formulate no-go theorems which make it possible to exclude certain alternatives as viable ontological interpretations of a given physical formalism. Exactly this way has been gone with respect to the particle interpretation of relativistic quantum mechanics where some sets of assumptions have been shown to be contradictory. These results are generally considered to be among the firmest foundations of ontological investigations about relativistic quantum physics. Reeh and Schlieder, Hegerfeldt, Malament and Redhead all gained mathematical results, or formalized their interpretation, which prove that certain sets of assumptions lead to contradictions. They all purport to be in the position to exclude certain interpretations which are, to say the least, closely connected with the particle concept.

Up to now it is a point of debate what exactly has been shown and how the different results relate to one another. The clarification of these questions is pivotal for an enquiry of the ontology of QFT since the particle interpretation is, besides the field interpretation, the most widespread one.

In the following study I will analyze the relation between two of the above-mentioned no-go theorems, namely the one resulting from Michael Redhead's interpretation of the Reeh-Schlieder theorem and a no-go theorem by David Malament. I will try to show that both theorems are equivalent from an interpretative point of view. Two reasons render the ensuing analysis particularly fruitful. Firstly, we have a firm mathematical ground to start from and, secondly, the detailed comparison will force us to have a very close look at the exact presuppositions and the legitimacy of the conclusions drawn by either author. A successful proof of their equivalence allows to transfer interpretive results about one theorem to the other theorem. That such a transfer would help stems from the fact that both theorems are often discussed separately with different issues

[1]The introduction in [1] is a good account of the first period of research about problems with localizability in quantum physics. References of more modern investigations will follow in the next sections.

being emphasized. I will finish my paper with some contrasting views which call the naturalness and tenability of different assumptions into question. It will turn out that a proper understanding and evaluation of the considered theorems is far more intricate than one might have expected.

2 Non-Localizability and Particles

2.1 Interpreting the Reeh-Schlieder Theorem

The Reeh-Schlieder theorem[2] is a central analyticity result from algebraic QFT (AQFT). AQFT is an axiomatic reformulation of QFT which stands out from similar attempts by its conceptually clear structure. The guiding principle of AQFT is to base QFT on a small set of physically intuitive assumptions which are imposed on algebras of 'local observables', i. e. observables that are assumed to be measurable in finite regions of space-time. The central assumption of AQFT is the locality principle which requires observables belonging to spacelike separated space-time regions to commute. For this reason AQFT is sometimes called 'local quantum physics'. The other assumptions of AQFT can be found in appendix A.

From a physical point of view the Reeh-Schlieder theorem is based on vacuum correlations. Although the theorem stems from an analysis of the vacuum state, the 0-particle state, it can easily be extended to other N-particle states with $N \neq 0$ and all states with bounded energy.[3] This already demonstrates the scope of its importance. As will be shown one consequence of the Reeh-Schlieder theorem is that a decision about the observation of an N-particle state cannot be based on local measurements. I will begin with the mathematical statement of the Reeh-Schlieder theorem. With Ω being the vacuum state and $R(O)$ the von Neumann algebra associated with the bounded open space-time region O the following result can be derived on the basis of the axioms of AQFT (see appendix A):

> *Reeh-Schlieder Theorem:* For *any* bounded open region O in space-time the set $\{A\Omega : A \in R(O)\}$ is dense in Hilbert space $\mathcal{H}$.

One set lies dense in another set if its elements are so finely distributed over the whole second set that for any given element in this second set and any given distance one can find an element in the first set which lies closer to this element in the second set than expressed by the given distance. The set $\{A\Omega : A \in R(O)\}$ is said to be *generated* from the vacuum state Ω by the von Neumann algebra $R(O)$ of local observables associated with O because it is the set which one gets when all operators A in $R(O)$ are applied to

[2] Although [2] is the original source later accounts render an easier access. For references confer the titles which are cited in the following footnotes.

[3] See [3], p. 102.

the vacuum state. With these explanations the content of the Reeh-Schlieder theorem can be expressed as follows: choosing suitable elements of $R(O)$ and acting with them on Ω any vector in $\mathcal{H}$ can be approximated arbitrarily closely.

Note that the statement of the Reeh-Schlieder theorem is different and much stronger than the well-known fact that the Hilbert space $\mathcal{H}$ can be spanned by eigenstates of the number operator which can be built up from the vacuum state by application of suitable creation operators. It can be shown that local algebras $R(O)$ - about which the Reeh-Schlieder theorem talks - do not contain creation (or annihilation) operators. The strength of the Reeh-Schlieder theorem becomes clearer when one considers that O can be a small neighborhood of a point in space-time. What the Reeh-Schlieder theorem now states is that acting on the vacuum state Ω with elements of $R(O)$ we can approximate as closely as we like any state in $\mathcal{H}$, in particular one that is very different from the vacuum in some space-like separated region O'. The Reeh-Schlieder theorem is thus clearly exploiting long distance correlations of the vacuum.

Even though the Reeh-Schlieder theorem is an astonishing result it is not immediately obvious what the conceptual consequences are. To this end we need the following

> *Corollary of the Reeh-Schlieder theorem:* Ω is a separating vector for $R(O)$, i. e. $A\Omega = 0 \Rightarrow A = 0$.

The reason for the expression "separating vector" can be seen by setting $A \equiv (A_1 - A_2)$ where $A_1, A_2 \in R(O)$. If Ω is a separating vector then two elements of $R(O)$ which yield the same result when acting on Ω must be one and the same operator. The use of this corollary yields an interesting interpretive result. Again, the assumptions are the standard axioms of AQFT. It will be shown that:

> *Redhead's Claim*: Local measurements can never decide whether we observe an N-particle state.[4]

Let us call this statement 'Redhead's claim' because it was he who expressed this conclusion more explicitly than anybody else, at least to my knowledge. The underlying mathematical result is the fact that a projection operator P_Ψ which corresponds to an N-particle state Ψ can never be an element of a local algebra $R(O)$. Since the proof of 'Redhead's claim' is very instructive and easy I will restate it here.

Proof of Redhead's Claim: Given the Reeh-Schlieder theorem and its above-mentioned corollary the proof is a straightforward reductio ad absurdum. Consider an arbitrary N-particle state Ψ (with $N \neq 0$). Since Ψ is orthogonal to the 0-particle state Ω the corresponding projector P_Ψ satisfies $P_\Psi \Omega = 0$. Let us now ask whether P_Ψ is an element of a local algebra $R(O)$ corresponding to a bounded region O. If this were the case then one could decide by a local measurement, restricted to the region O, whether we have an N-particle state Ψ or not, or, to be more cautious, whether we will find such a state

[4]Cf. [4].

or not when we perform such a measurement. Let us assume now as a trial that P_Ψ is an element of a local algebra $R(O)$. In this case our corollary is applicable so that $P_\Psi \Omega = 0$ would imply $P_\Psi = 0$. This, however, contradicts our assumption that P_Ψ is the projection operator corresponding to the N-particle state Ψ, which cannot be $\mathbf{0}$ unless Ψ itself is 0. Therefore, we are forced to drop our assumption that P_Ψ is an element of a local algebra $R(O)$. q. e. d.

There are two possible objections as regards the relevance of Redhead's claim that local measurements can never decide whether we observe an N-particle state. The first objection is that Redhead's argument is not only applicable to particle states. Since local algebras are of type III according to von Neumann's classification[5] they do not contain *any* one-dimensional projection operators so that the above claim applies to a huge class of states with N-particle-states being only one special kind. While this comment is correct the Redhead's argument is still of considerable interest since it has the most serious consequences in the case of particle states when particles are taken to be local objects.

The second objection to the relevance of Redhead's claim I have in mind is that its physical basis is no more surprising than EPR-correlations. On the one hand this is partly correct since vacuum correlations which underly the Reeh-Schlieder theorem are in fact EPR-like and lead to violations of Bell inequalities as was shown in [5]. However, on the other hand Redhead's argument applies to 1-particle states as well and this is different from the entanglement of subsystems in standard EPR case.

The exact meaning and range of Redhead's claim will become clearer when we compare it with the above-mentioned no-go theorem by David Malament which is the topic of the next subsection. The comparison itself will follow after the exposition of these two results to be compared.

2.2 Malament's No-go Theorem

Malament's no-go theorem [6] is another analyticity result which rests on a lemma of Borchers [7]. In short Malament's theorem says that a relativistic quantum theory of a fixed number of particles predicts a zero probability for finding a particle in any spatial set. The following account of Malament's no-go theorem orientates itself closely to Malament's own cristal clear exposition.

Malament's no-go theorem rests on four conditions. We assume the existence of projection operators P_Δ on Hilbert space $\mathcal{H}$ representing the proposition that a particle detector would respond if a position measurement were performed in the spatial set Δ. Δ is taken to be a bounded open subset of a spacelike hyperplane in Minkowski space-time M. Furthermore we will assume that there is a strongly continuous, unitary representation $U(\mathbf{a}), \mathbf{a} \in M$ (in $\mathcal{H}$) of the translation subgroup of the Poincaré group

[5]See e. g. [3].

in M. Malament stipulates the following four conditions as natural constraints for a relativistic quantum theory of particles.

(i) *Translation Covariance Condition:*

$$P_{\Delta+\mathbf{a}} = U(\mathbf{a})P_\Delta U(-\mathbf{a}) \tag{1}$$

for all $\mathbf{a}$ in M and all spatial sets Δ. $\Delta + \mathbf{a}$ denotes the set Δ after a translation by $\mathbf{a}$.

(ii) *Energy Condition:* The spectrum of the Hamiltonian operator $H(\mathbf{a})$ is bounded below provided that $H(\mathbf{a})$ satisfies $U(t\mathbf{a}) = e^{-itH(\mathbf{a})}$ for all unit vectors $\mathbf{a}$ in M which are future directed and timelike.

(iii) *Localizability Condition:* If Δ_1 and Δ_2 are disjoint spatial sets in the same hyperplane,

$$P_{\Delta_1}P_{\Delta_2} = P_{\Delta_2}P_{\Delta_1} = \mathbf{0}. \tag{2}$$

The Localizability Condition is the essential ingredient of the particle concept: A particle - in contrast to a field - cannot be found in two disjoint spatial sets at the same time. P_{Δ_1} and P_{Δ_2} must therefore be orthogonal. The condition is very weak since it does not set any finite limit to the travelling speed of a particle.

(iv) *Locality Condition:* If Δ_1 and Δ_2 are spatial sets (not necessarily in the same hyperplane) that are space-like related,

$$P_{\Delta_1}P_{\Delta_2} = P_{\Delta_2}P_{\Delta_1}. \tag{3}$$

The Locality Condition is the relativistic part of Malament's assumptions: Probabilities for measurements in Δ_1 must be statistically independent from probabilities for measurements in the space-like related Δ_2. Therefore, P_{Δ_1} and P_{Δ_2} must commute. This condition again is very weak since it does not require for P_{Δ_1} and P_{Δ_2} to be orthogonal which would rule out that the particle can travel at superluminal speed.

Using a lemma of Borchers and each of the four conditions above Malament derives that (the proof can be found in appendix B)

$$P_\Delta = 0 \qquad \text{for any spatial set } \Delta. \tag{4}$$

This means that the probability for finding a particle anywhere in space is 0 no matter how large Δ is. Since that is an unacceptable conclusion Malament's proof has the weight of a no-go theorem provided that we acccept his four conditions as natural assumptions for a particle interpretation. To put it another way, a relativistic quantum theory of one and thereby of any number of particles has to assume, in order not to contradict itself, a world in which particles can never be detected. Accepting this conclusion Malament's no-go theorem shows that there is no middle ground between QM and QFT, i. e. no theory with a fixed number of particles (like in QM) and which is relativistic (like QFT) without running into the non-localizability problem of the no-go theorem. One is forced towards QFT (without a fixed number of particles).

For now I will leave the question aside whether or not Malament's theorem actually has the intended bearing. I will come back to this point in the concluding discussion.

3 Two Theorems, One Result

Although both Redhead's interpretation of the Reeh-Schlieder theorem as well as Malament's no-go theorem are concerned with the non-localizability of relativistic N-particle states they seem to be two different results on a related matter. This is not to say that there is any contradiction between the two. It only seems that we have two more results which enrich our knowledge about the possibilities and restrictions for an ontological interpretation of QFT.

I wish to show that there is, in fact, just one result. Even though Redhead and Malament draw two different conclusions and use two different theorems for their considerations I shall demonstrate that the mathematical machinery employed is the only difference from an interpretive point of view. The assumptions made as well as the result one gets under these circumstances are exactely the same with respect to the question of localizability of relativistic N-particle states. In order to prove this claim I will compare Redhead's and Malament's work part by part using one common mathematical language. Since Malament's no-go theorem is already very clearly and explicitly structured I shall use his theorem as the standard for comparison.

Redhead starts from the Reeh-Schlieder theorem with its very general assumptions and content. It is only at the end of his considerations that Redhead gets more specific with respect to the question of the localizability of relativistic N-particle states. Malament, in contrast to this, addresses the question from the very start. Accordingly, his assumptions, in particular his four conditions, are more specific from the outset of his argumentation. In order to see this we will start from the end where the parallels can be seen most easily. To this end we will compare the two final results which have direct impact on the leading question of particle localization. In the second step we will examine whether the assumptions of the two results are identical so that one can legitimately say that we have effectively just one result from an interpretive standpoint.

The Proofs On the one hand Malament shows that, given his assumptions,

$$P_\Delta = 0 \qquad \text{for any spatial set } \Delta, \tag{5}$$

where P_Δ is the projection operator onto the set Δ which can be any subset of a spacelike hyperplane of Minkowski space-time. This means that the result applies to every possible inertial observer, i. e. for any given inertial observer and any given time we get $P_\Delta = 0$ no matter where and how large Δ is chosen.

Redhead, on the other hand, concludes that local measurements can never decide whether we observe an N-particle state[6]. In Redhead's consideration of the Reeh-Schlieder theorem there is no explicit reference to subsets of spacelike hyperplanes, however. His conclusion, therefore, seems to be more general in that respect already since it applies, due to the Reeh-Schlieder theorem, to any bounded open set in space-time. But let us consider whether there are any implicit restrictions to these space-time

[6]See section 2.1.

sets in the context of localizability. We are in this case concerned with position measurements which should be as general as possible as long as the area where the measurement takes place stays finite. Nevertheless, how ever general the position measurements we consider are, each single measurement has to take place at some time in some observers frame of reference. Or, to put the same thing in other words, the 'bounded open set in space-time' we consider has to be a bounded open subset of a spacelike hyperplane, just as in Malament's no-go theorem. So we passed the first check in the comparison of Redhead's interpretation of the Reeh-Schlieder theorem and Malament's no-go theorem with the result that they refer to the same set of measurement regions.

Now let us compare what Malament and Redhead actually say about position measurements in those space-time regions we discussed above. Malament on the one hand concludes that "there cannot be a relativistic quantum mechanics of (localizable) particles".[7] He derives his conclusion by showing that $P_\Delta = 0$ for any spatial set, given his assumptions. Since this result is a striking contradiction to all experimental facts Malament can claim to have a no-go theorem provided that one accepts his assumptions.

Redhead on the other hand concludes, given his assumptions in turn, that "it is not a local question to ask "are we in an N-particle state?"".[8] Formally he shows that P_Ψ can never be an element of a local algebra, where Ψ is an N-particle state. He proves this claim indirectly by showing that P_Ψ would have to be $\mathbf{0}$ if it were an element of a local algebra.

We can immediately see that there is at least a certain superficial similarity between Malament's and Redhead's results. In both cases some kind of localizability is assumed and shown to lead to contradictions by deriving that a certain class of projection operators, which are directly linked to localizability, would have to vanish in that case. Since, for different reasons, it is not acceptable for these projection operators to be zero we can exclude the possiblity of localizability for certain states under certain conditions.

In order to see whether there is more than just this superficial similarity between Malament's and Redhead's results we have to compare the respective classes of projection operators

$$P_\Delta,\ \Delta \text{ bounded open set in spacelike hyperplane} \quad (Malament)$$

and

$$P_\Psi \in R(O),\ O \text{ bounded open set} \quad (Redhead)$$

more closely. Whereas P_Δ is explicitly linked to localizability[9] this is true for P_Ψ only in an implicit way via the assumption that it is an element of a local algebra $R(O)$.

Let us compare P_Δ and P_Ψ for the same bounded open region Δ of space-time, i. e. P_Δ and $P_\Psi \in R(\Delta)$. Surely P_Δ should be an element of $R(\Delta)$ since it refers to

---

[7]See title of his paper [6].

[8]See [4], p. 127.

[9]The assumption of localizability is complete together with the localizability condition which we will discuss in the next paragraph.

measurements in space-time region Δ. As we have seen, however, P_Δ can only be the trivial element **0** since it vanishes under those assumptions we stated above.

We can thus take Malament's result

$$\underbrace{\begin{array}{l} P_\Delta,\ \Delta\text{: arbitrary spatial bounded open set} \\ \text{Conditions for translation covariance, energy, localizability and locality} \\ \text{("Mal.-assumptions") are fulfilled.} \\ \xRightarrow{\text{(Borchers' th.)}} P_\Delta = 0 \text{ in contradiction to experiments} \end{array}} \\ \Longrightarrow \text{Malament-assumptions cannot all be maintained.} \tag{6}$$

and reformulate it as follows:

$$\underbrace{\begin{array}{l} P_\Delta \in R(\Delta),\ \Delta\text{: arbitrary spatial bounded open set} \\ +\ \text{localizability assumption} \\ \xRightarrow{\text{(Borchers' th.)}} P_\Delta = 0 \text{ in contradiction to experiments} \end{array}} \\ \begin{array}{l} \Longrightarrow \text{Localizability assumption cannot be maintained for relativistic} \\ \quad \text{N-particle states. (Impossible since localizability is an indispen-} \\ \quad \text{sible ingredient of the particle concept)} \\ \quad \underline{\text{or}} \\ \quad P_\Delta \notin R(\Delta) \end{array} \tag{7}$$

where "+"denotes the logical conjunction of propositions. Redhead's result can now be formulated in an absolutely similar fashion as

$$\underbrace{\begin{array}{l} P_\Psi \in R(\Delta),\ \Delta\text{: arbitrary spatial bounded open set,} \\ \qquad\qquad \Psi\text{: N-particle state} \qquad\qquad (*) \\ \xRightarrow{\text{(Reeh-Schl. th.)}} P_\Psi = 0 \text{ in contradiction to } (*) \end{array}} \\ \Longrightarrow P_\Psi \notin R(\Delta). \tag{8}$$

Since in both cases we are dealing with N-particle states (implicitly in Malament's case, explicitly in Redhead's case) propositions about P_Δ and P_Ψ amount to the same thing: the impossibility to measure N-particle states when they are assumed to be localized.[10]

4 Discussion

The results about non-localizability which have been explored in this section may appear to be not very astonishing in the light of the following facts about ordinary QM:

[10]A more detailed and comprehensive proof of the equivalence of Malament's and Redhead's theorems will appear in a future publication.

Quantum mechanical wave functions (in position representation) are usually smeared out over all $\mathbb{R}^3$, so that everywhere in space there is a non-vanishing probability for finding a particle. This is even the case arbitrarily close after a sharp position measurement due to the instantaneous spreading of wave packets over all space. Note, however, that ordinary QM is non-relativistic. A conflict with special relativity theory would thus not be very surprising although it is not at all clear whether the above-mentioned quantum mechanical phenomena can actually be exploited to allow for superluminal signalling. QFT, on the other side, has been designed to be in accordance with special relativity theory. The local behaviour of phenomena is one of the leading principles upon which the theory was built. This makes non-localizability within the formalism of QFT a much severer problem for a particle interpretation.

Only very recently Malament's reasoning has come under attack in [8] and [9]. Both argue to the effect that there are alternatives to Malament's conclusion. The main line of thought in both criticisms is that Malament's 'mathematical result' might just as well be interpreted as evidence that the assumed concept of a sharp localization operator is flawed and has to be modified either by allowing for unsharp localization [9] or for so-called "hyperplane dependent localization" [8]. I fully agree to the extent to which the conclusiveness of Malament's interpretation is concerned. However, one problem is that the proposed alternatives are not sufficiently worked out to allow for a final evaluation.

In his article "A dissolution of the problem of locality" [10] Saunders draws a different conclusion from Malament's (as well as from similar) results. Rather than granting Malament's four conditions and deriving a problem for a particle interpretation Saunders takes Malament's proof as further evidence that one cannot hold on to all four conditions. According to Saunders it is the localizability condition which might not be a natural and necessary requirement on second thought.

A short word on terminology: Saunders calls Malament's localizability condition "weak-placing condition" since it does not, as Saunders "strong-placing condition" does, entail microcausality which Malament postulates separately in his locality condition. Saunders "strong-placing condition" reads as follows:

$$\Delta_1, \Delta_2 \text{ spacelike related } \Rightarrow P_{\Delta_1} P_{\Delta_2} = P_{\Delta_2} P_{\Delta_1} = \mathbf{0}.$$

Stressing that "relativity requires the language of events, not of things" Saunders argues that the localizability condition looses its plausibility when it is applied to events: It is not sensible to postulate that the same event can not occur at two disjoint spatial sets at the same time. One can only require for the same *kind* of event not to occur at both places.

For Saunders the particle interpretation as such is not at stake in Malament's argument. The question is rather whether QFT speaks about things in any physical sense. Saunders considers Malament's result to give a negative answer to this question.

One thing seems to be clear by now. Since Malament's 'mathematical result' appears to allow for various different conclusions it cannot be taken as conclusive evidence against the tenability of a particle interpretation of QFT. Since I could show that Redhead's interpretation of the Reeh-Schlieder theorem is equivalent to Malament's theorem the same conclusion applies to Redhead's results as well. All one can say at the current stage of research is that a particle interpretation of QFT is faced with problems. It is not clear, however, whether it will be possible to overcome these problems or not and, in the positive case, how far the usual concept of a particle has to be modified in order to cope with these problems.

Appendix A: Assumptions of AQFT

- $\{\mathcal{O}_i\}$ Bounded open regions in Minkowski space-time.
- **Local observables:** Hermitian elements in local (non-commutative) von Neumann-algebras $\mathcal{R}(\mathcal{O}_i)$ which are subsets of $\mathcal{B}(\mathcal{H})$, i. e. the set of bounded operators acting on Hilbertspace $\mathcal{H}$.

(i) **Locality:**

$$\text{If } \mathcal{O}_1 \subset \mathcal{O}_2', \text{ then } \mathcal{R}(\mathcal{O}_1) \subset [\mathcal{R}(\mathcal{O}_2)]',$$

where $\mathcal{O}'$ denotes the set of all points which are spacelike separated from all points in $\mathcal{O}$ and $[\mathcal{R}(\mathcal{O})]'$ the set of all operators (in $\mathcal{R}$) which commute with all operators in $\mathcal{R}(\mathcal{O})$.

(ii) **Covariance:** The Poincaré group $\mathcal{P}_+^\uparrow$ is represented by automorphisms on the net of local algebras, i. e. $\alpha_g[\mathcal{R}(O)] = \mathcal{R}[g(O)]$. The map $g \to \alpha_g(R)$ is assumed to be strongly continuous for any $R \in \mathcal{R}$ [$\leftarrow$ global algebra generated by all local algebras $\mathcal{R}(\mathcal{O})$]

(iii) **Isotony:**

$$\text{If } \mathcal{O}_1 \subseteq \mathcal{O}_2, \text{ then } \mathcal{R}(\mathcal{O}_1) \subseteq \mathcal{R}(\mathcal{O}_2).$$

(iv) **Diamond:** $\mathcal{R}(\mathcal{O}) = \mathcal{R}[D(\mathcal{O})]$, $D(\mathcal{O})$: 'causal shadow' of $\mathcal{O}$.

(v) **Weak additivity:** is essentially the assumption that the space-time continuum is homogeneous, so that no subquantal phenomena like minimal legth are present.

(vi) **Irreducibility:** $\mathcal{H}$ has no non-trivial subspace invariant under elements of all $\mathcal{R}(\mathcal{O})$.

(vii) **Spectrum:** Spectrum of translation subgroup of $\mathcal{P}$ is contained in closed forward lightcone.

(viii) **Vakuum:** There is a unique state Ω, invariant under all Poincaré transformations, i. e. $U(g)\Omega = \Omega \; \forall g \in \mathcal{P}$.

Appendix B: Proof of Malament's No-Go Theorem

I begin by restating and explaining the central lemma which Malament employs.

Borchers' Lemma: Let the spectrum of the generator H of the group of unitary time translation operators $U(t) = e^{-itH}$ be bounded below. For two projection operators P_1 and P_2 which satisfy

$$P_1 P_2 = \mathbf{0} \tag{9}$$

and

$$[U(t)P_1U^*(t), P_2] = \mathbf{0} \quad \text{for } |t| < \epsilon, \epsilon > 0 \tag{10}$$

it holds that

$$U(t)P_1U^*(t)P_2 = \mathbf{0} \quad \text{identically in } t, \text{ i. e. for all } t. \tag{11}$$

$U(t)P_1U^*(t)$ denotes the same proposition about a measurement result as P_1 only with respect to a later time t, let us therefore abbreviate $P_{1,t} \equiv U(t)P_1U^*(t)$. Borchers' lemma thus says that if P_1 and P_2 are orthogonal and if $P_{1,t}$ and P_2 commute for all sufficiently small $t < \epsilon$ then $P_{1,t}$ and P_2 are orthogonal for all times t. That analyticity is of vital importance in this lemma can be seen from the fact that a proposition (9) which holds for one particular value of t (viz. $t = 0$) can, when certain conditions (10) hold for a small range of values t, be extended to all values of t (note that $U(t)P_1U^*(t) = P_{1,t} = P_1$ for $t = 0$). Since $U(t)P_1U^*(t)P_2 = \mathbf{0}$ implies $P_2U(t)P_1U^*(t) = \mathbf{0}$ (take the c. c.) and thereby $[U(t)P_1U^*(t), P_2] = \mathbf{0}$ one can say equivalently that a proposition (10) which holds for a small range of values t can, when a certain condition (9) holds for one particular value of t (viz. $t = 0$), be extended to all values of t.

Using this lemma of Borchers and the above-mentioned conditions for translation covariance, energy, localizability and locality Malament derives that

$$P_\Delta = 0 \qquad \text{for any spatial set } \Delta \tag{12}$$

with the interpretation that the probability for finding a particle in any finite region of space is 0 no matter how large Δ is. Since this result is in contradiction to the fact that "particle detectors" do click sometimes and since the above conditions are meant to capture the minimal traits for a relativistic quantum theory of particles Malament concludes that such a theory is not viable.

In the following pages Malament's proof of (12) will be restated in three steps.

1.) Consider an arbitrary spatial set Δ and take any vector $\mathbf{a}$ which is parallel to Δ such that Δ and $\Delta + \mathbf{a}$ are disjoint (see figure 1). Under these circumstances it holds that:

$$P_\Delta P_{\Delta+\mathbf{a}} = P_{\Delta+\mathbf{a}} P_\Delta = \mathbf{0}. \tag{13}$$

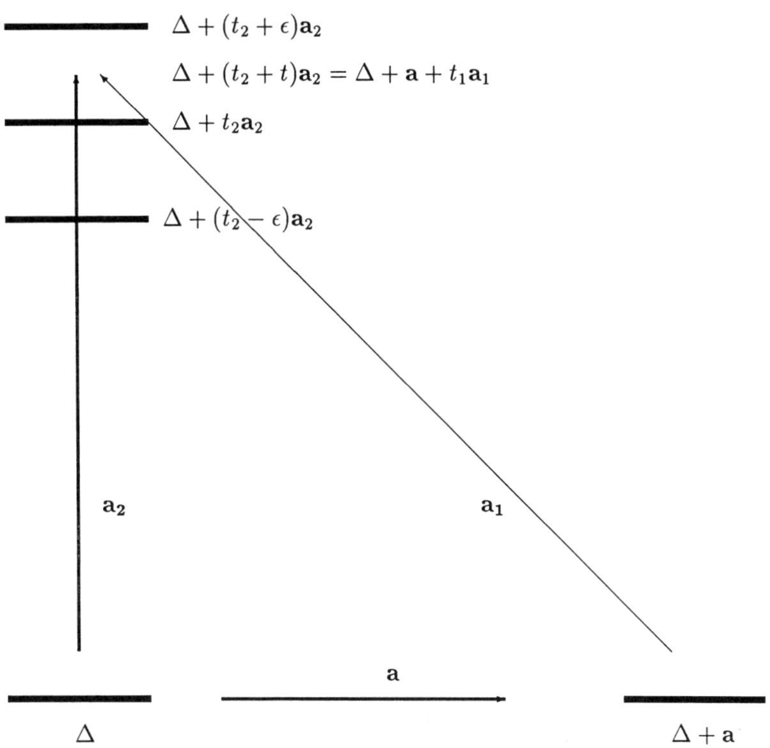

Figure 1: Slight modification of Malament's original figure.

2.) Take any future directed timelike unit vector $\mathbf{a_1}$ and a sufficiently small $|t|$ such that Δ and $\Delta + \mathbf{a} + t\mathbf{a_1}$ are spacelike related. With the Locality Condition from section 2.2 it follows that

$$[P_\Delta, P_{\Delta+\mathbf{a}+t\mathbf{a_1}}] = \mathbf{0}, \tag{14}$$

and because of the Translation Covariance Condition we get

$$[P_\Delta, U(t\mathbf{a_1})P_\Delta U(-t\mathbf{a_1})] = \mathbf{0}. \tag{15}$$

With equations (13) and (15) and under the assumption of the Energy Condition from section 2.2 we are now in the position to apply the lemma of Borchers for the first time to get

$$P_\Delta U(t\mathbf{a_1})P_\Delta U(-t\mathbf{a_1}) = U(t\mathbf{a_1})P_\Delta U(-t\mathbf{a_1})P_\Delta = \mathbf{0} \tag{16}$$

for all t! Using the Translation Covariance Condition once again, now "in the opposite direction", yields:

$$P_\Delta P_{\Delta+\mathbf{a}+t\mathbf{a_1}} = \mathbf{0}. \tag{17}$$

3.) Take any future directed timelike unit vector $\mathbf{a_2}$ and choose $t_2 > 0, \epsilon > 0$ such that $\Delta + (t_2 + t)\mathbf{a_2}$ and $\Delta + \mathbf{a}$ are timelike related for all t with $|t| < \epsilon$. Apply the translation covariance condition once, the lemma by Borchers and apply the translation covariance condition once again in order to get:

$$P_\Delta U[(t+t_2)\mathbf{a_2}]P_\Delta U[-(t+t_2)\mathbf{a_2}) = \mathbf{0} \tag{18}$$

Taking $t = -t_2$ yields:

$$P_\Delta P_\Delta = \mathbf{0} \tag{19}$$

which, since P_Δ is a projection operator, implies:

$$P_\Delta = \mathbf{0}. \tag{20}$$

q. e. d.

Acknowledgement

I would like to thank Jeff Barrett, Jeremy Butterfield, David Malament and Miklos Rédei for discussions on these issues.

References

[1] Arthur S. Wightman. On the localizabilbity of quantum mechanical systems. *Reviews of Modern Physics*, 34(4):845, October 1962.

[2] H. Reeh and S. Schlieder. Bemerkungen zur Unitäräquivalenz von Lorentzinvarianten Feldern. *Nuovo Cimento*, 22:1051–1068, 1961.

[3] Rudolf Haag. *Local Quantum Physics: Fields, Particles, Algebras.* Springer, Berlin, Heidelberg, New York, second edition, 1996.

[4] Michael L. G. Redhead. More ado about nothing. *Foundations of Physics*, 25:123–137, 1995.

[5] Michael L. G. Redhead and Fabian Wagner. Unified treatment of EPR and Bell arguments in Algebraic Quantum Field Theory. *Foundations of Physics Letters*, 11:111–125, 1998.

[6] David Malament. In defense of dogma: Why there cannot be a relativistic quantum mechanics of (localizable) particles. In Rob Clifton, editor, *Perspectives on Quantum Reality: Non-Relativistic, Relativistic, and Field-Theoretic*, The University of Western Ontario Series in Philosophy of Science, pages 1–10, Dordrecht, Boston, London, 1996. Kluwer Academic Publishers.

[7] Hans-Jürgen Borchers. A remark on a theorem of B. Misra. *Communications in Mathematical Physics*, 4:315–323, 1967.

[8] Gordon N. Fleming and Jeremy Butterfield. Strange positions. In Jeremy Butterfield, editor, *From Physics to Philosophy*, pages 108–165. Cambridge University Press, 1999.

[9] Paul Busch. Unsharp localization and causality in relativistic quantum theory. *Journal of Physics A: Mathematics General*, 32:6535, 1999.

[10] Simon Saunders. A dissolution of the problem of locality. In David Hull, Micky Forbes, and Richard M. Burian, editors, *PSA 1994: Proceedings of the 1994 Biennial Meeting of the Philosophy of Science Association, Vol. 2*, pages 88–98, 1995.

Decoherence and its Implications in Quantum Computation and Information Transfer
T. Gonis and P.E.A. Turchi (Eds.)
IOS Press, 2001

Quantum reference systems: reconciling locality with quantum mechanics

Gyula BENE

Institute for Theoretical Physics, Eötvös University, Pázmány Péter sétány 1/A, H-1117 Budapest, Hungary

Abstract. The status of locality in quantum mechanics is analyzed from a nonstandard point of view. It is assumed that quantum states are relative in the sense that they depend on and are defined with respect to some bigger physical system which contains the former system as a subsystem. Hence, the bigger system acts as a reference system. It is shown that quantum mechanics can be reformulated in accordance with this new physical assumption.

Additional laws express the (probabilistic) relation among states which refer to different quantum reference systems. They replace von Neumann's postulate about the measurement (collapse of the wave function). The dependence of the quantum states on the quantum reference systems resolves the apparent contradiction connected with the measurement (Schrödinger's cat paradox). There is another important consequence of this dependence: states may not be comparable, i.e., they cannot be checked by suitable measurements simultaneously. This special circumstance is fully reflected mathematically by the theory. Especially, it is shown that certain joint probabilities (or the corresponding combined events) which play a vital role in any proof of Bell's theorem do not exist. The conclusion is that the principle of locality holds true in quantum mechanics, and one has to give up instead of locality an intuitively natural-looking feature of realism, namely, the comparability of existing states.

1 Introduction

During the development of physics research is extended to phenomena at such spatial, time or energy scales that are far beyond the range of everyday experience. These phenomena sometimes force us to revise our previous concepts which have seemed to us natural or even indispensable, but are actually rooted in our limited previous experience and finally prove to be of approximate validity. This is reflected in the growing level of abstraction of physical theories. Quantum phenomena have forced already several such revisions. A primary example is the surrender of determinism, but the wavelike behaviour of particles, the existence of discrete energy levels and Heisenberg's uncertainty relations also imply substantial revisions of some basic features of classical physics. It is well known that quantum phenomena force us to surrender or revise at least one more very basic concept: Bell's theorem tells us that that we have to give up (or revise) either locality, or realism, or inductive inference[1]-[9]. On the other hand, these concepts are so deeply rooted in scientific thinking that one is reluctant to give up any of them. Many people think that perhaps locality is the weakest concept of the three (cf. arguments for nonlocality in Ref.[10] and counterarguments in Ref.[11]). Nevertheless, all the fundamental equations of physics satisfy the principle of locality, including e.g. the standard model of the elementary particles, so it is rather implausible that locality would be violated just in quantum measurements. The myth of nonlocality is much less attractive today than it has been in the fifties or sixties when people tried to build nonlocal quantum field theories.

There are at least two other myths related to quantum mechanics: one of them is the notorious belief that around some large particle number or mass quantum mechanics gradually becomes invalid and classical mechanics starts to be correct. Actually this very simple view has no experimental support at all. On the contrary, all the available experimental results confirm the validity of quantum mechanics, even in the macroscopic situation of superfluid He. The actual motivation of the above view is the paradox of quantum measurements, which involves the third myth, the fictitious collapse of the wave function[12]. This process was invented for the interpretation of quantum mechanics, i.e., for making contact between theoretical results and the experience. Nevertheless, if considered an actual physical process, it contradicts the Schrödinger equation and violates the principle of locality.

In the present paper we show that a consistent theory can be constructed without accepting any of the three myths above. As a guiding principle, we try to confine the number of assumptions to a minimum and base our considerations on logical clarity and consistency. Obviously, this is all we can safely do in a range of phenomena where our intuition does not work. As expected, one has to revise rather substantially some basic concepts. Namely, it will be shown that one has to give up the belief that a physical system at a given instant of time has a quantum state in an absolute sense, i.e., a state which depends only on the system to be described (apart from unitary transformations which may appear if another coordinate system is chosen). Instead, quantum states will express basicly a relation between a system to be described and another system containing the former one. The latter, bigger system acts as reference system[13]. This new kind of dependence on reference systems will be explained in detail in Section 2. The idea that quantum states are relational has been put forward (in mathematically and conceptually different ways) in Refs.[14],[15]. The relational nature of quantum mechanics is expressed also by the proposal that correlations (rather than states) are the basic entities ("correlations without correlata") [16].

In Section 3. the basic rules of quantum mechanics are formulated in a way that takes into account the dependence of the states on quantum reference systems. von Neumann's measurement postulate will be replaced by the rules which express the indeterministic relation among states defined with respect to different quantum reference systems. The most important of these rules, which has no counterpart in standard quantum mechanics, utilizes the eigenstates of the reduced density matrix. These eigenstates (Schmidt states[17]) play a central role also in the modal interpretations[1][18]. Their significance has also been emphasized within the framework of decoherence theory [19], [20]. According to the present

[1]One may even consider the present approach a modal interpretation in the sense that it also aims at giving a physical meaning to the quantum formalism ("properties attribution") like modal interpretations.

approach, Schrödinger's equation is universally valid and measurements are considered as any other interaction between two physical systems. If the dynamics of this interaction has the special form of a perfect (von Neumann type) measurement, the usual quantum mechanical expressions for the probabilities are recovered.

The dependence of quantum states on quantum reference systems (as defined in the present paper) leads to an even more radical consequence: some states which exist may not be compared. This is totally unusual for our intuition and classical experience, but if the states which we would like to compare are defined with respect to different reference systems, there is no logical necessity that one can indeed compare them. This basically influences the picture we have about reality. We cannot think of existing states as of letters in a big book. If two letters exist, we may compare them in principle, without changing them or their relationship. Thus (if we do not possess complete knowledge about them) it makes sense to speak about the joint probability of their simultaneous presence. In case of existing quantum states one may check any of them by a suitable measurement, without disturbing that particular state. But such a measurement usually does disturb the other states, therefore, these measurements usually cannot be performed simultaneously (or one by one) without influencing any of the relevant states. Thus one cannot give any operational definition of the joint probability of the simultaneous presence of states. This seems to be rather odd, and one might think that independently of this situation such probabilities must exist on the ground that each existing state somehow corresponds to an element of the reality. On the other hand, one should keep in mind that this set-like picture of the reality is based on the fact that in classical physics the property of comparability is always given. In the absence of that one has no reason to expect that the consequences remain intact. Let us emphasize that this modification of the concept of reality is completely independent of any influence of consciousness. Therefore, the resulting reality concept still preserves the idea that the world is made up of objects whose existence is independent of human consciousness. The non-comparability of states will be discussed in Section 4.

In Section 5. the derivation of Bell's inequality is analyzed from the point of view of the present modified quantum mechanics. It will be shown that the role of the "hidden variable" is played by a particular quantum state. It turns out that the usual "shadowing property" which would express local realism, does not follow, because the corresponding three-fold joint probability is physically meaningless due to the non-comparability of the relevant quantum states. On the other hand, all reasonable requirements of locality are fulfilled.

In Section 6. we summarize the results and conclude.

2 The basic new concept: quantum reference systems

Let us consider a simple example, namely, an idealized measurement of an $\hat{S}_z$ spin component of a spin-$\frac{1}{2}$ particle. Be the particle P initially in the state

$$\alpha|\uparrow> +\beta|\downarrow> , \tag{1}$$

where $|\alpha|^2 + |\beta|^2 = 1$ and the states $|\uparrow>$ and $|\downarrow>$ are the eigenstates of $\hat{S}_z$ corresponding to the eigenvalues $\frac{\hbar}{2}$ and $-\frac{\hbar}{2}$, respectively. The other quantum numbers and variables have been suppressed. The dynamics of the measurement is given by the relations $|\uparrow> |m_0> \rightarrow |\uparrow> |m_\uparrow>$ and $|\downarrow> |m_0> \rightarrow |\downarrow> |m_\downarrow>$, where $|m_0>$ stands for the state of the measuring device M (e.g. a Stern-Gerlach apparatus) before the measurement (no spot on the photographic plate), while $|m_\uparrow>$ ($|m_\downarrow>$) is the state of the measuring device after the measurement that corresponds to the measured spin value $\frac{\hbar}{2}$ ($-\frac{\hbar}{2}$). The shorthand notation $\rightarrow$ stands for the unitary time evolution during the measurement, which is assumed to fulfill the time dependent Schrödinger equation corresponding to the total Hamiltonian of the combined $P+M$ system. As the initial state of the particle is given by Eq.(1), the linearity of the Schrödinger equation implies that the measurement process can be written as

$$(\alpha|\uparrow> +\beta|\downarrow>)|m_0> \rightarrow |\Psi> = \alpha|\uparrow> |m_\uparrow> +\beta|\downarrow> |m_\downarrow> \quad . \tag{2}$$

This simplified dynamics is called a von Neumann type perfect measurement. Let us consider now the state of the measuring device M after the measurement. As the combined system

$P + M$ is in an entangled state, the measuring device has no own wave function and may be described by the *reduced density matrix*[21]

$$\hat{\rho}_M = Tr_P\left(|\Psi >< \Psi|\right) = |m_\uparrow > |\alpha|^2 < m_\uparrow| + |m_\downarrow > |\beta|^2 < m_\downarrow| \quad , \tag{3}$$

where Tr_P stands for the trace operation in the Hilbert space of the particle P. Nevertheless, if we look at the measuring device, we certainly see that either $\frac{\hbar}{2}$ or $-\frac{\hbar}{2}$ spin component has been measured, that correspond to the states $|m_\uparrow >$ and $|m_\downarrow >$, respectively. These are obviously not the same as the state (3). Indeed, $|m_\uparrow >$ and $|m_\downarrow >$ are pure states while $\hat{\rho}_M$ is not. As is well known by now, $\hat{\rho}_M$ cannot be considered a statistical mixture of $|m_\uparrow >$ and $|m_\downarrow >$, i.e., "ignorance interpretation" cannot solve the problem of objectification[22]. Why do we get different states? According to orthodox quantum mechanics, one may argue as follows. The reduced density matrix $\hat{\rho}_M$ has been calculated from the state $|\Psi >$ (cf. Eq.(2)) of the whole system $P+M$. A state is a result of a measurement (the preparation), so we may describe M by $\hat{\rho}_M$ if we have gained our information about M from a measurement done on $P + M$. On the other hand, looking at the measuring device directly is equivalent with a measurement done directly on M. In this case M is described by either $|m_\uparrow >$ or $|m_\downarrow >$. We may conclude that performing measurements on different systems (each containing the system we want to decribe) gives rise to different descriptions (in terms of different states). Let us call the system which has been measured (it is $P + M$ in the first case and M in the second case) the *quantum reference system*. Using this terminology, we may tell that we make a measurement on the quantum reference system R, thus we prepare its state $|\psi_R >$ and using this information we calculate the state $\hat{\rho}_S(R) = Tr_{R\setminus S}|\psi_R >< \psi_R|$ of a subsystem S. We shall call $\hat{\rho}_S(R)$ the state of S with respect to R. Obviously $\hat{\rho}_R(R) = |\psi_R >< \psi_R|$, thus $|\psi_R >$ may be identified with the state of the system R with respect to itself.

Let us emphasize that up to now, despite of the new terminology, there is nothing new in the discussion. We have merely considered some rather elementary consequences of basic quantum mechanics.

Let us return now to the question why the state of the system S (i.e., $\hat{\rho}_S(R)$) depends on the choice of the quantum reference system R. In the spirit of the Copenhagen interpretation one would answer that in quantum mechanics measurements unavoidably disturb the systems, therefore, if we perform measurements on different surroundings R, this disturbance is different, and this is reflected in the R-dependence of $\hat{\rho}_S(R)$. Nevertheless, this argument is not compelling. We may also take the realist point of view and assume that the states of the systems have already existed before the measurements, and appropriate measurements do not change these states. Then the R-dependence of $\hat{\rho}_S(R)$ becomes an inherent property of quantum mechanics, the states themselves represent actually existing properties (or correspond to some elements of the reality), and quantum formalism becomes a desciption of the reality rather than a calculational tool which relates consecutive measurement results. Let us leave at this decisive point the traditional framework of quantum mechanics and follow the new line just sketched. Below a comparison between the two approaches is given.

Standard QM:	**Present approach:**
Quantum reference system dependence is caused by the influence of the measurement. Measurement is a primary concept.	Quantum reference system dependence is a fundamental property. States exist and depend on quantum reference systems even in the absence of any measurement. Measurement is a derived concept.

The meaning of the quantum reference systems is now analogous to the classical coordinate systems. Choosing a classical coordinate system means that we imagine what we would experience if we were there. Similarly, choosing a quantum reference system R means that we imagine what we would experience if we performed a measurement on R that does not disturb $\hat{\rho}_R(R) = |\psi_R><\psi_R|$. In order to see that such a measurement exists, consider an operator $\hat{A}$ (which acts on the Hilbert space of R) whose eigenstates include $|\psi_R>$. The measurement of $\hat{A}$ will not disturb $|\psi_R>$. Let us emphasize that the possibility of nondisturbing measurements is an expression of realism: the state $\hat{\rho}_R(R)$ exists independently whether we measure it or not.

Despite the analogy, there are several differences between the concept of classical and quantum reference systems. These are summarized below:

Classical reference system (CRS):	**Quantum reference system (QRS):**
CRS is an abstraction of actual physical systems, as their detailed structure is not important.	QRS is a physical system. The detailed structure cannot be eliminated.
measurements are done on a system by devices attached to the CRS	measurement is done on the QRS
there is a one-to-one relationship (transformation) between descriptions with respect to two different CRS-s.	there is a stochastic relationship between descriptions with respect to two different QRS-s (indeterminism), or they may not even be compared

As the dependence of $\hat{\rho}_S(R)$ on R is a fundamental property now, one has to specify the relation of the different states and has to relate the formalism with experience. This can be done in terms of suitable postulates[13]. Below these postulates are listed and are applied to the theoretical description of measurements.

3 Rules of the new framework

Postulate 1. ***The system S to be described is a subsystem of the reference system R.***

Postulate 2. ***The state $\hat{\rho}_S(R)$ is a positive definite, Hermitian operator with unit trace, acting on the Hilbert space of S.***

Definition 1. ***$\hat{\rho}_S(S)$ is called the internal state of S.***

Postulate 3. ***The internal states $\hat{\rho}_S(S)$ are always projectors, i.e., $\hat{\rho}_S(S) = |\psi_S><\psi_S|$.***

In the following these projectors will be identified with the corresponding wave functions

$|\psi_S>$ (as they are uniquely related, apart from a phase factor).

Postulate 4. ***The state of a system S with respect to the reference system R (denoted by $\hat{\rho}_S(R)$) is the reduced density matrix of S calculated from the internal state of R, i.e. $\hat{\rho}_S(R) = Tr_{R\setminus S}(\hat{\rho}_R(R))$, where $R \setminus S$ stands for the subsystem of R complementer to S.***

Definition 2. ***An isolated system is such a system that has not been interacting with the outside world. A closed system is such a system that is not interacting with any other system at the given instant of time (but might have interacted in the past).***

Postulate 5. ***If I is an isolated system then its state is independent of the reference system R: $\hat{\rho}_I(R) = \hat{\rho}_I(I)$.***

Postulate 6. ***If the reference system $R = I$ is an isolated one then the state $\hat{\rho}_S(I)$ commutes with the internal state $\hat{\rho}_S(S)$.***

This means that the internal state of S coincides with one of the eigenstates of $\hat{\rho}_S(I)$.

Definition 3. ***The possible internal states are the eigenstates of $\hat{\rho}_S(I)$ provided that the reference system I is an isolated one.***

Postulate 7. ***If I is an isolated system, then the probability $P(S, j)$ that the eigenstate $|\phi_{S,j}>$ of $\hat{\rho}_S(I)$ coincides with $\hat{\rho}_S(S)$ is given by the corresponding eigenvalue λ_j.***

Postulate 8. ***The result of a measurement is contained unambigously in the internal state of the measuring device.***

Postulate 9. ***If there are n ($n = 2,\ 3,\ ...$) disjointed physical systems, denoted by $S_1, S_2, ...S_n$, all contained in the isolated reference system I and having the possible internal states $|\phi_{S_1,j}>, |\phi_{S_2,j}>, ..., |\phi_{S_n,j}>$, respectively, then the joint probability that $|\phi_{S_i,j_i}>$ coincides with the internal state of S_i ($i = 1, ..n$) is given by***

$$P(S_1, j_1, S_2, j_2, ..., S_n, j_n)$$
$$= Tr_{S_1+S_2+...+S_n}[\hat{\pi}_{S_1,j_1}\hat{\pi}_{S_2,j_2}...\hat{\pi}_{S_n,j_n}\hat{\rho}_{S_1+S_2+...+S_n}(I)], \tag{4}$$

where $\hat{\pi}_{S_i,j_i} = |\phi_{S_i,j_i}><\phi_{S_i,j_i}|$.

Postulate 10. ***The internal state $|\psi_C>$ of a closed system C satisfies the time dependent Schrödinger equation $i\hbar\partial_t|\psi_C> = \hat{H}|\psi_C>$.***

Here $\hat{H}$ stands for the Hamiltonian.

Let us emphasize that **Postulate 6** and **8** do not have any counterpart in standard quantum mechanics. They replace the measurement postulate but are not simple translations of that (as it will be shown below). The relation to the experience (that provides the whole construction with a physical meaning) is expressed by **Postulate 8**. On the other hand, **Postulate 6** and **Postulate 8** are mathematically equivalent with the proposals in Refs.[19],[18]. The present approach differs from these at the level of interpretation.

In order to demonstrate the working of the postulates let us consider again the measurement discussed in the previous section. A measurement is treated as a usual interaction between two systems, therefore, it is specified by a Hamiltonian or the corresponding unitary time evolution. For simplicity we assume that it is given by Eq.(2) and that the measuring device + measured object composite is an isolated system. Then **Postulate 5** and **Postulate 3** imply that $\hat{\rho}_{P+M}(P+M) = |\Psi><\Psi|$. According to **Postulate 4** the state of the measuring device with respect to the compound system $P+M$ is

$$\hat{\rho}_M(P+M) = |m_\uparrow> |\alpha|^2 < m_\uparrow| + |m_\downarrow> |\beta|^2 < m_\downarrow| \quad . \tag{5}$$

Applying **Postulate 6** we get that the internal state $\hat{\rho}_M(M)$ (cf. **Definition 1**) of the measuring device is either $|m_\uparrow><m_\uparrow|$ (with probability $|\alpha|^2$, according to **Postulate 7**) or $|m_\downarrow><m_\downarrow|$ (with probability $|\beta|^2$). Finally, **Postulate 8** tells us that the actual measurement results (the experience) correspond to this $\hat{\rho}_M(M)$. As one can see, we get the same result as in orthodox quantum mechanics (this time without assuming the collapse of the wave function). One reason for this was the special dynamics (2). In orthodox quantum mechanics one usually specifies only which observable has been measured. It tacitly assumes a simple approximation for the dynamics of the measurement like (2). In such cases our postulates lead to the same results. If the dynamics is different (e.g., a more detailed description is given), nothing ensures that one gets back the results of the traditional approach exactly. This means that the above postulates cannot be considered as mere translations of orthodox quantum mechanics.

One should also be aware that the present approach leaves much less flexibility than orthodox quantum mechanics. While in the traditional approach one may rather freely choose a bordering line between quantum and classical regime, according to the present approach such bordering line does not exist, and one should in principle always apply a quantum description. Based on this description one may justify the approximations which lead to the traditional approach.

4 Non-comparability of the states

It is an important feature of the present approach that the states defined with respect to different quantum reference systems are not necessarily comparable. Below we explain what it means.

The present approach works as follows. We assume that the internal state of an isolated system I is known. Then the states of all the subsystems S_j of I with respect to I (i.e., $\hat{\rho}_{S_j}(I)$) are also known together with their possible internal states (i.e., the eigenstates of $\hat{\rho}_{S_j}(I)$). The aim of a measurement performed on S_j is to learn which of the possible internal states is the actual one. This can be done without disturbing that state. It can be achieved if an operator commuting with $\hat{\rho}_{S_j}(I)$ is measured, i.e., if the dynamics of the measurement is approximately given by

$$|\phi_{S_j,k} > |m_0 > \rightarrow |\phi_{S_j,k} > |m_k > \quad , \tag{6}$$

where $|\phi_{S_j,k} >$ stands for the k-th possible internal state of S_j (cf. **Definition 3**).

If two subsystems S_1 and S_2 are disjoint ($S_1 \cap S_2 = \emptyset$) then one can perform such nondisturbing measurements on both systems simultaneously without disturbing either $\hat{\rho}_{S_1}(I)$ or $\hat{\rho}_{S_2}(I)$. Correspondingly, **Postulate 9** provides us with a positive definite expresssion for the joint probability that these states coincide with specified possible internal states.

There is a special situation if $S_2 = I \setminus S_1$. The internal state of the isolated system can be expressed in this case by the possible internal states of S_1 and S_2 as

$$|\psi_I >= \sum_k c_k |\phi_{S_1,k} > |\phi_{S_2,k} > \ . \tag{7}$$

This is the Schmidt representation[17]. It readily implies that there is a unique relationship between the internal state of S_1 and that of S_2. Indeed, Eq.(4) yields

$$P(S_1, j_1, S_2, j_2) = P(S_1, j_1)\delta_{j_1 j_2} \tag{8}$$

or

$$P(S_2, j_2|S_1, j_1) = \delta_{j_1 j_2} \ . \tag{9}$$

We can rewrite Eq.(4) in the general case as follows:

$$P(S_1, j_1, ...S_n, j_n) =< \psi_I|\phi_{S_1,j_1} >< \phi_{S_1,j_1}|...|\phi_{S_n,j_n} >< \phi_{S_1,j_1}|\psi_I > \ . \tag{10}$$

Eq.(7) implies

$$< \psi_I|\phi_{S_1,j_1} >< \phi_{S_1,j_1}| = \sum_k c_k^* < \phi_{I\backslash S_1,k}| < \phi_{S_1,k}|\phi_{S_1,j_1} >< \phi_{S_1,j_1}| \tag{11}$$

$$= c_{j_1}^* < \phi_{S_1,j_1}| < \phi_{I\backslash S_1,k}| =< \psi_I|\phi_{I\backslash S_1,j_1} >< \phi_{I\backslash S_1,j_1}| \tag{12}$$

i.e., one can replace $\hat{\pi}_{S_1,j_1} = |\phi_{S_1,j_1} >< \phi_{S_1,j_1}|$ by $\hat{\pi}_{I\backslash S_1,j_1} = |\phi_{I\backslash S_1,j_1} >< \phi_{I\backslash S_1,j_1}|$ in Eq.(4). This sometimes makes possible simultaneous nondisturbing check of internal states of non-disjoint systems.

Suppose, e.g., that $I = S_1 + S_2 + S_3$ (where $S_i \neq \emptyset$ and $S_i \cap S_j = \emptyset$ if $i \neq j$) and we would like to learn the internal state of $A = S_1 + S_2$ and of $B = S_2 + S_3$. The systems A and B are not disjoint, but their complementary systems S_3 and S_1 are. Therefore, one may perform nondisturbing measurements on S_3 and S_1. The knowledge of the internal states of these systems uniquely specifies the internal states of the original systems A and B (cf. Eq.(9)). Correspondingly, Eq.(4) yields a positive definite expression for $P(A, j, B, k)$.

There are, however, situations when neither the systems nor their complementary systems are disjoint. The simplest case is if $I = S_1 + S_2 + S_3 + S_4$ (where $S_i \neq \emptyset$ and $S_i \cap S_j = \emptyset$

if $i \neq j$) and we would like to learn the internal state of $A = S_1 + S_2$ and of $B = S_2 + S_3$. Obviously,

$$A \neq B \, , A \cap B = S_2 \neq \emptyset \tag{13}$$

$$(I \setminus A) \cap B = (S_3 + S_4) \cap (S_2 + S_3) = S_3 \neq \emptyset \tag{14}$$

$$A \cap (I \setminus B) = (S_1 + S_2) \cap (S_1 + S_4) = S_1 \neq \emptyset \tag{15}$$

$$(I \setminus A) \cap (I \setminus B) = (S_3 + S_4) \cap (S_1 + S_4) = S_4 \neq \emptyset \tag{16}$$

Thus, one cannot relate this situation to the case of disjoint subsystems. Indeed, a measurement done on A or on $I \setminus A$ (which does not disturb $\hat{\rho}_A(A)$) usually does disturb $\hat{\rho}_B(B)$ and $\hat{\rho}_{I\setminus B}(I \setminus B)$. Therefore, there is no way to perform a measurement which conveys information about both the internal state of A and that of B. This is reflected mathematically by the postulates, namely, the positivity of $P(A, i, B, k)$ (if Eq.(4) is formally applied) is not guaranteed (it is usually not even real). In such a situation the states $\hat{\rho}_A(A)$, $\hat{\rho}_B(B)$ cannot be compared, although both exist separately. Conclusions based on assumptions about the simultaneous existence of $\hat{\rho}_A(A)$ and $\hat{\rho}_B(B)$ (in the sense that one imagines that he knows both states) or, especially, about a related joint probability can lead to contradictions. Therefore, non-comparability of existing states is an essential feature of the present approach.

5 Bell's inequality and the principle of locality

Let us consider a Bell-type experiment performed on two spin-$1/2$ particles. In order to exhibit the mathematical structure we write the internal state of the two particle system before the measurements as

$$\sum_j c_j |\phi_{P_1,j} > |\phi_{P_2,j} > \tag{17}$$

where $c_1 = a$, $c_2 = -b$, $|\phi_{P_1,1} >= |1, \uparrow>$, $|\phi_{P_1,2} >= |1, \downarrow>$, $|\phi_{P_2,1} >= |2, \downarrow>$, $|\phi_{P_2,2} >= |2, \uparrow>$. Eq.(17) is just the Schmidt representation, thus $P(P_1, j, P_2, k) = |c_j|^2 \delta_{j,k}$.

Let us consider now a typical experimental situation, when measurements on both particles are performed. We shall show that according to the present theory the observed correlations are exclusively due to the previous interaction between the particles. Before the measurements the internal state of the isolated system $P_1 + M_1 + P_2 + M_2$ (P_1, P_2 stands for the particles and M_1, M_2 for the measuring devices, respectively) is given by $\left(\sum_j c_j|\phi_{P_1,j} > |\phi_{P_2,j} >\right) |m_0^{(1)} > |m_0^{(2)} >$, while it is

$$\sum_j c_j \hat{U}_t(P_1 + M_1) \left(|\phi_{P_1,j} > |m_0^{(1)} >\right) \hat{U}_t(P_2 + M_2) \left(|\phi_{P_2,j} > |m_0^{(2)} >\right) \quad , \tag{18}$$

a time t later, i.e. during and after the measurements. Here $\hat{U}_t(P_i + M_i)$ $(i = 1, 2)$ stands for the unitary time evolution operator of the closed system $P_i + M_i$.

Eq.(18) implies that the internal states of the closed systems $P_1 + M_1$ and $P_2 + M_2$ evolve unitarily and do not influence each other. This follows readily if one applies **Postulates 4, 6** to Eq.(18) and takes into account the unitarity of $\hat{U}_t(P_i + M_i)$. This time evolution can be given explicitly through the relations

$$|\xi(P_i, j) > |m_0^{(i)} > \rightarrow \ |\xi(P_i, j) > |m_j^{(i)} > \quad , \tag{19}$$

where $i, j = 1, 2$ and $|\xi(P_i, j) >$ is the j-th eigenstate of the spin measured on the i-th particle along an axis $z^{(i)}$ which closes an angle ϑ_i with the original z direction. The time evolution of the internal state of the closed systems $P_i + M_i$ is given explicitly by $|\psi_{P_i} > |m_0^{(i)} > \rightarrow \sum_j < \phi_{P_i,j}|\psi_{P_i} > |\phi_{P_i,j} > |m_j^{(i)} >$. As we see, the i-th measurement process is completely determined by the initial internal states of the particle P_i. Therefore, any correlation between the measurements may only stem from the initial correlation of the internal states of the particles.

For the calculation of the state $\hat{\rho}_{M_1}(M_1)$ (which corresponds to the measured value, cf. **Postulate 8**) one needs to know the state of the whole isolated system $P_1 + P_2 + M_1 + M_2$. Using Eq.(19) the final state (18) may be written as

$$\sum_{j,k} \left(\sum_l c_l < \xi(P_1, j)|\phi_{P_1,l} >< \xi(P_2, k)|\phi_{P_2,l} > \right) \times |m_j^{(1)} > |m_k^{(2)} > |\xi(P_1, j) > |\xi(P_2, k) > .$$

Direct calculation shows that

$$\hat{\rho}_{M_1}(P_1 + P_2 + M_1 + M_2)$$
$$= \sum_j \left(\sum_l |c_l|^2 | < \xi(P_1, j) | \phi_{P_1,l} > |^2 \right) |m_j^{(1)} >< m_j^{(1)}|.$$

Note that it is independent of the second measurement.

According to **Postulate 6** $|\psi_{M_1} >$ is one of the $|m_j^{(1)} >$-s. (Similarly one may derive that $|\psi_{M_2} >$ is one of the $|m_k^{(2)} >$-s.) The probability of the observation of the j-th result (up or down spin in a chosen direction) is

$$P(M_1, j) = \sum_l |c_l|^2 | < \xi(P_1, j) | \phi_{P_1,l} > |^2 . \tag{20}$$

This may be interpreted in conventional terms: $|c_l|^2$ is the probability that $|\psi_{P_1} >= |\phi_{P_1,l} >$, and $| < \xi(P_1, j) | \phi_{P_1,l} > |^2$ is the conditional probability that one gets the j-th result if $|\psi_{P_1} >= |\phi_{P_1,l} >$. Indeed, the initial internal state of P_1 uniquely determines the internal state of $P_1 + M_1$ (owing to the unitary time evolution), and the internal state of $P_1 + M_1$ uniquely determines the internal state of its complementary system $P_2 + M_2$ (and vica versa). Therefore, the joint probability that the initial internal state of P_1 coincides with its l-th possible internal state and the final internal state of M_1 coincides with its j-th possible internal state is the same as the joint probability $P(P_2{+}_2, l, M_1, j)$ that the final internal state of $P_2 + M_2$ coincides with its l-th possible internal state and the final internal state of M_1 coincides with its j-th possible internal state. As $P_2 + M_2$ and M_1 are disjoint systems, $P(P_2{+}_2, l, M_1, j)$ is positive definite, namely, if calculated according to Eq.(18), it is just the summand on the right hand side of Eq.(20).

Thus we see that the initial internal state of P_1 determines the outcome of the first measurement in the usual probabilistic sense. One may show quite similarly that the initial internal state of P_2 determines the outcome of the second measurement in the same way. In this sense the internal states of P_1 and P_2 play the role of local hidden variables. On the other hand, to hidden variables some supernatural features are attributed (e.g. that they cannot be measured in any way) and they are assumed to be comparable with the results of both measurements. None of these properties applies to the internal states of P_1 and P_2. As for the non-comparability we mean that the initial internal state of P_1 and P_2 is not comparable with both $|\psi M_1>$ and $|\psi_{M_2}>$ (while they can be compared with one of them). This means that in our theory there is no way to define the joint probability $P(P_1, l_1, P_2, l_2, (0); M_1, j, M_2, k, (t))$, i.e., the probability that initially $|\psi_{P_1} >= |\phi_{P_1,l_1}>$ and $|\psi_{P_2} >= |\phi_{P_2,l_2}>$ *and* finally $|\psi_{M_1} >= |m_j^{(1)}>$ and $|\psi_{M_2} >= |m_k^{(2)}>$. Intuitively we would write

$$
\begin{aligned}
&P(P_1, l_1, P_2, l_2, (0); M_1, j, M_2, k, (t)) \\
&= |c_{l_1}|^2 \delta_{l_1,l_2} | < \xi(P_1, j)|\phi_{P_1,l_1} > |^2 | < \xi(P_2, k)|\phi_{P_2,l_2} > |^2 ,
\end{aligned}
\tag{21}
$$

as $|c_{l_1}|^2 \delta_{l_1,l_2}$ is the joint probability that $|\psi_{P_1} >= |\phi_{P_1,l}>$ and $|\psi_{P_2} >= |\phi_{P_2,l}>$, and $| < \xi(P_i, j)|\phi_{P_i,l_i} > |^2$ is the conditional probability that one gets the j-th result in the $i - th$ measurement if initially $|\psi_{P_i} >= |\phi_{P_i,l_i}>$ $(i = 1, 2)$. Certainly the existence of such a joint probability would immediately imply the validity of Bell's inequality, thus it is absolutely important to understand why this probability does not exist.

As our postulates provide us with equal time joint probabilities, we should try to express $P(P_1, l_1, P_2, l_2, (0); M_1, j, M_2, k, (t))$ with them. As above, we may note that the initial internal state of P_1 is uniquely related to the final internal state of $P_1 + M_1$, the initial internal state of P_2 is uniquely related to the final internal state of $P_2 + M_2$, and the final internal state of $P_1 + M_1$ is uniquely related to the final internal state of $P_2 + M_2$. Therefore, if $P(P_1, l_1, P_2, l_2, (0); M_1, j, M_2, k, (t))$ exists, it coincides with $P(P_1+M_1, l_1, M_1, j, M_2, k)\delta_{l_1,l_2}$. But the systems $P_1 + M_2$, M_1 and M_2 are not disjoint, neither are their complementary systems. As a result, if one tries to apply Eq.(4), one obtains

$$
\begin{aligned}
P(P_1 + M_1, l_1, M_1, j, M_2, k) &= \sum_{j'} < \xi(P_1, j')|\phi_{P_1,l_1} >< \phi_{P_1,l_1}|\xi(P_1, j) > \\
&\times \left(\sum_{l'} c_{l'} < \xi(P_1, j)|\phi_{P_1,l'} >< \xi(P_2, k)|\phi_{P_2,l'} > \right) \\
&\times \left(\sum_{l''} c^*_{l''} < \xi(P_1, j')|\phi_{P_1,l''} >^* < \xi(P_2, k)|\phi_{P_2,l''} >^* \right)
\end{aligned}
\tag{22}
$$

This expression fails to be real and positive when Bell's inequality is violated. This can be seen because summing Eq.(22) over l_1 one gets the correct joint probability

$$P(M_1, j, M_2, k) = \left| \sum_l c_l < \xi(P_1, j)|\phi_{P_1,l} >< \xi(P_2, k)|\phi_{P_2,l} > \right|^2 \quad . \tag{23}$$

which can also be obtained directly from Eq.(4). This is the usual quantum mechanical expression which violates Bell's inequality and whose correctness is experimentally proven.

The other side of the non-comparability is that a nondisturbing measurement of $|\psi_{P_1+M_1} >$ inevitably disturbs $|\psi_{P_1+M_1+P_2+M_2} >$, and thus $P(M_1, j, M_2, k)$, too. Therefore, if one measures $|\psi_{P_1+M_1} >$, $|\psi_{M_1}$ and $|\psi_{M_2}$ (in order to give an operational definition for $P(P_1 + M_1, l_1, M_1, j, M_2, k)$) $P(M_1, j, M_2, k)$ will be changed. Indeed, after having measured $|\psi_{P_1+M_1}$ [2] by a further measuring device M_3 we get for the internal state of the whole system

$$\sum_l c_l \left(\sum_j < \xi(P_1, j)|\phi_{P_1,l} > |\xi(P_1, j) > |m_j^{(1)} > \right) \\ \times \left(\sum_k < \xi(P_2, k)|\phi_{P_2,l} > |\xi(P_2, k) > |m_k^{(2)} > \right) |m_l^{(3)} > \ . \tag{24}$$

As the systems $M_1,\ M_2,\ M_3$ are disjointed, we may apply **Postulate 9** for $n = 3$ and we get for $P(M_3, l_1, M_1, j, M_2, k)$ the positive definite expression

$$P(M_3, l_1, M_1, j, M_2, k) = |c_{l_1}|^2 | < \xi(P_1, j)|\phi_{P_1,l_1} > |^2 | < \xi(P_2, k)|\phi_{P_2,l_1} > |^2 \ . \tag{25}$$

This readily implies Bell's inequality. Indeed, if **Postulate 9** is applied for Eq.(24) one gets

$$P(M_1, j, M_2, k) = \sum_l |c_l|^2 | < \xi(P_1, j)|\phi_{P_1,l} > |^2 | < \xi(P_2, k)|\phi_{P_2,l} > |^2 \tag{26}$$

which satisfies Bell's inequality and differs from Eq.(23). This is due to the extra measurement which is equivalent with a measurement of the "hidden variable".

Summarizing, we have seen that the initial internal state of P_1 (P_2) determines the first (second) measurement process, therefore, these states 'carry' the initial correlations and 'transfer' them to the measuring devices. As the measurement processes do not influence each other, the observed correlations may stem only from the 'common past' of the particles. On the other hand, any attempt to compare the initial internal states of P_1 and P_2 with

[2] This is equivalent by recording the initial internal state of P_1.

the results of both measurements changes the correlations, thus a joint probability for the simultaneous existence of these states cannot be defined. This means that the reason for the violation of Bell's inequality is that the usual derivations always assume that the states (or "stable properties", "hidden variables" etc.) which carry the initial correlations can be freely compared with the results of the measurements. This comparability is usually thought to be a consequence of realism. According to the present theory, the above assumption goes beyond the requirements of realism and proves to be wrong, because each of the states $|\psi_{P_1+M_1}>$, $|\psi_{P_2+M_2}>$, $|\psi_{M_1}>$ and $|\psi_{M_2}>$ exists individually, but they cannot be compared without changing the correlations.

6 Summary and conclusion

A new approach to quantum mechanics was presented. The measurement postulates were abandoned and replaced by different ones to get a self-consistent, universally valid quantum mechanics. The present approach was based on the idea of quantum reference systems and the relational nature of quantum states, and, from the mathematical point of view the Schmidt representation (eigenstates of density matrices) was utilized. It was briefly shown how this idea solved the objectification problem, then the concept of non-comparability was discussed. Finally, it was shown that locality was maintained in a Bell-type experiment and the violation of Bell's inequality was rendered possible by non-comparability of existing states.

According to the present approach the principle of locality is valid in quantum mechanics, and it is the concept of realism that should be modified. This modification amounts to surrendering the overall comparability of existing things (states) which are defined with respect to different (quantum) reference systems.

Finally, let us mention that the present approach is self-contained, i.e., it does not rely upon the concept of *a priori* classical objects. This means that (if correct) it should give account of classical behavior. This is a great challenge but it has not been discussed here.

7 Acknowledgements

The author is indebted to D.Dieks, N.D.Mermin, P.Mittelstaedt, L.E.Szabó and P.Vermaas for useful discussions and remarks. This work has been supported by the Hungarian Academy of Sciences under Grant No. OTKA T 029752 and the János Bolyai Research Fellowship.

References

[1] J.S.Bell, Physics **1** (1964) 195, reprinted in *Proc. of Int. School of Physics "Enrico Fermi", Course 49* ed.: B.d'Espagnat, (Academic Press, New York, 1974).

[2] J.F.Clauser and M.A.Horne, Phys.Rev.D **10**, 526 (1974).

[3] J.F.Clauser, M.A.Horne, A.Shimony, and R.A.Holt, Phys.Rev.Lett. **23**, 880 (1969).

[4] N.D.Mermin, Am.J.Phys. **58**, 731 (1990).

[5] D.M.Greenberger, M.A.Horne, A.Shimony, and A.Zeilinger, Am.J.Phys. **58**, 1131 (1990).

[6] B. d'Espagnat, Phys.Rep. **110**, 201 (1984).

[7] B. d'Espagnat, Sci.Am. **241** (1979) 128., J.Stat.Phys. **56** (1989) 747.

[8] A.Aspect, P.Grangier, and G.Roger, Phys.Rev.Lett. **49** (1982) 91, A.Aspect, J.Dalibard, and G.Roger, Phys.Rev.Lett. **49** (1982) 1804.

[9] N.D.Mermin, Phys.Rev.Lett. **65** (1990) 3373.

[10] H.P.Stapp, Am.J.Phys.**65** (1997), 300.

[11] N.D.Mermin, Am. J. Phys. **66**, 920 (1998)

[12] J.von Neumann, *Mathematische Grundlagen der Quantenmechanik* (Springer, Berlin, 1932).

[13] Gy.Bene, *Quantum reference systems: a new framework for quantum mechanics*, Physica A **242** (1997) 529.

[14] H.Everett Rev.Mod.Phys. **29** (1957) 454.

[15] C.Rovelli, Int.J.Theor.Phys. **35** 1637 (1996).

[16] N.D.Mermin, *The Ithaca Interpretation of Quantum Mechanics*, quant-ph/9609013, *What Is Quantum Mechanics Trying to Tell Us?*, quant-ph/9801057, Am. J. Phys. **66**, 753 (1998).

[17] E.Schmidt, Math. Annalen **. 63** (1907) 433; A.C.Zaanen, *Linear Analysis* (North Holland, Amsterdam, 1960) pp.432-440; O.Kübler and H.D.Zeh, Ann.Phys. (NY) **76** (1973) 405.

[18] D.Dieks, Found.Phys. **19** (1989) 1395, Phys.Rev.A **49** (1994) 2290 and references therein, Phys.Lett. A **197** (1995) 367.

[19] H.D.Zeh, Found.Phys. **1** (1971) 69, E.Joos and H.D.Zeh Z.Phys.B **59** (1985) 223, D.Giuliani, E.Joos, C.Kiefer, J.Kupsch, I.-O.Stamatescu and H.D.Zeh, *Decoherence and the Appearence of a Classical World*, (Springer, 1995).

[20] A.Albrecht, Phys.Rev. **D 48** (1993) 3768.

[21] L.D.Landau and E.M.Lifshitz *Quantum mechanics* *[Course on theoretical physics, Vol.3]* (Pergamon Press, New York, 1965) pp.21-24.

[22] P.Bush, P.J.Lahti, and P.Mittelstaedt, *The Quantum Theory of Measurement*, (Springer, 1991).

Decoherence: Logic and Interpretation

Decoherence and its Implications in Quantum Computation and Information Transfer
T. Gonis and P.E.A. Turchi (Eds.)
IOS Press, 2001

Quantum Logic and the Problem of Decoherence

Peter Mittelstaedt, *University of Cologne, Cologne, Germany*

Abstract: Quantum logic and in particular time dependent sequential quantum logic can be used for formulating the quantum measurement process and the history approach to quantum mechanics. There are interesting proposals to explain the decoherence of pointer values by means of histories. However, these attempts are exposed to the objection, that quantum logic as well as the history approach always presuppose the objectification of measurement results.

1. Introduction

For the Hilbert space H_S of a proper quantum system S the algebra of projection operators P_A, P_B, ...which project onto closed linear manifolds M_A, M_B, ... is given by the Hilbert lattice L_H. Except from more complicated properties [1], L_H is a complete, orthomodular, and irreducible lattice L_Q with elements 0 and 1. In addition L_H is atomic and fulfils the *covering law* [2]. If these properties are included we denote the lattice by L_Q*.

It is a most remarkable result that the lattice L_Q can be reconstructed as the lattice of propositions which are elementary, value definite, and restrictedly available. If the propositions attribute properties to an individual system, then one can reconstruct even the lattice L_Q*. This approach to quantum mechanics, called *quantum logic*, can be extended by additional properties leading to the Hilbert lattice L_H [1] and finally to the Hilbert space, making use of a theorem by *Piron* [3].

The quantum logical way of reasoning presupposes some conditions which cannot be justified within the framework of quantum mechanics in Hilbert space. The conditions mentioned are objectification in the measurement process and decoherence of the macroscopic pointer properties. These problems are well known as the *measurement problem*. We are not able to solve this problem but we are going to clarify the interrelations between the problem of decoherence and the quantum logical approach to quantum mechanics.

There are, of course, several attempts to avoid the inconsistency between the quantum logical approach to quantum mechanics and the lack of decoherence induced by quantum physics. We mention here the quantum logical formulation of the history approach by *Isham* [4], and the attempt to solve the measurement problem by means of gravity [5]. We find that in both cases the objectification in the measurement problem must be presupposed and can thus not be explained by one of these approaches.

2. Quantum language and quantum logic

a) Language, semantics, and pragmatics

Let S be a proper quantum system and A, B, ... elementary propositions which attribute predicates P(A), P(B), ... to system S at times t_1, t_2, ...Hence, we write for the elementary propositions A(S, t_1), B(S, t_2), ...We will assume here, that for every elementary proposition A there exists a finite testing procedure which shows whether or not P(A) pertains to S. If P(A) pertains to S at time t_1 , then the proposition A(S, t_1) is called to be „true", otherwise A(S, t_1) is said to be „false". The assumption, that for every elementary proposition there is a testing procedure which decides between „true" and „false" means, that these propositions are „value definite". Hence, an elementary proposition can either be proved (with result A) or disproved (with result $\bar{A}$), where $\bar{A}$ is the counter proposition of A.

Furthermore, we assume that after a successful proof of A a new proof attempt leads with certainty to the same result, provided the time interval between the two proof attempts is sufficiently small. This assumption means that there are repeatable measurement processes, which can be applied to the testing procedures. However, if after a successful proof of A, say, another proposition B is proved, then a new proof attempt will in general not lead to the previous positive result. Hence, we will not assume that two propositions A and B are in general simultaneously decidable. If accidentally two propositions A and B are always jointly decidable, we will call A and B to be „commensurable". In this case, after the proof attempt of B the result of the previous A-test is still available. However, in the general case the result of a previous test is only restrictedly available.

On the basis of the set $\mathcal{S}_Q^e$ of elementary propositions we introduce the logical connectives by the possibilities to attack or to defence them, i.e. by the possibilities to prove or to disprove the connective. First, we consider in table 1 the two-place operation called „sequential conjunction", [6], [7], [8], [9].

Table 1. Possibilities of attacks and defences of the sequential conjunction

connective	denotation	attacks	defences
$A \sqcap B$	A and then B	1. A?, 2.B?	1. A!, 2. B!

where „A?" means the challenge to prove A and „A!" the successful proof. This attack- and defence schema can be illustrated most conveniently by a proof-tree which is chronologically ordered. The first branching point corresponds to a A-test at t_1, the second one corresponds to the B-test at t_2.

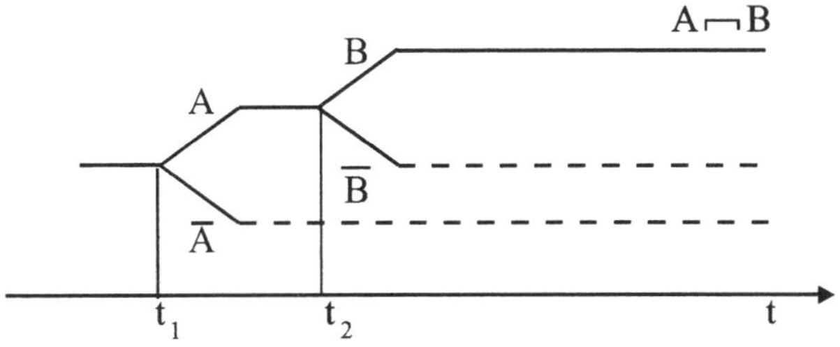

Fig. 1. Proof-tree for the sequential conjunction $A \sqcap B$

The temporal order is fixed here, but the time difference $\delta t = t_2 - t_1 > 0$ may assume arbitrary positive values. Note, that for the truth of $A(S, t_1) \sqcap B(S, t_2)$ the commensurability of A and B does not matter.

In a similar way the one-place operation „negation" may be introduced by the proof-tree for $\neg A$ (not A) with one branch for success and one branch for loss.

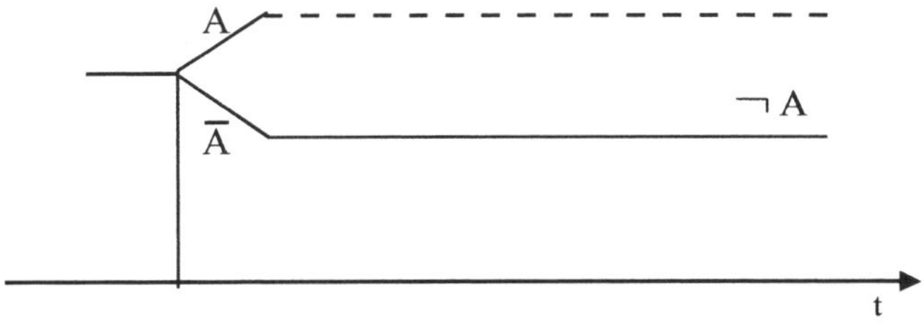

Fig. 2. Proof-tree for the negation $\neg A$

Obviously, there is no need for commensurability considerations.

The sequential conjunction (and other sequential connectives) refer to two times t_1 and t_2 with $t_1 < t_2$. The logical connectives A∧B (A and B), A∨B (A or B), and A→B (if A then B) refer to one common instant of time. The logical conjunction which corresponds to the sequential conjunction A⊓B is defined by the attack- and defence schema in table 2.

Table 2. Possibilities of attacks and defences of the logical conjunction

connective	denotation	attacks	defences
A∧B	A and B	A?; B?	A!, B!

In contrast to the sequential conjunction we have here an arbitrary number of attacks and defences in arbitrary order. In this way, it is guaranteed that the results of A- and B-tests can be attributed to a common time value. If A∧B is proved, then A and B are simultaneously true.

If the propositions A and B were commensurable and thus unrestrictedly available, then after a test of B the result of a preceding test of A would still be valid, and for a proof of A∧B two tests would be sufficient. However, since unrestricted availability will not be presupposed in general, the proof tree for A∧B consists of an infinite number of steps and cannot be reduced to two steps. In order to achieve generally a finite proof-tree we make use of the commensurability proposition k(A, B) which is defined to be true if and only if A and B are commensurable. The counter propositions is denoted here by $\bar{k}(A,B)$. The logical conjunction is then true if in addition to A and B also k(A, B) is shown to be true. Hence, for A∧B we have a proof-tree with three subsequent tests at times t_1, t_2, and t_3. Since the logical conjunction is understood as a simultaneous connective the time differences t_3-t_2 and t_2-t_1 must be sufficiently small.

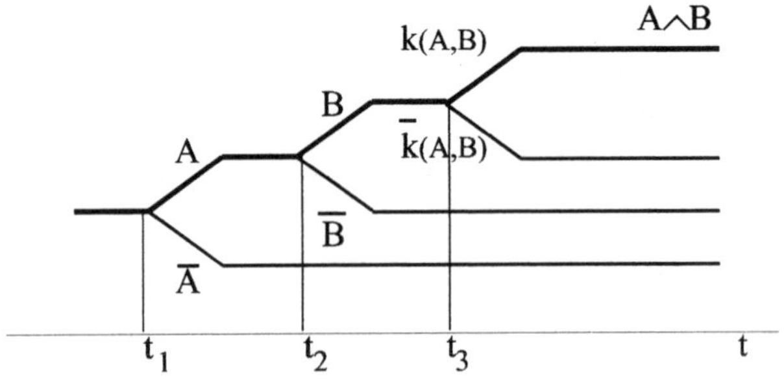

Fig. 3. Proof-tree for the logical conjunction A∧B

The commensurability propositions k(A. B) and $\bar{k}$(A, B) are contingent propositions whose truth must be shown by a convenient sequence of measurements. We will not go into detail here.By means of the commensurability propositions one can define also the logical disjunction A∨B (A or B) and the material implication A→B (if A then B) by proof trees with a finite numbers of steps. E.g. the material implication can be defined by the attack- and defence schema in table 3,

Table 3. Attack- and defence schema for the material implication

connective	denotation	attacks	defences
A→B	if A then B	1.A, 2.k(A,B)?	1.B, 2.k(A,B)!

which corresponds to a proof-tree with three branching points, fig.4.

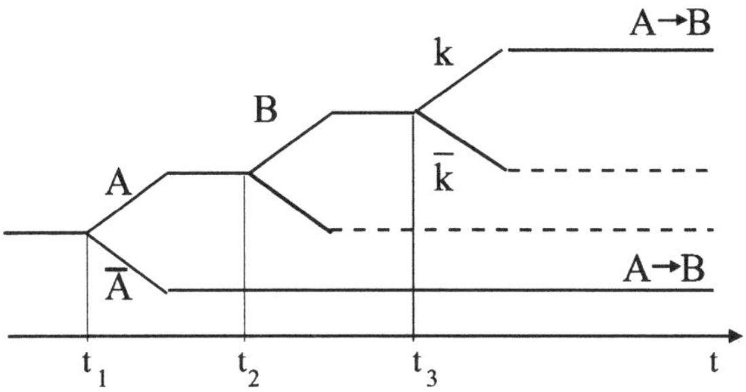

Fig. 4. Proof-tree for the material implication A→B

Furthermore, we can define here binary relations between propositions. The *proof equivalence* A ≡ B means that A can be replaced in any proof tree by B without thereby changing the result of the proof-tree. The relation of *value equivalence* A = B means that A is true if and only if B is true. The relation of implication A ≤ B can then be defined by A ≡ A∧B. Finally, we mention that A→B is true if and only if A ≤ B holds and k(A, B) is true if and only if the A ≤ (A∧B)∨(A∧¬B) holds.

The full quantum language $\mathcal{S}_Q$ can then inductively be defined by the set $\mathcal{S}_Q^e$ of elementary propositions, commensurability propositions, and the connectives mentioned. Together with the three relations the language $\mathcal{S}_Q$ reads

$$\mathcal{S}_Q = \{\mathcal{S}_Q^e; k, \bar{k}, \sqcap, \wedge, \vee, \rightarrow, \neg; \equiv, =, \leq\}.$$

b) Quantum logic

The semantics described here is a combination of a realistic semantics (with respect to elementary and commensurability propositions) and a proof semantics (with respect to the connectives). Hence, the truth of a compound proposition depends on the connectives contained in it as well as on the elementary and commensurability propositions and their truth values. There are, however, finitely connected propositions which are true in the sense of the semantics mentioned, irrespective of the truth values of the elementary and commensurability propositions. These propositions are called *formally true.*

The precondition that measurements are repeatable implies that $k(A,A)$ is always true and thus $A \rightarrow A$, the *law of identity*, is formally true. The value definiteness of elementary propositions implies that also finitely connected propositions are value definite and thus $A \vee \neg A$, the *tertium non datur law*, is formally true. In a similar way, it follows that $\neg(A \wedge \neg A)$, the *law of contradiction*, and $(A \wedge (A \rightarrow B)) \rightarrow B$, the *modus ponens law*, are formally true. - These propositions are true irrespective of the truth or falsity of elementary and commensurability propositions contained.

Formally true propositions can also be expressed by „formally true implications". E.g. the modus ponens law reads $A \wedge (A \rightarrow B) \leq B$. In addition, we make use of two special propositions, the true proposition V (verum) and the false proposition Λ (falsum), such that for all propositions $A \in \mathcal{S}_Q$ the relations $A \leq V$ and $\Lambda \leq A$ hold. Hence, we have $\Lambda = \neg V$. The formal truth of a proposition A can then be expressed by $V \leq A$. E. g. the tertium non datur law reads $V \leq A \vee \neg A$ and the law of contradiction $A \wedge \neg A \leq \Lambda$.

There are two kinds of propositions $A \in \mathcal{S}_Q$. If a compound proposition contains in addition to elementary and commensurability propositions only the *logical* connectives $\wedge$, $\vee$, $\wedge$, and $\rightarrow$, then it is called a „logical proposition". In the more general case, when the proposition contains also *sequential* connectives, in particular the sequential conjunction $\sqcap$, then it is called a „sequential proposition". In addition to the formally true logical propositions mentioned above, there are also formally true sequential propositions. If A and B are logical propositions then $A \wedge B \leq A \sqcap B$ is a formally true implication. Furthermore, the implications

$$A \vee (\neg A \wedge B) \leq A \sqcap B \leq (A \vee \neg B) \wedge B$$

are formally true, where $A \vee (\neg A \wedge B)$ is the logical infimum and $(A \vee \neg B) \wedge B$ the logical supremum of the sequential proposition $A \sqcap B$.

The totality of formally true implications can be summarised in a calculus which contains „beginnings" $\Rightarrow A \leq B$ and rules $A \leq B \Rightarrow C \leq D$. Here, we distinguish the calculus $\mathbf{L}_Q$ of formally true *logical* propositions and the calculus $\mathbf{S}_Q$ of formally true sequential propositions. For the explicit form of these formal systems we refer to the literature [6], [7]. More important for the characterisation of quantum logic $\mathbf{L}_Q$ are propositions which are formally true in classical logic but not in quantum logic. The shortest proposition of this kind is $A \rightarrow (B \rightarrow A)$. For the proof of $A \rightarrow (B \rightarrow A)$ the truth values of A and B do not matter. However the truth value of the commensurability proposition k(A,B) does matter and thus proposition $A \rightarrow (B \rightarrow A)$ is not formally true in quantum logic. Since in the context of classical logic k(A,B) is always true, the proposition $A \rightarrow (B \rightarrow A)$ is formally true in classical logic.

For an algebraic characterisation of the calculi $\mathbf{L}_Q$ and $\mathbf{S}_Q$ we consider the corresponding Lindenbaum-Tarski algebras. The Lindenbaum-Tarski algebra of the calculus $\mathbf{L}_Q$ is given by a complete, orthomodular lattice L_Q. Subsets of mutually commensurable propositions constitute a Boolean sublattice $L_B \subseteq L_Q$ of the lattice L_Q [8]. Moreover, if the entire quantum language $\mathcal{S}_Q$ refers to one individual quantum system, then the lattice L_Q is atomic (where the atoms correspond to pure states) and fulfils the covering law [10]. In this case we denote the lattice by L_Q^*. The Hilbert lattice L_H of projection operators in Hilbert space [11] can be obtained from the lattice L_Q^* by adding the Solér law, the operational meaning of which is, however, still open [1].

Analogously, the Lindenbaum-Tarski algebra of the calculus $\mathbf{S}_Q$ of sequential quantum logic can be constructed. This algebra is no longer given by a lattice but by a Baer* semigroup. For the explicit formulation of this formal system one needs the operations $\neg$ and $\sqcap$ and an additional 1-place operation *, an involution. The system $< \mathcal{S}_Q, \sqcap >$ is a semigroup, $< \mathcal{S}_Q, \sqcap . * >$ an involution semigroup, and $< \mathcal{S}_Q, \sqcap, *, ' >$ a Baer* semigroup, where the operation ′ is called *projection.* We will not go into details here and refer to the literature [7], [12].

3. Quantum theory of measurement

a) Quantum logical description of the measurement process

The quantum theory of measurement in the version of J. von Neumann which presupposes the objectification of the measurement results as the *projection postulate* can be reformulated in terms of quantum logic. Let S = S(W) be a proper quantum system S with Hilbert space H_S and with the preparation W ∈ L_Q, where the atomic proposition W describes a pure state corresponding to a projection operator P[φ], $\varphi \in H_S$. For the experimental test of another elementary proposition A a measurement process must be performed which has two possible outcomes, A and ¬A. This mans that after a measurement process one of the two sequential propositions W⌐¬A or W ⌐¬ ¬A is true for system S. If the conditional probabilities for A and ¬A are given by p(W, A) and p(W, ¬A), respectively, after this first step of the measurement process the system is described by the ensemble Γ(W; A) = { p(W, A), p(W, ¬A); W ⌐¬A, W ⌐¬ ¬A } of two weighted alternatives. The ensemble Γ(W; A) represents the observers knowledge about S after step (I) of the measurement process.

In a second step (II) this ensemble can be reduced merely by reading one of the possible outcomes W ⌐¬A or W ⌐¬ ¬A. Hence, the whole measurement process with result A, say, reads W $\Rightarrow_{(I)}$ Γ(W; A) $\Rightarrow_{(II)}$ W ⌐¬A. In step (I) the system S(W) is changed such that proposition A is objectified, whereas in step (II) the observers ignorance about S is removed.

For the various proof trees in quantum logic there is no need to consider the full measurement process, since for proof trees *probabilities* do not matter but only the possible alternatives. Hence, in the proof tree of the elementary proposition A, as shown in (Fig.5),

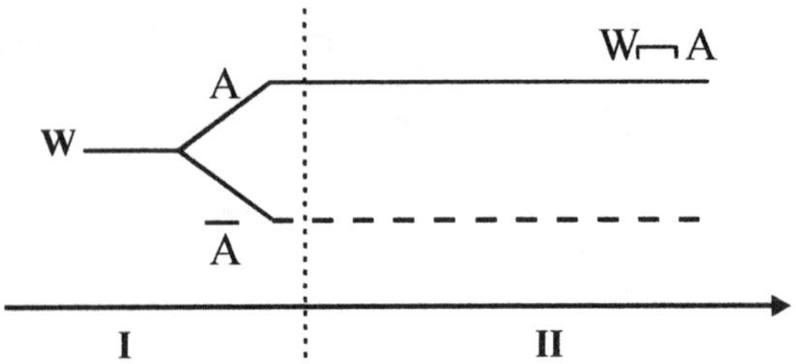

Fig. 5. Proof-tree of proposition A in quantum logic

step (I) corresponds to the transition from W to the mixture $\{ W \sqcap A, W \sqcap \neg A \}$ of sequential propositions, whereas the step (II) describes the reduction of this mixture of propositions to $W \sqcap A$ or to $W \sqcap \neg A$.

The sequential propositions $W \sqcap A$ and $W \sqcap \neg A$ describe some part of the history of the system S but not its state after the measurement. States in the sense of subspaces of Hilbert space are given by logical propositions which contain only the logical connectives. In order to find the state of the system S whose history is given by $W \sqcap A$ we make use of the result mentioned above:

For any sequential proposition A there exists a logically connected proposition [A] such that $A \leq [A]$ and $\forall B, A \leq B \Rightarrow [A] \leq B$, i.e. [A] is the logical supremum of A.

For the sequential conjunction $W \sqcap A$ the logical supremum $[W \sqcap A]$ is given explicitly by $[W \sqcap A] = (W \vee \neg A) \wedge A$. The step from the sequential conjunction $W \sqcap A$ to the final state $[W \sqcap A]$ is step (III) of the measurement process. In this step the description of the history is reduced to a description of the final state. Obviously, this reduction is connected with some loss of information. We call step (III) „the abstraction from the history". Since $(W \vee \neg A) \wedge A$ contains only the information which we need for predictions of future measurements, we call $[W \sqcap A]$ „predictive content" of the sequential proposition $W \sqcap A$.

b) Decoherence and the measurement problem

The quantum theory of measurement does not reproduce the description of the measurement process given in the previous subsection. In order to explain this surprising statement, let us briefly recall the main steps of the contemporary quantum theory of measurement [13], [14]. The measurement process consists of three steps, 1. *preparation*, 2. *premeasurement*, and 3. *objectification* and *reading*. If the object system is prepared in a pure state $\varphi(S)$ and the apparatus M in a neutral state $\Phi(M)$, then the compound system S+M is prepared in the state $\Psi(S+M) = \varphi(S) \otimes \Phi(M)$. Here we consider a repeatable unitary premeasurement of a discrete observable $A = \sum a_i P[\varphi_i]$ with eigenstates φ_i and eigenvalues a_i . In a second step a unitary operator U_A is applied top the state $\Psi(S+M)$ such that

$$U_A \Psi(S+M) = \Psi'(S+M) = \sum c_i \varphi_i \otimes \Phi_i$$

with $c_i = (\varphi_i, \varphi)$ and eigenstates of the pointer observable $Z = \sum Z_i P[\Phi_i]$. In order to achieve this special biorthogonal decomposition of the state Ψ' the unitary operator U_A must fulfil the following calibration condition: If the system S is prepared in an eigenstate φ_i of A, then the post-premeasurement state reads $\Psi' = \varphi_i \otimes \Phi_i$.

If after the premeasurement (when the interaction is turned off and the compound system is in the state $\Psi'(S+M)$) the system S and the apparatus M are considered as separate objects, then these systems have to be described by the correlated mixed states

$$W_S' = \Sigma\, |c_i|^2 P[\varphi_i] \qquad \text{and} \qquad W_M' = \Sigma\, |c_i|^2 P[\Phi_i]\,,$$

respectively. However, it is a most important theoretical result that these mixed states do not admit ignorance interpretation. This means that the states W_S' and W_M' must not be considered as weighted mixtures of states $\Gamma_S' := \{\varphi_i, |c_i|^2\}$ and $\Gamma_M' := \{\Phi_i, |c_i|^2\}$, such that S is actually in state φ_n, say, and the observer doesn't know this state but only its probability $p(\varphi, a_n) = |c_n|^2$. Analogously, the apparatus M is not actually in a state Φ_n and the observer knows only its probability. Hence, after the premeasurement the values a_i and Z_i of the object observable A and of the pointer observable Z are objectively undetermined and not merely subjectively unknown. (Fig.6)

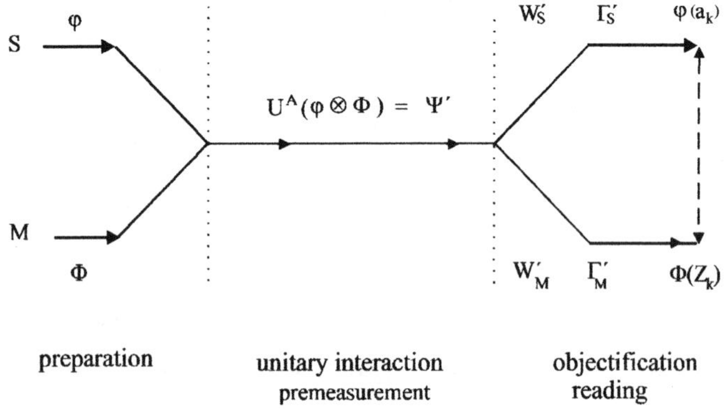

Fig.6. Schematic representation of the measurement process in quantum mechanics

However, the final step of the measurement process, reading the results a_i and Z_i , presupposes that these values are objectively determined. Hence, there must be some real process that transforms the mixed states W_S' and W_M' into the weighted mixtures of states Γ_S' and Γ_M', respectively. This process, the *objectification*, is not contained in the quantum theory of measurement mentioned. Instead, it is added to this theory as a new postulate or as a hypothetical assumption. J. von Neumann called it „projection postulate". With respect to the objectification of the values Z_i of the *macroscopic* pointer, the transition from W_M' to Γ_M' is called „decoherence".

These considerations show that the quantum logical description of the measurement process is not reproduced by the quantum theory of unitary premeasurements. It is neither clear how the transition from the preparation $W = P[\varphi]$ to the mixture $\Gamma(W,A) = \{W \sqcap\!\!\!\!\rightarrow A, W \sqcap\!\!\!\!\rightarrow \neg A\}$ of sequential propositions really works, nor do we know a quantum mechanical process that transforms a sequential proposition $W \sqcap\!\!\!\!\rightarrow A$ into the final result $[W \sqcap\!\!\!\!\rightarrow A]$. Hence, neither step (I) , the „objectification" nor step (III), the „abstraction from the history" correspond to a realisable quantum mechanical process. In particular, this means that elementary propositions cannot be tested in the described way and are thus not value definite. Hence, the nonobjectifiability of pointer values, or the lack of decoherence precludes the basic assumptions of value definitness and finite decidability of quantum logical propositions.

4. The history approach

There is an interesting attempt to overcome the problem of objectification in the measurement process which goes back to Wigner [15]. Since quantum mechanics does not lead to an adequate description of the measurement outcomes, von Neumann introduced the „projection postulate" as an additional requirement. Wigner tried to present a formulation of quantum mechanics without the *ad hoc* introduced projection postulate. For this reason he admits that quantum mechanics gives only probability connections between successive observations on quantum systems. Obviously, this formulation gives some primacy to the act of observation which is considered as an irreducible basic concept in quantum mechanics.

Let $W = W(t_0)$ with $W = W^+$, $tr\{W\} = 1$ be the preparation state of a system S at time $t = t_0$, and P_A, P_B, and P_C observables (projection operators) with eigenvalues $(A, \neg A)$, $(B, \neg B)$, and $(C, \neg C)$ which are measured successively at times $t_1 < t_2 < t_3$, $(t_1 > t_0)$, respectively. If the successive measurements are not carried out immediately but at times t_1, t_2, t_3 with finite intervals the projection operators have to be transformed by some unitary development operator $U(t, t') = \exp[-iH(t - t')/\hbar]$, e.g.

$$P_B(t_2) = \exp[-iH(t_2 - t_1)/\hbar]P_B(t_1)\exp[iH(t_2 - t_1)/\hbar].$$

The probability to find the eigenvalue A after the first measurement is given by

$$p(W; P_A(t_1)) = tr\{W(t_0)P_A(t_1)\}$$

and the post-measurement (Lüders) state reads

$$W'(t_1) = P_A(t_1)W(t_0)P_A(t_1)/tr\{W(t_0)P_A(t_1)\}.$$

The conditional probability for $B(t_2)$ given that $A(t_1)$ was found then reads

$$p(W'; P_B(t_2)) = tr\{P_B(t_2)P_A(t_1)W(t_0)P_A(t_1)P_B(t_2)\}/tr\{W(t_0)P_A(t_1)\}.$$

Finally, we find the probability to get first $A(t_1)$ and then $B(t_2)$ by multiplying this expression with $p(W; P_A(t_1))$

$$p(W(t_0); A(t_1) \sqcap B(t_2)) = tr\{P_B(t_2)P_A(t_1)W(t_0)P_A(t_1)P_B(t_2)\}$$

where we have used the notation $A \sqcap B$ known from sequential quantum logic. The generalisation of this formula to three and more measurements leads to

$$p(W(t_0); A(t_1) \sqcap B(t_2) \sqcap C(t_3)) = tr\{P_C(t_3)P_B(t_2)P_A(t_1)W(t_0)P_A(t_1)P_B(t_2)P_C(t_3)\}$$

etc. and is known as „Wigners formula". Wigner hoped to avoid the problems connected with the collapse of the state vector, if quantum mechanics is restricted to probability statements of this kind. We will comment this conjecture at the end of this section.

It is obvious, that the probability of „Wigners formula" refers to the preparation $W(t_0)$ and to the sequential proposition $Q = A(t_1) \sqcap B(t_2) \sqcap C(t_3)$ which is given by the product $P_C(t_3) \cdot P_B(t_2) \cdot P_A(t_1)$ of projection operators in Hilbert space. Hence, the subject of Wigners probabilities are products of time ordered projection operators, or of time ordered sequences of propositions. These expressions are also called „histories"

The intimate relation between sequential propositions and „histories" was first elaborated by Isham [4] in 1994. It must be emphasised that most studies in quantum logic consider propositions at a single time and are therefore not applicable to the present problem. However, quantum logic is applicable to the history approach, if in addition to the theory mentioned the logic of sequential propositions is used. From a mathematical point of view quantum logic and sequential quantum logic are quite different. Standard quantum logic corresponds to an orthomodular lattice, or to the weaker structure of an orthomodular partially ordered set whereas the appropriate mathematical tool for describing sequential propositions is a Baer*-semigroup [7].

Up to now we have considered histories as objects that are constructed within the framework of ordinary quantum mechanics or of sequential quantum logic. In this context a history is a time-ordered sequence

$$h(A) = A_1(t_1) \sqcap A_2(t_2) \sqcap \cdots A_n(t_n), \quad t_1 \leq t_2 \leq \cdots t_n$$

of elementary propositions which correspond to projection operators. However, in the history-interpretation of quantum mechanics a history is considered as an irreducible entity in its own right that is not necessarily constructed by a time ordered sequence of propositions. Histories in this sense can be characterised by the following axioms: [4]

On the set $\bar{h} = \{ h(A), h(B), ...\}$ of histories there is

1. - a complex-valued function D_W: $\bar{h} \times \bar{h} \rightarrow \mathbf{C}$, the „decoherence functional" $D_W(h(A), h(B))$. In terms of Hilbert space W is a state operator and $D_W(h(A), h(B)) = tr\{h(A)\cdot W \cdot h^{+}(B)\}$.

2. - a partial ordering relation $R_C \subseteq \bar{h} \times \bar{h}$, „h(B) is coarser than h(A)", denoted by $h(A) \leq h(B)$ and called „coarse graining relation".

3. There is a unit history h(E) (which is always realised) with $h(A) \leq h(E)$ for all h(A), and a null history h(N) (which is never realised) with $h(N) \leq h(A)$ for all h(A).

4. Two histories h(A) and h(B) are called to be disjoint, denoted $h(A) \perp h(B)$, if either h(A) or h(B) is realised.

5. For two disjoint histories h(A) and h(B) the two-place operation $O_{\vee}$: $\bar{h} \times \bar{h} \rightarrow \bar{h}$, denoted by $h(A) \vee h(B)$, is defined as the least upper bound of h(A) and h(B) with respect to the partial ordering relation R_C. (Heuristically it means the history „h(A) or h(B)".)

Histories were used by many authors as a means for describing the phenomenon of decoherence in large quantum systems without thereby being in conflict with quantum mechanics. We mention here in particular the work of Gell-Mann and Hartle [16] [17], Omnes [18], and Kiefer [19]. However, histories provide merely a description of the decoherence phenomenon but not its justification. Indeed, if histories are not used in the abstract form given by the axioms mentioned but written explicitly as a time-ordered sequence

$$h(A) = A_1(t_1) \sqcap A_2(t_2) \sqcap \cdots \qquad ; t_1 \leq t_2 \leq \cdots$$

of elementary propositions, then it becomes clear that the objectification in the measurement is already presupposed. The sequential conjunction $A(t_A) \sqcap B(t_B)$ is defined by two successive measurements at times t_A and t_B, respectively. For every measurement of this kind, the objectification of the final results must be presupposed. Hence, histories cannot be used for justifying the objectification and decoherence.

A different way of reasoning refers to quantum gravity. There is a new interest in sequential quantum logic and its extension to the history approach since some authors consider it as a basis for constructing quantum gravity [4],[16], [17]. For this program quantum logic must first be generalised to relativistic quantum logic [20]. We will not go here into details. Of course, quantum gravity has its own interest. For the present problem of objectification and decoherence it becomes relevant, since quantum gravity could perhaps help understanding the measurement problem. Penrose [5],[21] and others argue in favour of gravity induced superselection rules. An experiment, recently proposed by Penrose [21] could provide perhaps a clear-cut test of whether or not objectification is a gravitational phenomenon.

However, sequential quantum logic as well as the history approach presuppose objectification, since sequential conjunctions are time-ordered sequences of *objective* facts. Since quantum mechanics leads to the nonobjectifiability of the measurement outcomes, this result obviously invalidates also sequential quantum logic, the history approach, and quantum gravity based on it. Hence, quantum gravity cannot be used for explaining the objectification by gravity induced superselection rules, simply since objectification was already presupposed. Obviously, this way of reasoning is a vicious circle.

5. Concluding remarks

Is quantum logic fundamental? We have seen that the most general pragmatic preconditions of a scientific language imply quantum logic and that quantum logic leads - together with some additional formal conditions - to the Hilbert space and to quantum mechanics. Quantum mechanics is *consistent* with the pragmatic preconditions of quantum logic in the sense in the sense that there is no general commensurability of quantum mechanical propositions. However, quantum mechanics is *inconsistent* with the pragmatics in the sense that it does not lead to the objectification of system values after premeasurement. Hence, quantum mechanics cannot justify the value definiteness of elementary propositions which was presupposed in the pragmatics.

The quantum theory of unitary premeasurements can not explain the decoherence of the pointer values. The attempt to explain decoherence by means of consistent histories fails if histories are built up of sequential propositions which always presuppose the objectification in the measurement process. In addition, if histories are used as a means to formulate quantum gravity, the influence of gravity on the measurement process can not be used for the derivation of some gravity induced pointer objectification. Hence, within the framework of the present considerations we cannot find out where the decoherence in the measurement process comes from.

References

[1] M. P. Solèr, *Characterisation of Hilberst Spaces by Orthomodular Lattices,* Communications in Algebra, **23(1)**, 1995, 219-243.

[2] J. M Jauch, *Foundations of Quantum Mechanics*, Addison-Wesley Publ. Co., Reading, Mass. 1968

[3] C. Piron, *Foundations of Quantum Physics*, W.A. Benjamin, Readin, Mass. 1976.

[4] C. J. Isham, *Quantum logic and the history approach to quantum theory,* J. Math. Phys. **35**, 1994, 2157-2185.

[5] R. Penrose, *The large, the small and the human mind*, Paperback edition, Cambridge University Press, Cambridge, 1999, Appendix 2.

[6] P. Mittelstaedt, *Quantum Logic*, D.Reidel Publ. Co., Dordrecht, 1978.

[7] E. W. Stachow, *Logical Foundation of Quantum Mechanics,* Inter. Journ.of Theor. Phys., **19**,1980, 251-304.

[8] P. Mittelstaedt, *Sprache und Realität in der modernen Physik,* BI-Wissenschaftsverlag, Mannheim, 1986.

[9] P. Mittelstaedt, *Language and Reality in Quantum Physics*, in: Symposium on the Foundations of modern Physics, 1987, World Scientific, Singapore, 1987, 229-250.

[10] E. W. Stachow, *Structures of quantum language for individual systems,* in: Recent Developments in Quantum Logic, P. Mittelstaedt and E.-W. Stachow, eds., BI-Wissenschaftsverlag, Mannheim, 1984, 129-145.

[11] G. Birkhoff and J. von Neumann, *The Logic of Quantum Mechanics*, Ann. of Math., **37**, 1936, 823-843.

[12] D. J. Foulis, *Baer*- Semigroups*, Proceedings of the American Mathematical Society, **11**, 1960, 648-654.

[13] P. Busch, P. Lahti, and P. Mittelstaedt, *The Quantum Theory of Measurement,* 2nd. ed.1996, Springer-Verlag, Heidelberg.

[14] P. Mittelstaedt, *The Interpretation of Quantum Mechanics and the Measurement Process*, Cambridge University Press, Cambridge 1998.

[15] E. Wigner, *The Subject of Our Discussion*, in: Foundations of Quantum Mechanics, ed. B. D'Espagnat, Proceedings of the International School of Physics „Enrico Fermi", Course IL, 1971, Academic Press, New York.

[16] M. Gell-Mann and J. B. Hartle, *Classical equations for quantum systems,* Phys. Rev. **47**, 1993, 3345-82.

[17] J. B. Hartle, *Spacetime quantum mechanics and the quantum mechanics of spacetime,* in: Les Houches 1993, Session LVII „Gravitation and Quantisation", eds. B. Julia and J. Zinn-Justin, 1995, Elsevier, 285-480.

[18] R. Omnes, *The Interpretation of Quantum Mechanics*, Princeton University Press, Princeton, 1994, Ch. 4.

[19] C. Kiefer, *Consistent Histories and Decoherence,* in: Decoherence and the Appearence of a Classical World in Quantum Theory, eds. D.Giulini et al., Springer-Verlag, Berlin 1996, 157-186.

[20] P. Mittelstaedt, *Relativistic Quantum Logic*, Int. J. Theor. Phys. **22**, 1983, 293-314.

[21] R. Penrose, *Quantum computation, entanglement and state reduction,* Phil. Trans. Roy. Soc. (London) **A356**, 1999, 1927-39.

Decoherence and its Implications in Quantum Computation and Information Transfer
T. Gonis and P.E.A. Turchi (Eds.)
IOS Press, 2001

Decoherence, Histories and Interpretation

Roland OMNES
Laboratoire de Physique Théorique, B. 210, Université de Paris-Sud, 91405 Orsay, France

Abstract. The interpretation of quantum mechanics has been significantly revised in the last two decades in view of three new ideas (or older ideas brought to completion): decoherence, full inclusion of classical physics in the quantum framework and a logical language using consistent histories. They follow directly from the basic principles of quantum physics and, when used together, they change interpretation from a somewhat philosophical matter to a standard deductive theory. The present paper is a brief account of these ideas, their main features and how they combine to provide an interpretation. The main outcome is a theory of measurements where the usual Copenhagen rules are replaced by so many theorems, with a few extensions covering recent important experiments. Alhough wave packet reduction is found to be correct as a practical recipe, there is no other physical effect behind it than decoherence.

1. Interpretation

Interpretation is a sub-theory linking the basic laws of quantum mechanics with the practice of experiments. It is necessary because of the abstractness of the laws and of the conflict between their probabilism with the causal working of a laboratory device.

The Copenhagen interpretation (by Bohr, Heisenberg and Pauli) fulfilled this task by means of some philosophical directions for understanding quantum physics and some important rules for using it (the so-called "measurement theory"). These rules have withstood time remarkably well and they are strongly justified, more than ever after so many recent experiments dealing with foundations. Some of the experiments require however an extension of the Copenhagen rules to new fields and the problem of how to do it directly from the basic principles had to be solved.

Von Neumann's approach to interpretation was somewhat different from Bohr's [1]. It was much less inclined towards philosophical matters and it started with the ordinary pattern of a standard theory with well-defined principles or axioms, proofs and theorems, at least in principle. The principles included the Hilbert space formalism for observables and the Schrödinger equation for dynamics. The main philosophical difference with Bohr was to refuse splitting physics into a classical domain and a quantum one with different basic laws.

Von Neumann succeeded in finding a theoretical expression for a quantum property (which he called an "elementary predicate"), by means of a projection operator in Hilbert space. That was an important step towards founding a specific language for interpretation, but he also recognized three non-trivial problems blocking the way towards a fully deductive interpretation:

1. The first difficulty was known later as the Schrödinger cat problem: a quantum superposition is amplified to the macroscopic level by a measurement and the consequences are so unwieldy that they seem to conflict with experimental facts, and even their existence.

2. The second problem arose when trying to derive a consistent classical physics from quantum principles. Even trying to express a classical property in a quantum language by means of projection operators turned out to be too difficult for the methods that were then available.

3. The third difficulty was a matter of logic. If the propositions describing quantum events consist of elementary predicates, they cannot obey the rules of standard logic.

The three obstacles looked so forbidding that the prospect of a frontal attack of that type was more or less dismissed for a long time. I consider on the contrary that von Neumann's work was essential because he recognized a few well-defined problems to which one had "only" to find an answer, whereas Bohr's more philosophical considerations led to a much more perplexing and fuzzy situation. Now that the three main problems have been solved, one can rely on a deductive interpretation, which has no definite name yet though "decohering histories" catches most of its significance.

I will now review the answers to Questions 1 to 3 and the results one obtains when putting them together [2].

2. Decoherence

The most worrying difficulty was certainly the problem of macroscopic interferences. They looked unavoidable because of the linear character of quantum mechanics and were never observed (except in interferences with light). The idea of decoherence was proposed to solve this problem and it is certainly the most important advance in the field of interpretation. Since it has been discussed by Wojciech Zurek from the standpoint of theory and by Serge Haroche for experiments, I will consider it only briefly.

The problem is well known. When measuring a quantum observable A on a microscopic superposition of two eigenstates of A with eigenvalues a_1 and a_2, the superposition is amplified to the whole measuring device. Let the initial state be $c_1|a_1> +$ $c_2\ |a_2>$ and assume for convenience that the measured system is destroyed during the measurement (like a photon when it is detected). It was first thought sufficient to modelize the measuring apparatus by a pointer sliding along a ruler, with position x. The final wave function for the pointer turned out to be a sum

$$\psi\ (x) = c_1\ \phi_1(x) + c_2\ \phi_2(x)\ , \qquad (1)$$

where $\phi_1(x)$ is a wave function centered at a point x_1 whereas $\phi_2(x)$ is centered at a point x_2 macroscopically distant from x_1.

If one thinks however of a realistic (macroscopic) measuring device, the position of a pointer is only a collective coordinate and there are many other coordinates describing all the atoms in the apparatus as well as atmospheric molecules and photons in the surroundings. Denoting the collection of these "environment" coordinates by y, Equation (1) is more properly written as

$$\psi\ (x, y) = c_1\ \phi_1(x, y) + c_2\ \phi_2(x, y)\ . \qquad (2)$$

When trying to guess how a function such as $\phi_1(x, y)$ may depend on y, one expects it to be very complicated, its phase changing very rapidly for some value of y to a slightly different one. The motion of a pointer, when reaching a final position to indicate a datum, looks like some sort of a cataclysm in the little world of the y's. Many atoms rub against each other in neighboring parts of the apparatus for instance, with drastic changes in the state of the environment. These changes are not seen at a macroscopic level but they strongly affect the fine details of the wave function. One can thus expect that two final functions such as $\phi_1(x, y)$ and $\phi_2(x, y)$ have a very different dependence on y or, for the same value of y, one expects ϕ_1 and ϕ_2 to have very different phases.

Remembering that the phase itself is a very rapidly varying function of y, one expects then a complete lack or loss of coherence between $\phi_1(x, y)$ and $\phi_2(x, y)$ so that

$$\int \phi_1^*(x, y)\ \phi_2(x', y)\ dy = 0, \qquad (3)$$

for any couple of values x and x'. I leave it to other speakers to show that this decoherence effect can be investigated theoretically in a more careful way [3-8], its existence can be shown experimentally [9] and, finally, it can explain why macroscopic interferences are not seen.

3. Quantum properties

As already mentioned, an important contribution by von Neumann was his introduction of "elementary predicates". I prefer to call them "properties" in so far as these sentences assert a definite property of a quantum system and the property itself is more important than the sentence stating it.

Consider an observable A, *i.e.*, a self-adjoint operator, and a set Δ of real numbers in the spectrum of A. Many interesting properties of the underlying system can be expressed by a sentence saying that "the value of A is in Δ". Von Neumann associated a definite projection operator E with the property, namely

$$E = \int_{a \in \Delta} |a \rangle\langle a|, \tag{4}$$

with an unwritten summation over degeneracy indices (if any).

This relation between predicates and projection operators was strongly inspired by considerations from logic (one may remember that in addition to being a mathematician and a physicist, von Neumann was also a logician). It is generally agreed that the most basic feature of a logical proposition is to be either true or false (principle of "excluded middle"). If one uses the number 1 to symbolise "true" (as done in computers) and 0 for "false", these two numbers are also the only possible eigenvalues of a projection operator.

After playing a bit with this kind of operators, one becomes easily convinced that they are much closer to the meaning of a property than any sentence using words could be (even the one we started with). They can express any physically relevant property (stating for instance that some reaction has taken place at some time) and, by the way, time can be introduced by considering time-dependent observables $A(t)$ and time-dependent projection operators $E(t)$.

4. Classical properties

Von Neumann tried to extend his version of quantum properties to classical ones. The idea was bold because it meant that one expects classical physics to result completely from quantum principles, including not only dynamical features but also logical ones. He did not succeed however because the necessary techniques were not yet available.

The idea is nonetheless essential. Why should one use different principles for describing a preparation process and a measurement in one way (classical) whereas the emission and absorption of a photon would be described in another (quantum) way ? Bridging this gap of "correspondence", *i.e.*, understanding how classical concepts can emerge from quantum laws is moreover an essential step towards the unity of physics.

Gell-Mann and Hartle gave arguments for embedding classical properties in a quantum framework [10, 11]. They introduced a coarse-grained description of matter and made a strong use of decoherence. I am personally in favor of a more direct approach relying on well-tried mathematical techniques, as will be now explained [12, 13].

4.1 Classical properties

The most important feature of classical physics is to consider simultaneous values of position and momentum. It is at variance with the uncertainty relations, but Planck's constant is so small that it turns out most often negligible when compared with the uncertainties in the working of an apparatus. One does not need therefore to make sense of perfect classical properties, which would state that "the position and momentum have exactly the values (x_0, p_0)" with an infinite number of digits for x_0 and p_0. It is enough to state that "the particle position and momentum have the values (x_0, p_0) up to upper error bounds $(\Delta x, \Delta p)$" and this last statement will be our basic example of a classical property, assuming $\Delta x.\Delta p >> \hbar$.

Considering for simplicity a unique coordinate x, one can associate the classical property with a rectangle R in phase space with center coordinates (x_0, p_0) and sides $(2\Delta x, 2\Delta p)$. The key question is then: can one associate a projection operator in Hilbert space with the rectangle R ?

4.2 Weyl calculus

Years ago, Hermann Weyl extended a remark by Wigner [14] to get a simple method for associating a function $a(x, p)$ on classical phase space with a quantum operator A [15]. It is given by the formula

$$a(x,p) = \int < x + y/2 \mid A \mid x - y/2 > \exp(ip.\, y /\hbar)\, dy, \tag{5}$$

Taking into account the well-known expression of plane waves,

$$< x \mid p > = (2\pi\hbar)^{-1/2} \exp(ip.\, x /\hbar)\,, \tag{6}$$

one checks easily the correspondence between the observables ($I, X, P, XP_y - YP_x, P^2/2m + V(X)$) and the functions ($1, x, p, xp_y - yp_x, p^2/2m + V(x)$) for the most familiar quantum quantities and their classical expressions (for more arguments justifying Eq.(5), its extension to arbitrary dimension and some properties, see [2] and [13). Conversely, using Fourier transforms, one has :

$$< x' \mid A \mid x" > = \int a\left((x' + x")/2, p\right) \exp(-ip.(x" - x')/\hbar)\, dp\, (2\pi\hbar)^{-1}. \quad (7)$$

The correspondence between operators and functions in phase space (often called "symbols") is central in modern mathematics and more properly tin microlocal analysis, or pseudo-differential calculus [16]. One can rely on it and the first question to be asked is then: How can one define the symbol $e(x, p)$ of a projection operator which would be associated with the previous classical property ?

4.3 Projection operators and classical properties

An obvious first try would be to propose a symbol of a projection $e(x, p)$ as equal to 1 in the domain R and zero outside; but it does not work. The reason is rather obvious: using referring to the sharp boundaries of R is almost as much at variance with the uncertainty relations as giving exactly the values of two coordinates (x, p). Moreover, a slight modification of the boundary on a scale smaller than h cannot be physically significant and, conversely, a well-defined classical domain cannot specify a unique projection operator.

Physical remarks of that sort boil down to a recipe which had long been known to mathematicians. One can associate a *family* of projection operators with a classical domain R. All these operators are equivalent, *i.e.* close each other in some convenient way (the so-called trace-norm topology). One can write them by means of an operator whose symbol $f(x, p)$ is equal to 1 in the most part of R, zero outside R and very regular near the boundary (where it is infinitely differentiable).

There are many interesting features of this expression of classical properties, but they will be left aside. The main point is its existence: classical properties can be expressed in the quantum language of projections !

4.4 Classical dynamics

Bohr's "correspondence principle" was never stated very precisely but it can be. Its best expression is found in a theorem by the Russian mathematician Yuri Egorov, which goes essentially as follows [16, 17]

One considers a macroscopic system with a *collective* hamiltonian H (ignoring the environment hamiltonian and also the coupling between collective and environment dynamics, which is responsible for friction; its effect can be dealt with later together with decoherence). Let $h(q, p)$ be the Weyl symbol for H. It can be used as a classical Hamilton function to define classical motion by means of standard Hamilton equations.

When given a cell C_0 in phase space (*i.e.* a connected and simply connected domain), the corresponding classical property is expressed by a family of projection operators if and only if the cell is big enough and its shape regular enough (more precise details are given in references 2, 13 and 22). Let E_0 be one of these operators. Denoting by $C(t)$ the cell generated from C_0 through classical motion during a time t one would expect that under an exact correspondence between classical and quantum dynamics, one had

$$E(t) = U^{-1}(t)E_0U(t), \qquad (8)$$

$U(t) = exp(-iH/\hbar)$ being the quantum evolution operator. Egorov's theorem says by what amount the right-hand and left-hand sides of Eq.(8) differ. Details are given in the previously mentioned references but the physical meaning of the result is clear.

Equation (8) is essentially correct in most practical cases for a macroscopic system, the error being very small when the cells are big enough and regular in shape. The error is essentially controlled by quantum fluctuations and wave-packet spreading; it increases with time as might be expected. One of the most interesting aspect of Egorov's theorem is provided by the cases when it does not hold (or, more properly, when the difference between the two sides of Eq.(8) is significant).

There are two main cases of exception. The first one occurs when classical motion is chaotic (this is by the way a nice way for introducing chaos). Eq.(8) ceases to hold, even approximately, when chaos has grown strong enough for giving the cell C a distorted wiggly shape down to the scale of Planck's constant. There is then no more a quantum-classical correspondence.

The other case of exception occurs when there are narrow potential barriers. It has been investigated experimentally by using SQUIDs (superconducting quantum interference devices) with a nice confirmation : SQUIDs having the shape and size of a hairpin, *i.e.* macroscopic, can exhibit typical quantum effects and particularly tunnel effects [18, 19].

5. Consistent histories

The third difficulty von Neumann encountered had to do with logic. Bohr had also met it in another guise when he proposed his principle of "complementarity". Logical problems are many in interpretation. Feynman for instance worried about the fact that for discussing interferences in a Young device, there is no way but saying that the wave function is split into two parts at some time while later no wave but a particle is hitting the detector at a specific location. Moreover, why is it wrong to believe that a particle goes through a unique hole in that experiment ?

A convenient way to clear up every logical problem once and for all is to use "consistent histories" [20, 10]. I am aware that a few people claim that histories are not necessary and one can understand quantum mechanics without using them. I can only agree with them since these people make no blunder if they are already good physicists. It would be a pity if after some time one had not developed a hunch for knowing which statements about an experiment can make sense or not ! Consistent histories, which can explicitly select the meaningful statements, are certainly not necessary for every purpose but their main effect is to turn interpretation into a fully deductive theory. (It should be stressed however that they do not claim to reintroduce some philosophical realism into quantum mechanics, as was mistakenly assumed by some critics; realism is a matter of philosophy, which is best left to philosophers).

5.1 Histories

A history a consists of a sequence of properties with projection operators $E_1(t_1)$, $E_2(t_2)$, $E_3(t_3)$, $\ldots E_n(t_n)$ at times $t_1 < t_2 < \ldots < t_n$. It may involve classical properties (describing experimental devices: preparation, detection and so on) and quantum events. Because different quantum events are always competing (for instance a collision of two particles can give end up in scattering or reaction), one considers complete families of mutually exclusive histories. Two histories are exclusive if they involve respectively two orthogonal projections at a common time. Completeness is best expressed by introducing a history operator

$$C_a = E_n^{(j_n)}(t_n) \;\ldots\ldots\; E_2^{(j_2)}(t_2)\, E_1^{(j_1)}(t_1), \tag{9}$$

completeness being then given by

$$\sum_a C_a = I, \tag{10}$$

where the sum is over all histories in the family.

One can define a probability on histories by

$$p(a) = Tr\,(C_a \rho\; C_a^\dagger\;), \tag{11}$$

where ρ is the state operator (density operator) at a time $0 \leq t_1$. The probabilities (11) are positive but not generally additive (One may trace that back to the fact that additivity of amplitudes is not generally compatible with additive probabilities). The probabilities of histories are additive if the *consistency conditions*

$$Tr(C_a \rho\ C_b^\dagger) = 0, \tag{12}$$

are satisfied for any couple of mutually exclusive histories a and b. The normalisation condition

$$\sum_a p(a) = 1,$$

follows.

5.2 Logic

Within a complete family of consistent histories, there are many possible "propositions", *i.e.* statements about the physical system at hand. A history is such a proposition and any property $E_j(t_j)$ is another. It is rather obvious how one may introduce the logical operations "and, or, not" among these propositions, for instance by plotting each elementary property as a point in n-dimensional space and taking intersections, unions and complements in the resulting set of points.

Elementary logic would then be complete if one knew how to define inference, namely the relation "if α then β " (to be written $\alpha \Rightarrow \beta$) between two propositions α and β. This is done by means of the conditional probability for β, given α, namely

$$p(\beta|\alpha) = p(\alpha \text{ and } \beta) / p(\alpha),$$

through the condition

$$p(\beta|\alpha) = 1. \tag{13}$$

A nice point about these logical definitions is that they automatically satisfy the basic axioms of standard logic [21, 22], thus solving von Neumann's Problem 3. An inference is considered as approximate if

$$p(\beta|\alpha) \geq 1 - e,$$

with ε a small number.

I can only mention a few consequences without entering in their details, which can be found elsewhere [2]:

• It does not make sense to say that a particle went through only one hole in an interference device because the consistency conditions and thereby the rules of logic are not satisfied.

• Bohr's principle of complementarity is nothing but the fact that several families of histories can describe the same system without being mutually consistent. One must therefore stick to a unique family of relevant histories (relevant for the sake of understanding clearly what is going on) [2, 22, 23].

• Feynman's trouble is removed : one can speak of the wave function ψ just after crossing the holes in the interference device as a quantum property (through the projection $|\psi\rangle\langle\psi|$) and leter speak of the position of a particle in some domain V on the screen through the projection

$$\int_V |x\rangle\langle x|\, dx.$$

• When applying these rules to classical physics, one first notices that determinism is a logical equivalence ($\alpha \Rightarrow \beta$ and $\beta \Rightarrow \alpha$) between two classical properties at different times. The long-searched relation between classical determinism and quantum probabilism amounts then to the validity of these two inferences with a *probability of error* which is controlled by Egorov's theorem [2, 22]. Determinism is an (approximate) logical consequence of the (probabilistic) quantum principles !

• Fortunately, consistency conditions and inferences are easily checked in the majority of applications without any calculation, either because of decoherence or by means of classical considerations (one keeps in mind Egorov's theorem though not having to make it explicit).

6. Measurements

One can finally obtain a deductive interpretation using only the basic axioms of quantum theory (the Hilbert space setup and Schrödinger's dynamics) and three of their consequences: decoherence, semiclassical logic and consistent histories. Let us sketch it in the case of measurements [2, 22].

In the simplest case, a measurement is an interaction between a measured system Q and a measuring system M. This interaction must have two features :

• If Q is initially in an eigenstate $|a\rangle$ of the "measured" observable A, the measuring apparatus will show a datum which is a classical property (*e.g.* the position of a pointer) having a one-to-one relation with a. This is due to the specific dynamics of a measurement interaction.

• Because of decoherence, the various possible data are mutually exclusive with no superposition !

The analysis of a measurement consists then in introducing histories in which every relevant property is included (preparation, results through the properties $|a><a|$, and data), checking consistency and drawing inferences. The results are as follows:

• A result (namely the eigenvalue a of A) and the corresponding datum are logically equivalent. Though usually assumed, this equivalence between (for instance) the value of a spin component and the datum from a detector behind a Stern-Gerlach device is most satisfactory.

• The statistical frequency of a specific result is given by Born's rule in the limit of many measurements.

• When two successive measurements of different observables are performed on the same quantum system, the result of the second measurement can be computed *as if* there had been reduction (wave function collapse) from the first measurement (though only decoherence occurred in the first measuring device).

One may say in a nutshell that the usual rules of measurement theory have become so many theorems. The extension to more tricky situations is easy (to delayed-choice measurements or the continuous observation of a unique atom for instance).

Of course, there always remain some controversies after an advance has been made and new ideas have occurred. Here is therefore the right place for using the first tense "I", standing for a personal opinion with no special weight. Some questions one can hear here and there are:

• Is there a problem of actualization ? (or why is there a unique datum at the end of a measurement ?) I think there is no problem of a physical nature and only some habit to be got in quantum logic, but some people disagree (for more details see [2].

• Is it possible to turn decoherence backwards ? Certainly, there is a non-zero probability for doing it, but it is outrageously small, with no empirical consequence. By the way, one expects no measurement nor any kind of classical physics in a toy universe containing very few particles, so one is clearly dealing with a quantitative matter of confidence level as when one says "the Sun will raise to-morrow" though the Earth might have then reached Sirius through a tunnel effect.

• Does one need some sort of many-worlds reality ? I think not.

• What is the meaning of the word "reality" when applied to the quantum world ? I would say it does not matter as far as one can account for everything observed.

• Are there hidden variables ? Perhaps they exist and are unobservable or they stand behind supersymmetric strings, but presently one can do without them.

Finally, if understood as a link between the basic principles and what is really observed, this interpretation is consistent and complete. It is most useful for non-trivial applications (quantum computers, perhaps).

References

[1] J. von Neumann, *Mathematische Grundlagen der Quantenmechanik*, Springer, Berlin (1932). English translation by R. T. Beyer, *Mathematical Foundations of Quantum Mechanics*, Princeton University Press (1955).

[2] For more details, see R. Omnès, *Understanding Quantum Mechanics*, Princeton University Press (1999).

[3] K. Hepp, E. H. Lieb, Helv. Phys. Acta, **46**, 573 (1974).

[4] W. H. Zurek, Phys. Rev. **D24**, 1516 (1981), *ibid.* **D26**, 1862 (1982),Physics Today **44** (10), 36 (1991).

[5] A. O. Caldeira, A. J. Leggett, Physica **A 121**, 587 (1983).

[6] E. Joos, H. D. Zeh, Z. Phys. **B 59**, 229 (1985).

[7] W. G. Unruh, W. H. Zurek, Phys. Rev. **D 40**, 1071 (1989).

[8] R. Omnès, Phys. Rev. **A 56**, 3383 (1997).

[9] M. Brune, M., E. Hagley, J. Dreyer, X. Maître, A. Maali, C. Wunderlich, J. M. Raimond, S. Haroche, Phys. Rev. Letters, **77**, 4887 (1996).

[10] M. Gell-Mann, J. B. Hartle, in *Complexity, Entropy, and the Physics of Information*, W. H. Zurek, edit. Addison-Wesley, Redwood City, CA (1991).

[11] M. Gell-Mann, J. B. Hartle, Phys. Rev. **D 47**, 3345 (1993).

[12] R. Omnès, J. Stat. Phys. **57**, 357 (1989).

[13] R. Omnès, J. Math. Phys. **38**, 697 (1997).

[14] E. P. Wigner, Phys. Rev. **40**, 749 (1932).

[15] H. Weyl, Bull. Amer. Math. Soc. **56**, 115 (1950).

[16] L. Hörmander, *The Analysis of Partial Differential Operators*, 3 volumes, Springer, Berlin (1985).

[17] Yu. V. Egorov, Uspehi Mat. Nauk. **24**, 5, 235 (1969).

[18] A. J. Leggett, in *Chance and Matter*, Les Houches Session XLVI, 1986, J. Souletie, J. Vannimenus, R. Stora, eds., North-Holland, Amsterdam (1987).

[19] J. Clarke, A. N. Cleland, M. H. Devoret, D. Estève, J. M. Martinis, Science **239**, 992 (1988).

[20] R. G. Griffiths, J. Stat. Phys. **36**, 219 (1984).

[21] R. Omnès, J. Stat. Phys. **53**, 893 (1988).

[22] R. Omnès, *The Interpretation of Quantum Mechanics*, Princeton University Press (1994).

[23] R. G. Griffiths, Phys. Rev. **A 54**, 2759 (1996).

Decoherence and its Implications in Quantum Computation and Information Transfer
T. Gonis and P.E.A. Turchi (Eds.)
IOS Press, 2001

Mathematics of Decoherence, Causality and Prediction of Future.

V P BELAVKIN
School of Mathematics, Nottingham University, NG7 2RD, UK

ABSTRACT. It is reminded that before the rise of quantum mechanics 75 years ago, the quantum theory had appeared first is the decoherent form of the statistics of quantum thermal noise and quantum spontaneous jumps which have never been explained by quantum mechanics. Moreover, the only reasonable probabilistic interpretation of quantum theory put forward by Max Born was in fact in irreconcilable contradiction with traditional mechanical reality and causality. This led to numerous quantum paradoxes, some of them, related to the great inventors of quantum theory such as Einstein and Schrödinger, are reconsidered in the paper. The development of quantum measurement theory, initiated by von Neumann, indicated a possibility for resolution of the decoherence problem by a divorce of the algebra of the dynamical generators and a subalgebra of the actual observables. It is shown that within this approach quantum causality can be rehabilitated in the form of a superselection rule for compatibility of past observables with the potential future. We call this rule, together with the self-compatibility of measurements insuring the consitency of histories, the nondemolition principle. The application of this rule in the form of the dynamical commutation relations leads to the derivation of the von Neumann projection postulate, as well as of more general reductions, instantaneous, spontaneous, and even continuous in time. This gives a quantum probabilistic solution in the form of the dynamical filtering equations to the notorious decoherence and measurement problem which was tackled unsuccessfully by many famous physicists starting from Schrödinger and Bohr. The simplest Markovian quantum stochastic model for the time continuous decoherence and measurement involves a boundary-value problem in second quantization for input "offer" waves in one extra dimension, and a reduction of the algebra of "actual" observables to an Abelian subalgebra for the output waves.

1. INTRODUCTION: INTERPRETATIONS OF QUANTUM MECHANICS

In 1923 de Broglie, ispired by the works of Einstein and Planck, extended the wave-corpuscular duality also to material particles. He used the Hamilton-Jacobi theory which had been applied both to particles and waves.

In 1925, Schrödinger gave a seminar on de Broglie's material waves, and a member of the audience suggested that there should be a wave equation. Within a few weeks Schrödinger found his celebrated wave equation, first in a relativistic, and then in the non-relativistic form [1]. Instead of seeking the classical solutions to the Hamilton-Jacobi equation

$$H\left(q, \frac{\hbar}{i}\frac{\partial}{\partial q}\ln\psi\right) = E$$

he suggested finding those wave functions ψ which satisfy the linear equation

$$H\left(q, \frac{\hbar}{i}\frac{\partial}{\partial q}\right)\psi = E\psi$$

(It coinsides with the former equation only if H is linear with respect to p). He also obtained the non-stationary wave equation written in terms of the Hamiltonian operator H as

$$i\hbar\frac{\partial}{\partial t}\psi(t) = H\psi(t).$$

The creators of the rival matrix quantum mechanics, Born and Jordan, who discovered the "Heisenberg form" of this equation a year earlier, were forced to accept the simplicity and beauty of Schrödinger's approach. In 1926 Max Born put forward the statistical interpretation of the wave function. This was developed in Copenhagen by Niels Bohr and gradually was accepted by almost all physicists as the "Copenhagen interpretation". Bohr, Born and Heisenberg considered electrons and quanta as unpredictable particles which cannot be visualized in the real space and time.

Schrödinger was a champion of the idea that the most fundamental laws of the microscopic world are absolutely random before he discovered wave mechanics. While he was preparing his Inaugural lecture for December 9, 1922 at the University of Zürich, he wrote to Pauli:

> *I for my part believe, horribile dictu, that the energy-momentum law is violated in the process of radiation.*

(quoted on page 152 in [3]).

Following de Broglie, Schrödinger initially thought that the wave function corresponds to a physical vibration process in a real continuous space-time because it was not stochastic, but he was puzzled by the failure to explain the blackbody radiation and photoelectric effect from this wave point of view. In fact the wave interpretation applied to light quanta leads back to classical electrodynamics, as his relativistic wave equation for a single photon coincides mathematically with the classical wave equation. However after realizing that the time-dependent ψ is a complex function in his fourth 1926 paper [2], submitted just a few days before Born's interpretational paper, he admitted that the wave function ψ cannot be given a direct interpretation, and described the wave density $\psi\bar{\psi}$ as a sort of weight function for superposition of point-mechanical configurations.

Bohr invited Schrödinger to Copenhagen and tried to convince him of the particle-probabilistic interpretation of quantum mechanics. The discussion between them went on day and night, without reaching any agreement. The conversation, however deeply affected both men. Schrödinger recognized the necessity of admitting both wave and particles, but he never devised a comprehensive interpretation rival to Copenhagen orthodoxy. Bohr ventured more deeply into philosophical waters and emerged with his concept of complementarity:

> *Evidence obtained under different experimental conditions cannot be comprehended within a single picture, but must be regarded as complementary in the sense that only the totality of the phenomena exhausts the possible information about the objects.*

Ten years later Schrödinger accepted the probabilistic interpretation of $\psi\bar{\psi}$, although he did not consider these probabilities classically, but as the strength of our belief or anticipation of an experimental result. In this sense his probabilities are closer to propensities than to the frequencies of the statistical interpretation of Born and Heisenberg. Schrödinger had never accepted the subjective positivism of Bohr and Heisenberg, and his philosophy is closer to that called representational realism. He was content to remain a critical unbeliever.

The most outspoken opponent of a/the probabilistic interpretation was Einstein. Albert Einstein admired the new development of quantum theory but with suspicion, and rejected its acausality and probabilistic interpretation. It was against his scientific instinct to accept quantum mechanics with its statistical interpretation as a complete description of physical reality. There are famous sayings of him on that account:

> *'God is subtle but he is not malicious'. 'God doesn't play dice'.*

In the famous debates on the probabilistic interpretation of quantum mechanics of Einstein and Niels Bohr, Schrödinger was often taking the side of his friend Einstein, and this may explain why he was distancing himself from the statistical interpretation of his wave function.

There have been many other attempts to retain the deterministic realism in the quantum world, the most extravagant among these being the many world interpretation. Certainly there are some advantages living in many worlds. From the philosophical and practical point of view, however, to have an infinite (continuum?) number of real worlds at the same time seems not better than to have none.

2. Propensities, Paradoxes, Predictions

How wonderful we have met with a paradox, now we have some hope of making progress - Niels Bohr.

In 1932 von Neumann put quantum theory on firm theoretical basis by setting the mathematical foundation for new, quantum, probability theory, the quantitative theory for counting non commuting events of quantum logics. This noncommutative probability theory is based on essentially more general axioms than the classical (Kolmogorovian) probability of commuting events, which form common sense Boolean logic, first formalized by Aristotle. It has been under extensive development during the last 30 years since the introduction of algebraic and operational approaches for treatment of noncommutative probabilities, and currently serves as the mathematical basis for quantum information and measurement theory.

2.1. Uncertainties and New Logics. Bohr was concerned with the paradox of spontaneous emission. He addressed the question: How does the electron know when to emit radiation? Bohr, Born and Heisenberg abandoned causality of traditional physics in the most positivistic way. Max Born said:-

If God made the world a perfect mechanism, ... we need not solve innumerable differential equations, but can use dice with fair success.

Heisenberg uncertainty relations. In 1927 Heisenberg derived [4] his famous uncertainty relation

$$\Delta Q \Delta P \geq \hbar/2, \quad \Delta T \Delta E \geq \hbar/2$$

which gave mathematical support to the revolutionary complementary principle of Bohr. As Dirac stated:-

Now when Heisenberg noticed that, he was really scared.

Einstein launched an attack on the uncertainty relation at the Solvay Congress in 1927, and then again in 1930, by proposing cleverly devised thought experiments which would violate this relation. Most of these imaginary experiments were designed to show that interaction between the microphysical object and the measuring instrument is not so inscrutable as Heisenberg and Bohr maintained. He suggested, for example, a box filled with radiation with a clock. The clock is designed to open a shutter and allow one photon to escape. By weighing the box the photon energy and the time of escape can both be measured with arbitrary accuracy.

After proposing this argument Einstein is reported to have spent a happy evening, and Niels Bohr an unhappy one. After a sleepless night he showed next morning that

Einstein was not right. Mathematically his solution can be expressed by the following formula

$$X = t + Q, \quad Y = i\hbar\frac{\partial}{\partial t} + P$$

for the measuring quantity X, the pointer coordinate of the clock, and the observable Y for indirect measurement of photon energy $E = i\hbar\frac{\partial}{\partial t}$ in the Einstein experiment, where Q and P are the position and momentum operators of the compensation weight under the box. Due to the initial independence of the weight, the commuting observables X and Y have even greater uncertainty

$$\Delta X \Delta Y \geq \Delta T \Delta E + \Delta Q \Delta P \geq \hbar$$

than that predicted by Heisenberg uncertainty $\Delta T \Delta E \geq \hbar/2$.

Nonexistence of hidden variables. Since then there were innumerable attempts to explain the uncertainty relations by introducing hidden variables ever in more sophisticated forms, perhaps not yet discovered, which would determine the complementary variables if the hidden variables were measured precisely.

Von Neumann's monumental book [5] on the mathematical foundations of quantum theory was therefore a timely contribution, clarifying, as it did, this point. Inspired by Lev Landau, he introduced, for the unique characterization of the statistics of a quantum ensemble, the statistical density operator ρ which eventually, under the name state, became a major tool in quantum statistics. He considered the set $\mathcal{L}$ of all possible events in a quantum system, described by the observables E with eigen-values valued 0 or 1 only, and defined the quantum probability of each event E in a state ρ as its expectation

$$\Pr(E) = \mathrm{Tr} E\rho \equiv \langle E \rangle .$$

Then he proved that in order to have positive probabilities for all events, and the probability one for the identity event $E = I$, i.e.

$$\Pr(E) \geq 0, \quad \Pr(I) = 1,$$

the statistical operator must be positive-definite and normalized. He applied this technique to the analysis of the completeness problem of quantum theory, whether it constitutes a logically closed theory or whether it could be reformulated as an entirely deterministic theory through the introduction of hidden parameters, that is, additional variables which, unlike ordinary observables, are inaccessible to measurements. He came to the conclusion that

> "*the present system of quantum mechanics would have to be objectively false, in order that another description of the elementary process than the statistical one may be possible.*"

To prove this theorem, von Neumann showed that there is no such state ρ which would be dispersion-free simultaneously for all quantum events $E \in \mathcal{L}$. For each such state, he argued,

$$\langle E^2 \rangle = \langle E \rangle = \langle E \rangle^2$$

would imply that $\rho = 0$ or $\rho = I$, the zero or identity operator, which can't be statistical operators as 0 and I are not normalized. Thus no state can be considered as a mixture of dispersion-free states, each of them associated with a definite value of hidden parameters. There are simply no such states, and thus, no hidden parameters. This in particular implies that the statistical nature of pure states, which are described by one-dimensional projectors $\rho = P_\psi$ corresponding to wave functions ψ, cannot be removed by supposing them to be a mixture of dispersion-free substates: each substate of a pure state coincides with this state, and thus is not dispersion-free.

Complementarity and common sense. In view of the decisive importance of this analysis for the foundations of quantum theory, Birkhoff and von Neumann [9] setup a system of formal axioms for the calculus of logicotheoretical propositions concerning results of possible measurements in a quantum system. They started with formalizing the calculus of quantum events $E \in \mathcal{L}$ described by orthoprojectors E, and showed that this calculus coincides with the calculus of the linear subspaces given by the ranges of the orthoprojectors E of the Hilbert space, in the same sense as the propositional calculus of classical events coincides with the calculus of subsets of a Boolean algebra. In this calculus the logical ordering $E \leq F$ implemented by the algebraic relation $EF = E$ coincides with

$$\mathrm{range}\,(E) \subseteq \mathrm{range}\,(E)\,,$$

the conjunction $E \wedge F$ corresponds to the intersection,

$$\mathrm{range}\,(E \wedge F) = \mathrm{range}\,(E) \cap \mathrm{range}\,(F)\,,$$

however the disjunction $E \vee F$ is represented by the linear sum of the corresponding subspaces but not their union,

$$\mathrm{range}\,(E \vee F) \neq \mathrm{range}\,(E) \cup \mathrm{range}\,(F)$$

This implies that the distributive law characteristic for propositional calculus of classical logics no longer holds, but it still holds for compatible events, described by commutative orthoprojectors.

Two events E, F are called complementary if $E \vee F = I$, orthocomplementary if $E + F = I$, incompatible or disjunctive if $E \wedge F = 0$, and contradictive or orthogonal if $EF = 0$. As in the classical, common sense, logic contradictive events are incompatible. However

incompatible propositions or events are not necessary contradictive as can be easily seen for any two nonorthogonal but not coinciding one-dimensional subspaces. In particular, in quantum logics there exist complementary incompatible pairs E, F which are not ortho-complementary in the sense $E + F \neq I$ (it would be impossible in the classical case). This is a rigorous logico-mathematical formulation of Bohr's complementarity.

As an example, we can consider the statement that a quantum system is in a stable energy state E, and an incompatible proposition F, that it collapses at a given time t, say. The incompatibility $E \wedge F = 0$ follows from the fact that there is no state in which the system would collapse preserving its energy, however these two propositions are not contradictory (i.e. not orthogonal, $EF \neq 0$): the system might not collapse if it is in other than E stable state (remember the Schrödinger's earlier belief that the energy law is valid only on average, and is violated in the process of radiation).

In 1952 Wick, Wightman, and Wigner [10] showed that there are physical systems for which not every orthoprojector corresponds to an observable event, so that not every orthoprojector P_ψ corresponding to a wave function ψ is a pure state. This is equivalent to the admission of some selective events which are dispersion-free in all pure states. Jauch and Piron [11] incorporated this situation into quantum logics and proved in the context of this most general approach that the hidden variable interpretation is only possible if the theory is observably wrong.

2.2. Measurement and Decoherence. Heisenberg derived from the uncertainty relation that 'the nonvalidity of rigorous causality is necessary and not just consistently possible'. Max Born even stated:-

> *One does not get an answer to the question, what is the state after collision? but only to the question, how probable is a given effect of the collision?*

Spooky action at distance. After his defeat on uncertainty relations Einstein seemed to have become resigned to statistical interpretation of quantum theory, and at the 1933 Solvay Congress he listened to Bohr's paper on completeness of quantum theory without objections. Then, in 1935, he launched a brilliant and subtle new attack in a paper [6] with two young co-authors, Podolski and Rosen, which is known as the EPR paradox that has become of major importance to the world view of physics. They stated the following requirement for a complete theory as a seemingly necessary one:

> *Every element of physical reality must have a counterpart in the physical theory.*

The question of completeness is thus easily answered as soon as soon as we are able to decide what are the elements of the physical reality. EPR then proposed a sufficient condition for an element of physical reality:

> *If, without in any way disturbing the system, we can predict with certainty the value of a physical quantity, then there exists an element of physical reality corresponding to this quantity.*

Then they designed a thought experiment the essence of which is that two quantum "bits", particles spins of two electrons say, are brought together to interact, and after separation an experiment is made to measure the spin orientation of one of them. The state after interaction is such that the measurement result $\tau = \pm\frac{1}{2}$ of one particle uniquely determination the spin z-orientation $\sigma = \mp\frac{1}{2}$ of the other particle. EPR apply their criterion of local reality: since the value of σ can be predicted by measuring τ without in any way disturbing σ, it must correspond to an existing element of physical reality. Yet the conclusion contradicts a fundamental postulate of quantum mechanics, according to which the sign of spin is not an intrinsic property of a complete description of the spin but is evoked only by a process of measurement. Therefore, EPR conclude, quantum mechanics must be incomplete, there must be hidden variables not yet discovered, which determine the spin as an intrinsic property. It seams Einstein was unaware of the von Neumann's theorem, although they both had the positions at the same Institute for Advanced Studies at Princeton being among the original six mathematics professors appointed there in 1933.

Bohr carefully replied to this challenge by rejecting the assumption of local physical realism as stated by EPR [7]: 'There is no question of a mechanical disturbance of the system under investigation during the last critical stage of the measuring procedure. But even at this stage there is essentially a question of *an influence on the very conditions which define the possible types of predictions regarding the future behavior of the system*'. This influence became notoriously famous as Bohr's *spooky action at a distance.* He had obviously meant the semi-classical model of measurement, when one can statistically infer the state of one (quantum) part of a system immediately after observing the other (classical) part, whatever the distance between them. In fact, there is no paradox of "spooky action at distance" in the classical case, the statistical inference, playing the role of such immediate action, is simply based on the Bayesian selection rule of a posterior state from the prior mixture of all such states, corresponding to the possible results of the measurement. Bohr always emphasized that one must treat the measuring instrument classically (the measured spin, or another bit interacting with this spin, as a classical bit), although the classical-quantum interaction should be regarded as purely quantum. The latter follows from non-existence of semi-classical Poisson bracket (i.e. classical-quantum potential interaction). Schrödinger clarified this point more precisely then Bohr, and he followed in fact the mathematical pattern of von Neumann measurement theory.

Releasing Schrödinger's cat. Motivated by EPR paper, in 1935 Schrödinger published a three part essay [8] on 'The Present Situation in Quantum mechanics'. He turns to EPR paradox and analyses completeness of the description by the wave function for the entangled parts of the system. He notes that if one has pure states for each of two completely separated bodies, one has maximal knowledge for two taken together. But the converse is not true:

> *Maximal knowledge of a total system does not necessary imply maximal knowledge of all its parts, not even when these are completely separated one from another, and at the time can not influence one another at all.*

To make absurdity of the EPR argument even more evident he constructed his famous burlesque example in quite a sardonic style. A cat is shut up in a steel chamber equipped with a camera, with an atomic mechanism in a pure state $\rho_0 = P_\psi$ which triggers the release of a phial of cyanide if an atom disintegrates spontaneously (it is assumed that it might not disintegrate in a course of an hour with probability $\mathrm{Tr}\,(EP_\psi) = 1/2$). If the cyanide is released, the cat dies, if not, the cat lives. Because the entire system is regarded as quantum and closed, after one hour, without looking into the camera, one can say that the entire system is still in a pure state in which the living and the dead cat are smeared out in equal parts.

Schrödinger resolves this paradox by noting that the cat is a macroscopic object, the states of which (lived or dead) could be told apart by a macroscopic observation are distinct from each other whether observed or not. He calls this 'the principle of state distinction' for macroscopic objects, which is in fact the postulate that the directly measurable system (consisting of cat) must be classical:

> *It is typical in such a case that an uncertainty initially restricted to an atomic domain has become transformed into a macroscopic uncertainty which can be resolved through direct observation.*

In order to make this idea clear, let us give the solution of this elementary quantum decoherence problem in the purely mathematical way. Consider the atomic mechanism as a quantum bit, described by a ψ-function of $\sigma \in \{0,1\}$ with scalar (complex) values $\psi(\sigma)$ (if atom is disintegrated, $\sigma = 1$, if not, $\sigma = 0$). The Schrödinger's cat is a classical bit with only two pure states $\tau \in \{0,1\}$, alive ($\tau = 0$) or dead ($\tau = 1$). Other (mixed) states can be also described by the amplitudes $\varphi(\tau)$ of the probabilities $|\varphi(\tau)|^2$. Initially the cat is alive, so its amplitude (uniquely defined up to the phase factors by the square root of probability distribution on $\{0,1\}$) is $\delta_0(\tau)$ that is 1 if

$\tau = 0$, and 0 if $\tau = 1$. The only meaningful classical-quantum reversible interaction affecting not atom but cat as it is said after the hour, is described as both bits were quantum, by

$$U[\psi \otimes \varphi](\sigma, \tau) = \psi(\sigma)\varphi((|\tau - \sigma|)).$$

Applied to the initial product-state $\psi \otimes \delta_0$ it has the resulting probability amplitude

$$\chi(\sigma, \tau) = \psi(\sigma)\delta_0(|\tau - \sigma|) = 0 \quad \text{if} \quad \sigma \neq \tau.$$

Despite the fact that the initial state was pure, corresponding to the Cartesian product $(\psi, 0)$ of ψ and $\tau = 0$, the reversible evolution induces the mixed state for the quantum-classical system atom+cat.

Indeed, the observables of such a system are operator-functions of τ with values $X(\tau)$ in σ-matrices, represented as block-diagonal (σ, τ)-matrices $\hat{X}$ of the multiplication operator by $X(\tau)$. This means that the amplitude χ induces the same expectations

$$\left\langle \hat{X} \right\rangle = \sum_{\tau} \chi(\tau)^{\dagger} X(\tau)\chi(\tau) = \sum_{\tau} \mathrm{Tr}X(\tau)\varrho(\tau) = \mathrm{Tr}\hat{X}\hat{\varrho}$$

as the density matrix $\hat{\varrho}$ of the multiplication operator by

$$\varrho(\tau) = E(\tau)P_{\psi}E(\tau) = \Pr(\tau)P_{E(\tau)\psi}$$

where $\Pr(\tau) = |\psi(\tau)|^2$, $E(\tau)$ is the projection operator

$$E(\tau)\psi(\sigma) = \delta_0(|\tau - \sigma|)\psi(\sigma),$$

and $P_{E(\tau)\psi} = P_{\tau}$ projects onto $\psi_{\tau}(\sigma) = \delta_{\tau}(\sigma)$. The density 4×4-matrix $\hat{\varrho}$ is a mixture of two orthogonal projectors $P_{\tau} \otimes P_{\tau}$, $\tau = 0, 1$:

$$\hat{\varrho} = \sum_{\tau} \Pr(\tau) P_{\tau} \otimes P_{\tau}.$$

The spooky action at distance, affecting the atomic state by measuring τ, is simply the result of the statistical inference (prediction after the measurement) of the atomic posterior state $\rho_{\tau} = P_{\tau}$: the atom disintegrated if and only if the cat is dead.

Von Neumann's projection postulate. Inspired by Bohr's complementarity principle, von Neumann proposed even earlier the idea that every quantum measuring process involves an unanalysable element. He postulated [5] that, in addition to the continuous causal propagation of the wave function generated by the Schrödinger equation, during a measurement, due to an action of the observer on the object, the function undergoes a discontinuous, acausal, and instantaneous change. Thus, just prior to a measurement of the event F, disintegration of atom, say, the quantum pure state P_ψ changes to the mixed one

$$\rho = \lambda P_{E\psi} + \mu P_{F\psi} = E\rho E + F\rho F,$$

where $E = I - F$ is the orthocomplement any event, and $\lambda = \mathrm{Tr}E\rho$, $\mu = \mathrm{Tr}F\rho$ are the probabilities of E and F. This linear irreversible decoherence process should be completed by the nonlinear, acausal jump to one of the pure states

$$\rho \mapsto P_{E\psi}, \text{ or } \rho \mapsto P_{F\psi}$$

depending on whether F is false (the cat is alive, $\psi_0 = \lambda^{-1/2}E\psi$), or true (the cat is dead, $\psi_1 = \mu^{-1/2}F\psi$). The last step is the posterior prediction, filtering the decoherent mixture of ψ_0 and ψ_1 by selection of only one result of the measurement, is an unavoidable element in every measurement process relating the state of the pointer of the measurement (in this case the cat) to the state of the whole system. This assures that the same result would be obtained in case of immediate subsequent measurement of the same event F. The resulting change is described up to normalization by one of the projections

$$\psi \mapsto E\psi, \quad \psi \mapsto F\psi$$

and is called the von Neumann projection postulate.

Although unobjectionable from the purely logical point of view the von Neumann theory of measurement soon became the target of severe criticisms. Firstly it seams radically subjective, postulating the spooky action at distance (the filtering) in a purely quantum system instead of deriving it. Secondly the analysis is applicable to only the idealized situation of discrete instantaneous measurements.

However as we already mentioned when discussing the EPR paradox, the process of filtering is free from conceptual difficulty if it is understood as the statistical inference about a mixed state in an extended stochastic representation of the quantum system as a part of a semiclassical system, based upon the results of observation in its classical part. In order to demonstrate this, we can return to the dynamical model of Schrödinger's cat, identifying the quantum system in question with the Schrödinger's atom. The event E (the atom exists) corresponds then to $\tau = 0$ (the cat is alive), $E = E(0)$, and $F = E(1)$. This model explains that the origin of the von Neumann irreversible decoherence $P_\psi \mapsto \rho$ of the atomic state is in the ignorance

$$\rho = \varrho(0) + \varrho(1)$$

of the result of the measurement, and it interprets the filtering simply as the conditioning

$$\rho_\tau = \varrho(\tau) \,/\, \Pr(\tau) = P_{E(\tau)\psi}$$

of the joint classical-quantum state $\varrho(\cdot)$. Thus the atomic decoherence is derived from the unitary interaction of the quantum atom with the classical cat. A formal derivation of the projection postulate and the decoherence in the general case by explicit construction of unitary transformation in the extended semi-classical system in given in [12, 13].

3. Causality and Decoherent Dynamics

In mathematics you don't understand things. You just get used to them - John von Neumann.

The quantum probability approach resolves the famous paradoxes of quantum measurement theory in a constructive way by giving exact nontrivial models for the statistical analysis of the quantum observation processes underlying these paradoxes. Conceptually it is based upon a new idea of quantum causality called the Nondemolition Principle which divides the world into the classical past, forming the consistent histories, and the quantum future, the state of which is predictable for each such history. The differential analysis of these models is based on Itô stochastic calculus. The most recent mathematical development of these methods leads to a profound quantum filtering and control theory in quantum open systems which has found numerous applications in quantum statistics, optics and spectroscopy, and is an appropriate tool for the solution of the dynamical decoherence problem for quantum communications and computations.

3.1. Reality and Nondemolition Principle.

Schrödinger like Einstein was deeply concerned with the loss of reality and causality in the positivistic treatment of quantum measuring process by Heisenberg and Born. Schrödinger's remained unhappy with Bohr's reply to the EPR paradox, Schrödinger's own analysis was:

It is pretty clear, if reality does not determine the measured value, at least the measurable value determines reality.

Compatibility and nondemolition. Von Neumann's projection postulate and its dynamical realization can be generalized to include cases with continuous spectrum of values. In fact there many such developments, I will only mention here the most general operational approach to quantum measurements of Ludwig [14], and its mathematical implementation by Davies and Lewis [15] in the "instrumental" form. The stochastic realization of the corresponding completely positive reduction map $\rho \mapsto \varrho(\cdot)$, resolving the corresponding instantaneous quantum measurement problem, is given in [16, 12]. Because of the crucial importance of these realizations for developing understanding of the mathematical structure and interpretation of modern quantum theory, analyze the mathematical consequences which can be drawn from such schemes we need to.

The generalized reduction of the wave function $\psi(x)$, corresponding to a complete measurement with possibly continuous data y, is described by a function $V(y)$ whose values are linear operators $\psi \mapsto V(y)\psi$ but not unitary with the following normalization condition. The resulting wave-function

$$\chi(x, y) = [V(y)\psi](x)$$

is normalized with respect to a given measure μ on y in the sense

$$\iint |\chi(x, y)|^2 \,\mathrm{d}\mu \mathrm{d}\lambda = \int |\psi(x)|^2 \,\mathrm{d}\lambda$$

for any wave function ψ (normalized w.r.t. a measure λ). In the discrete case such as the case of two-point variables $y = \tau$ (EPR paradox, or Schrödinger cat with the projection-valued $V(\tau) = E(\tau)$) this is usually satisfied by taking μ to be the counting measure:

$$\int_y V(y)^\dagger V(y)\,\mathrm{d}\mu = I, \quad \text{or} \quad \sum_y V(y)^\dagger V(y) = I.$$

As in that simple case the realization can be always constructed [12] in terms of a unitary transformation U and a normalized wave function φ_0 such that

$$U[\psi \otimes \varphi_0](x, y) = \chi(x, y)$$

for any ψ. The additional system described by "the pointer coordinate y of the measurement apparatus" can be regarded as classical (like the cat) as the actual observables in question are the measurable functions $g(y)$ represented by commuting operators $\hat{g}$ of multiplication by these functions. Note that such observables, extended to the quantum part as $I \otimes \hat{g}$, are compatible with any possible a future event, represented by an orthoprojector $F \otimes I$. The probabilities (or, it is better to say, the propensities) of all such events are the same in all states whether an observable $\hat{g}$ was measured but the result is not read, or it was not measured at all. In this sense the measurements of $\hat{g}$ are called *nondemolition* with respect to the observables F, they do not demolish the picture of the future possibilities, or propensities of F. But they are not necessary compatible with the past representation of the initial operators $F \otimes I$ of the quantum system with respect to the initial states $\psi \otimes \varphi_0$. Indeed, the Heisenberg operators

$$G = U^* (I \otimes \hat{g}) U$$

in general do not commute with $F \otimes I$. One can see this from the example of the Schrödinger cat. The "cat observables" in Heisenberg picture are represented by commuting operators of multiplication by $g(\sigma + \tau)$, where the sum $\sigma + \tau = 0, 1$ is modulo 2. They do not commute with $F \otimes I$ unless F is also a multiplication operator $\hat{f}$ by a function $f(\sigma)$.

Transition from possible to actual. The analysis above shows that as soon as dynamics is taking into consideration even in the form of just a single unitary transformation, the measurement process needs to specify the arrow of time, what is the predictable future and what is the reduced past, what is possible and what is actual with respect to this measurement. As soon as a measured observable Y is specified, the algebra of possible observables (not the state which stays invariant in the Schrödinger picture unless the selection due to an inference has taken place!) reduces to the subalgebra commuting with Y, and *this reduction doesn't change the reality* (the wave function remains the same and induces the same but now mixed state on the smaller - reduced algebra!). Possible observables in an individual system are only those which are compatible with the actual observables. This is another formulation of Bohr's complementarity. More specifically this can be rephrased in the form of a dynamical postulate of quantum causality called the Nondemolition Principle [12] which we first formulate for a single instant of time t in a quite obvious form:

> In a given representation of a quantum system by an algebra $\mathcal{A}$ of (necessarily not all) operators on a Hilbert space $\mathcal{H}$, causal, or nondemolition observables are represented only by those operators Y on $\mathcal{H}$ which are compatible with $\mathcal{A}$:
>
> $$[X, Y] := XY - YX = 0, \quad \forall X \in \mathcal{A}$$
>
> (this is usually written as $Y \in \mathcal{A}'$, where $\mathcal{A}'$, called the commutant of $\mathcal{A}$, in this formulation is not necessarily contained in $\mathcal{A}$). Each measurement process of the history for a quantum system $\mathcal{A}$ can be represented as nondemolition by the causal observables in the appropriate representation of $\mathcal{A}$ on a Hilbert space $\mathcal{H}$.

As was proved in [17], this causality condition is necessary and sufficient for the existence of a conditional expectation for any state on the total algebra $\mathcal{A} \vee \mathcal{B}$ with respect to a commutative subalgebra $\mathcal{B}$ of nondemolition observables Y. Thus the nondemolition causality condition amounts exactly to the predictability of a posterior state on the algebra $\mathcal{A}$ upon the measurement results of the observables in $\mathcal{B}$. The transition from a prior ρ to the posterior state $\rho_y = P_{V(y)\psi}$ is simply the result of gaining a knowledge y defining the actual state in the decoherent mixture

$$\rho = \int V(y) P_\psi V(y)^* \, d\mu = \int P_{V(y)\psi} f(y) \, d\mu,$$

where $f(y) = \|V(y)\psi\|^2$ is the probability density of all possible y defining the output measure $d\nu = f d\mu$. As Heisenberg always emphasized, "quantum jump" is contained in the transitions from possible to actual.

If an algebra $\mathcal{B}$ of actual observables is specified at a time t, there must be a causal representation $\mathcal{B}_t$ of $\mathcal{B}$ on $\mathcal{H}$ with respect to the present $\mathcal{A}_t$ and all future possible representations $\mathcal{A}_s$, $s > t$ of the quantum system on the same Hilbert space $\mathcal{H}$ (they might not coincide with $\mathcal{A}_t$ if the system is open [18]). The past representations $\mathcal{A}_r$, $r < t$ which are incompatible with a $G \in \mathcal{B}_t$ are meaningless as noncausal for the observation at the time t, they should be replaced by the causal histories $\mathcal{B}_r$, $r < t$ of the actual observables on $\mathcal{H}$ which must be consistent in the sense of compatibility of all $\mathcal{B}_t$. Thus the dynamical formulation of the nondemolition principle of quantum causality and the consistency of histories reads as

$$\mathcal{B}_r \subseteq \mathcal{A}'_s, \quad \mathcal{B}_r \subseteq \mathcal{B}'_s, \quad \forall r \leq s.$$

These are the only possible conditions when the posterior states always exist as results of inference (filtering and prediction) of future quantum states upon the measurement results of the classical (i.e. commutative) past of process of observation. The act of measurement transforms quantum propensities into classical realities. As Lawrence Bragg, another Nobel prize winner, once said, everything in the future is a wave, everything in the past is a particle.

The true Heisenberg principle. The time continuous solution of the quantum measurement problem was motivated by analogy with the classical stochastic filtering problem which obtains the prediction of future for an unobservable dynamical process $x(t)$ by time-continuous measuring of another, observable process $y(t)$. Such a problem was first considered by Wiener and Kolmogorov who found its solution in the form of causal spectral filter but only for the stationary Gaussian case. The differential solution in the form of a stochastic filtering equation was then obtained by Stratonovich [22] in 1958 for an arbitrary Markovian pair (x, y). This was really a break through in the statistics of stochastic processes which soon found many applications, in particular for solving the problems of stochastic control under incomplete information (it is possible that this was one of the reasons why the Russians were so successful in launching the rockets to the Moon and other planets of the Solar system in 60s).

If $X(t)$ is the unobservable process, a Heisenberg coordinate process of a quantum particle, say, and $Y(t)$ is an observable quantum process, describing the trajectories $y(t)$ of the particle in a cloud chamber, say, why don't we find a filtering equation for the a posterior expectation $q(t)$ of $X(t)$ in the same way as we do it in the classical case if we know a history, i.e. a particular trajectory $y(r)$ up to the time t? This problem was first considered and solved for the case of quantum Markovian Gaussian pair (X, Y) corresponding to a quantum open linear system with linear output channel, in particular for a quantum oscillator matched to a quantum transmission line [23]. By studying this example, the nondemolition condition

$$[X(s), Y(r)] = 0, \qquad [Y(s), Y(r)] = 0 \quad \forall r \leq s$$

was first found, and this allowed to get the solution in the form of the causal equation for $q(t) = \langle X(t) \rangle_y$.

Let us describe this exact dynamical model of the causal nondemolition measurement first in terms of quantum white noise analysis for a one-dimensional quantum nonrelativistic particle of mass m which is conservative if not observed, in a potential field ϕ. But we shall assume that the particle is under indirect observation by measuring of its Heisenberg position operator $X(t)$ with an additive random error $e(t)$:

$$Y(t) = X(t) + e(t).$$

We take the simplest statistical model for the error process $e(t)$, the white noise model (the worst, completely chaotic error), assuming that it is a classical (i.e. commutative) Gaussian white noise given by the first momenta

$$\langle e(t) \rangle = 0, \quad \langle e(s) e(r) \rangle = \sigma^2 \delta(s - r).$$

The measurement process $Y(t)$ should be commutative, satisfying the causal nondemolition condition with respect to the noncommutative process $X(t)$ (and any other Heisenberg operator-process of the particle), what can be achieved by perturbing the particle Newton-Erenfest equation:

$$m \frac{\mathrm{d}^2}{\mathrm{d}t^2} X(t) + \nabla \phi(X(t)) = f(t).$$

Here $f(t)$ is a Langevin force perturbing the dynamics due to the measurement, which is assumed to be another classical (commutative) white noise.

$$\langle f(t) \rangle = 0, \quad \langle f(s) f(r) \rangle = \tau^2 \delta(s - r).$$

In classical measurement and filtering theory the white noises $e(t), f(t)$ are usually considered independent, and the intensities σ^2 and τ^2 can be arbitrary, even zeros, corresponding to the ideal case of the direct unperturbing observation of the particle trajectory $X(t)$. However in quantum theory corresponding to the standard commutation relations

$$X(0) = \hat{x}, \quad \frac{\mathrm{d}}{\mathrm{d}t} X(0) = \frac{1}{m}\hat{p}, \quad [\hat{x}, \hat{p}] = i\hbar\hat{1}$$

the particle trajectories do not exist, and it was always understood that the measurement error $e(t)$ and perturbation force $f(t)$ should satisfy a sort of uncertainty relation. This "true Heisenberg principle" had never been mathematically formulated and proved before the discovery [23] of quantum causality and nondemolition condition in the above form of commutativity of $X(s)$ and $Y(r)$ for $r \le s$. As we showed first in the linear case [23], and later even in the most general case [17], these conditions are fulfilled if and only if $e(t)$ and $f(t)$ satisfy the canonical commutation relations

$$[e(r), e(s)] = 0, \ [e(r), f(s)] = \frac{\hbar}{i}\delta(r - s), \ [f(r), f(s)] = 0.$$

This proves that the pair (e, f) must satisfy the uncertainty relation $\sigma\tau \ge \hbar/2$ similar to the pair $(\hat{x}, \hat{p})$, which constitutes the precise formulation of the true Heisenberg principle for the square roots σ and τ of the intensities of error e and perturbation f. The canonical pair (e, f) called quantum white noise cannot be considered classically despite of the possibility of the classical realizations of each process e and f separately due to the commutativity of the families e and f. Thus, a generalized matrix mechanics for the treatment of quantum open systems under continuous nondemolition observation and the true Heisenberg principle was invented exactly 20 years ago in [23].

3.2. Consistent Histories and Filtering. Schrödinger believed that all quantum problems including the interpretation of measurement should be formulated in continuous time in the form of differential equations. He thought that the measurement problem would have been resolved if quantum mechanics had been made consistent with relativity theory and the time had been treated appropriately. However Einstein and Heisenberg did not believe this, each for to his own reasons. While Einstein thought that the probabilistic interpretation of quantum mechanics was wrong, Heisenberg simply stated:-

> *Quantum mechanics itself, whatever its interpretation, does not account for the transition from 'possible to the actual'*

Perhaps the closest to the truth was Bohr when he said that it 'must be possible so to describe the extraphysical process of the subjective perception as if it were in reality in the physical world', extending the reality beyond the closed quantum mechanical form by including a subjective observer into a semiclassical system.

Mathematics of the decoherence. In order to formulate the differential non-demolition causality condition and to derive a filtering equation for the posterior states in the time-continuous case we need quantum stochastic calculus.

The classical differential calculus for the infinitesimal increments

$$\mathrm{d}x = x(t + \mathrm{d}t) - x(t)$$

became generally accepted only after Newton gave a simple algebraic rule $(\mathrm{d}t)^2 = 0$ for the formal computations of the differentials $\mathrm{d}x$ for smooth trajectories $t \mapsto x(t)$. In the complex plane $\mathbb{C}$ of phase space it can be represented by a one-dimensional algebra $\mathfrak{a} = \mathbb{C}\mathrm{d}$ of the elements $a = \alpha \mathrm{d}$ with involution $a^\star = \bar{\alpha}\mathrm{d}$. Here

$$\mathrm{d} = \begin{bmatrix} 0 & 1 \\ 0 & 0 \end{bmatrix} = \frac{1}{2}(\sigma_1 + i\sigma_2)$$

for $\mathrm{d}t$ is the nilpotent matrix, which can be regarded as Hermitian $\mathrm{d}^\star = \mathrm{d}$ with respect to the Minkowski metrics $(\mathbf{z}|\mathbf{z}) = 2\,\mathrm{Re}\, z_- \bar{z}_+$ in $\mathbb{C}^2$.

This formal rule was generalized to non-smooth paths early in the last century in order to include the calculus of forward differentials $\mathrm{d}w \simeq (\mathrm{d}t)^{1/2}$ for continuous diffusions w_t which have no derivative at any t, and the forward differentials $\mathrm{d}n \in \{0, 1\}$ for left continuous counting trajectories n_t which have zero derivative for almost all t (except the points of discontinuity where $\mathrm{d}n = 1$). The first is usually done by adding the rules

$$(\mathrm{d}w)^2 = \mathrm{d}t, \quad \mathrm{d}w\mathrm{d}t = 0 = \mathrm{d}t\mathrm{d}w$$

in formal computations of continuous trajectories having the first order forward differentials $\mathrm{d}x = \alpha\mathrm{d}t + \beta\mathrm{d}w$ with the diffusive part given by the increments of standard Brownian paths w. The second can be done by adding the rules

$$(\mathrm{d}n)^2 = \mathrm{d}n, \quad \mathrm{d}n\mathrm{d}t = 0 = \mathrm{d}t\mathrm{d}n$$

in formal computations of left continuous and smooth for almost all t trajectories having the forward differentials $\mathrm{d}x = \alpha\mathrm{d}t + \gamma\mathrm{d}m$ with jumping part given by the increments of standard compensated Poisson paths $m_t = n_t - t$. These rules were developed by Itô [19] into the form of a stochastic calculus.

The linear span of $\mathrm{d}t$ and $\mathrm{d}w$ forms the Wiener-Itô algebra $\mathfrak{b} = \mathbb{C}\mathrm{d} + \mathbb{C}d_w$, while the linear span of $\mathrm{d}t$ and $\mathrm{d}n$ forms the Poisson-Itô algebra $\mathfrak{c} = \mathbb{C}\mathrm{d} + \mathbb{C}d_m$, with the second order nilpotent $d_w = d_w^\star$ and the idempotent $d_m = d_m^\star$. They are represented together with d by the triangular Hermitian matrices

$$\mathrm{d} = \begin{bmatrix} 0 & 0 & 1 \\ 0 & 0 & 0 \\ 0 & 0 & 0 \end{bmatrix}, \quad d_w = \begin{bmatrix} 0 & 1 & 0 \\ 0 & 0 & 1 \\ 0 & 0 & 0 \end{bmatrix}, \quad d_m = \begin{bmatrix} 0 & 1 & 0 \\ 0 & 1 & 1 \\ 0 & 0 & 0 \end{bmatrix},$$

on the Minkowski space $\mathbb{C}^3$ with respect to the inner Minkowski product $(\mathbf{z}|\mathbf{z}) = z_- z^- + z_\circ z^\circ + z_+ z^+$, where $z^\mu = \bar{z}_{-\mu}$, $-(-, \circ, +) = (+, \circ, -)$.

Although both algebras $\mathfrak{b}$ and $\mathfrak{c}$ are commutative, the matrix algebra $\mathfrak{a}$ generated by $\mathfrak{b}$ and $\mathfrak{c}$ on $\mathbb{C}^3$ is not:

$$d_w d_m = \begin{bmatrix} 0 & 1 & 1 \\ 0 & 0 & 0 \\ 0 & 0 & 0 \end{bmatrix} \neq \begin{bmatrix} 0 & 0 & 1 \\ 0 & 0 & 1 \\ 0 & 0 & 0 \end{bmatrix} = d_m d_w .$$

The four-dimensional $\star$-algebra $\mathfrak{a} = \mathbb{C}\text{đ} + \mathbb{C}e_- + \mathbb{C}e^+ + \mathbb{C}e$ of triangular matrices with the canonical basis

$$e_- = \begin{bmatrix} 0 & 1 & 0 \\ 0 & 0 & 0 \\ 0 & 0 & 0 \end{bmatrix}, e^+ = \begin{bmatrix} 0 & 0 & 0 \\ 0 & 0 & 1 \\ 0 & 0 & 0 \end{bmatrix}, e = \begin{bmatrix} 0 & 0 & 0 \\ 0 & 1 & 0 \\ 0 & 0 & 0 \end{bmatrix},$$

given by the algebraic combinations

$$e_- = d_w d_m - \text{đ},\ e^+ = d_m d_w - \text{đ},\ e = e - d_w - \text{đ}$$

is the canonical representation of the differential $\star$-algebra for one-dimensional vacuum noise in quantum stochastic calculus [17, 20]. It realizes the HP (Hudson-Parthasarathy) table [21]

$$\mathrm{d}\Lambda_- \mathrm{d}\Lambda^+ = \mathrm{d}t, \quad \mathrm{d}\Lambda_- \mathrm{d}\Lambda = \mathrm{d}\Lambda_-, \quad \mathrm{d}\Lambda \mathrm{d}\Lambda^+ = \mathrm{d}\Lambda^+, \quad (\mathrm{d}\Lambda)^2 = \mathrm{d}\Lambda,$$

with zero products for all other pairs, for the multiplication of the canonical counting $\mathrm{d}\Lambda = \Lambda(e)$, creation $\mathrm{d}\Lambda^+ = \Lambda(e^+)$, annihilation $\mathrm{d}\Lambda_- = \Lambda(e^+)$, and preservation $\mathrm{d}t = \Lambda(\text{đ})$ quantum stochastic integrators in Fock space over $L^2(\mathbb{R}_+)$. As was proved in [20], any generalized Itô algebra can be represented as a $\star$-subalgebra of a multi-dimensional quantum vacuum algebra, similar to the representation of every semi-classical system with a given state as a subsystem of quantum system with a pure state. In particular, any real-valued process y_t with zero mean value $\langle y_t \rangle = 0$ and independent increments generating a two-dimensional Itô algebra has the differential $\mathrm{d}y$ in the form of a commutative combination of $\mathrm{d}\Lambda, \mathrm{d}\Lambda_-, \mathrm{d}\Lambda^+$. We shall call it standard if it has the stationary incremants with the standard variance $\langle y_t^2 \rangle = t$. In this case

$$y_t = \left(\Lambda^+ + \Lambda_- + \varepsilon\Lambda\right)_t . = \varepsilon m_t + (1 - \varepsilon) w_t,$$

where $\varepsilon \geq 0$ is defined by the equation $(\mathrm{d}y)^2 - \mathrm{d}t = \varepsilon \mathrm{d}y$.

Stochastic decoherence equations. The generalized wave mechanics which enables us to treat the quantum processes of continual in time observation, or in other words, quantum mechanics with trajectories $\omega = (y_t)$, was discovered only quite recently, in [24, 25, 27]. The basic idea of the theory is to replace the deterministic unitary Schrödinger propagation $\psi \mapsto \psi(t)$ by a linear causal stochastic one $\psi \mapsto \chi(t,\omega)$ which is not necessarily unitary for each history ω, but unitary in the mean square sense with respect to a standard probability measure $\mu(\mathrm{d}\omega)$ for $\mathrm{d}\omega$. Due to this positive measure

$$\Pr(t, \mathrm{d}\omega) = \|\chi(t,\omega)\|^2 \mu(\mathrm{d}\omega)$$

is normalized (if $\|\psi\| = 1$) for each t, and is interpreted as the probability measure for the histories $\omega_t = (y_r)_{r<t}$.of an output stochastic process y_t. In the same way as the abstract Schrödinger equation can be derived from only unitarity of propagation, the abstract decoherence wave equation can be derived from the mean square unitarity in the form of a linear stochastic differential equation. The reason that Bohr and Schrödinger didn't derive such equation despite their firm belief that the measurement process can be described 'as if it were in reality in the physical world' is that the appropriate (stochastic) differential calculus had not been yet developed early in that century. As Newton had to invent the differential calculus in order to formulate the equations of classical dynamics, we had to develop the quantum nondemolition calculus [24, 17] in order to derive the generalized wave equation for quantum dynamics with continual observation.

The abstract stochastic wave equation can be written as [25, 26]

$$\mathrm{d}\chi(t) + K\chi(t)\,\mathrm{d}t = L\chi(t)\,\mathrm{d}y_t, \quad \chi(0) = \psi$$

where $y_t(\omega)$ is the output process, which is assumed to be a martingale (e.g. the independent increment process with zero expectations) representing a measurement noise with respect to the input probability measure $\mu(\mathrm{d}\omega) = \Pr(0, \mathrm{d}\omega)$ (but not with respect to the output probability measure $\Pr(\infty, \mathrm{d}\omega)$). The stochastic process $\chi(t,\omega)$ normalized in the mean square sense for each t, is a wave function $\chi(t)$ in an extended Hilbert space, describing the process of continual decoherence of the initial pure state $\rho(0) = P_\psi$ into the mixture

$$\rho(t) = \int P_{\psi_\omega(t)} \Pr(t, \mathrm{d}\omega)$$

of the posterior states corresponding to $\psi_\omega(t) = \chi(t,\omega) / \|\chi(t,\omega)\|$. Using Itô multiplication rule

$$\mathrm{d}(\chi\chi^\dagger) = \chi^\dagger \mathrm{d}\chi + \mathrm{d}\chi^\dagger \chi + \mathrm{d}\chi^\dagger \mathrm{d}\chi$$

in the case of independent increments dy_t, the mean square normalization can be expressed as $K + K^* = L^* L$, i.e.

$$K = \frac{1}{2} L^* L + \frac{i}{\hbar} H,$$

where $H = H^*$ is the Schrödinger Hamiltonian such that this is the Schrödinger equation if $L = 0$. One can also derive the corresponding Master equation

$$\frac{\mathrm{d}}{\mathrm{d}t} \rho(t) + K\rho(t) + \rho(t) K^* = L\rho(t) L^*$$

for mixing the initially pure state $\rho(0) = \psi\psi^\dagger$, as well as a stochastic nonlinear wave equation for the dynamical prediction of the posterior state vector $\psi_\omega(t)$ which is the normalization of $\chi(t, \omega)$ at each ω.

Actually, there are two basic standard forms [27, 28] of such stochastic wave equations, corresponding to two basic types of stochastic integrators: the Brownian standard type, $\varepsilon = 0$, $y_t \simeq w_t$, and the Poisson standard type $\varepsilon = 1$, $y_t \simeq n_t - t$ with respect to the measure μ. To get these we shall assume that y_t is standard with respect to μ, with the multiplication table

$$(\mathrm{d}y)^2 = \mathrm{d}t + \nu^{-1/2} \mathrm{d}y, \quad \mathrm{d}y\mathrm{d}t = 0 = \mathrm{d}t\mathrm{d}y,$$

where $\nu > 0$ is the intensity of the Poisson process $n_t = \nu^{1/2} y_t + t$ with respect to the standard measure μ, and

$$L = \nu^{1/2}(C - I), \quad H = E + i\frac{\nu}{2}\left(C - C^\dagger\right).$$

This corresponds to the stochastic decoherence equation for the counting observation initially derived in the form [25, 29]

$$\mathrm{d}\chi(t) + \left(\frac{\nu}{2}\left(C^\dagger C - I\right) + \frac{i}{\hbar} E\right) \chi(t)\, \mathrm{d}t = (C - I)\, \chi(t)\, \mathrm{d}n_t.$$

The nonlinear filtering equation for $\psi_\omega(t)$ in this case has the form [28]

$$\mathrm{d}\psi_\omega + \left(\frac{\nu}{2}\left(C^\dagger C - \|C\psi_\omega\|^2\right) + \frac{i}{\hbar} E\right) \psi_\omega \mathrm{d}t = \left(C / \|C\psi_\omega\| - I\right) \psi_\omega \mathrm{d}n_t^{\psi_\omega},$$

where n_t^ψ is the output counting process which is described by the probability measure $\Pr(\infty, \mathrm{d}\omega)$ with the increment $\mathrm{d}n_t^\psi$ independent of n_t^ψ under the condition $\psi_\omega(t) = \psi$, having the conditional expectation $\nu \|C\psi\|^2 \mathrm{d}t$. This can be written also in the quasi-linear form [27, 28]

$$\mathrm{d}\psi(t) + \tilde{K}(t)\, \psi_\omega(t)\, \mathrm{d}t = \tilde{L}(t)\, \psi(t)\, \mathrm{d}\tilde{y}_t^{\psi(t)},$$

where $\mathrm{d}\tilde{y}_t^{\psi} = \nu^{-1/2} \|C\psi\|^{-1} \mathrm{d}n_t^{\psi} - \nu^{1/2} \|C\psi\| \,\mathrm{d}t$ is the martingale with respect to the output measure,

$$\tilde{K}(t) = \frac{1}{2}\tilde{L}(t)^* \tilde{L}(t) + \frac{i}{\hbar}\tilde{H}(t),$$

and $\tilde{H}(t), \tilde{L}(t)$ depend on t and ω through the dependence on $\psi = \psi_\omega(t)$:

$$\tilde{L} = \nu^{1/2}(C - \|C\psi\|), \quad \tilde{H} = E + i\frac{\nu}{2}\left(C - C^{\dagger}\right)\|C\psi\|.$$

The last form of the nonlinear filtering equation admits the limit $\nu \to \infty$ corresponding to the standard Wiener case $\varepsilon = 0$ when $y_t = w_t$ with respect to the input Wiener measure μ. If L and H do not depend on ν, i.e. C and E depend on ν as

$$C = I + \nu^{-1/2}L, \; E = H + \frac{\nu^{1/2}}{2i}\left(L - L^{\dagger}\right),$$

then $\tilde{y}_t^{\psi} \to \tilde{y}_t$ as $\varepsilon^2 = \nu^{-1} \to 0$, where $\tilde{y}_t$ with $\mathrm{d}\tilde{y}_t = \mathrm{d}y_t - 2\,\mathrm{Re}\langle\psi|L\psi\rangle \,\mathrm{d}t$ with respect to the output measure is another standard Wiener process $\tilde{w}_t$ which does not depend on ψ. If $\|\psi_\omega(t)\| = 1$ (which follows from the initial condition $\|\psi\| = 1$), the operator-functions $\tilde{L}(t)$, $\tilde{H}(t)$ defining the nonlinear filtering equation have the limits

$$\tilde{L} = L - \mathrm{Re}\langle\psi|L\psi\rangle, \quad \tilde{H} = H + \frac{i}{2}\left(L - L^{\dagger}\right)\mathrm{Re}\langle\psi|L\psi\rangle.$$

Quantum trajectories and filtering. Let us give the exact model based on quantum stochastic calculus for a quantum particle of mass m in a potential ϕ under indirect observation of its position given by the equation

$$\mathrm{d}Y_t = (2\lambda)^{1/2} X(t)\,\mathrm{d}t + \mathrm{d}y_t,$$

where $y_t = w_t$ is the standard Wiener process. This model coincides with the signal plus noise model $Y(t) = X(t) + y(t)$ in terms of the generalized derivatives

$$y(t) = (2\lambda)^{-1/2}\,\mathrm{d}y_t/\mathrm{d}t \equiv e(t),$$

where $e(t)$ is quantum white noise of intensity $\sigma^2 = (2\lambda)^{-1}$. Here $\lambda > 0$ is the accuracy of the measurement of $X(t)$ with respect to the standard Wiener process represented as $w_t = \left(\Lambda_- + \Lambda^+\right)_t$ on the vacuum in Fock space. It was proved in [24, 17] that Y_t is a commutative nondemolition process with respect to the coordinate and momentum Heisenberg processes if they are perturbed by a Langevin force $f(t)$ of intensity $\tau^2 = \lambda\hbar^2/2$, the generalized derivative of $(\lambda/2)^{1/2} f_t$, where $f_t = i\hbar\left(\Lambda_- - \Lambda^+\right)_t$:

$$\mathrm{d}P(t) + \nabla\phi(X(t))\,\mathrm{d}t = \left(\frac{\lambda}{2}\right)^{1/2}\mathrm{d}f_t, \quad P(t) = m\frac{\mathrm{d}}{\mathrm{d}t}X(t).$$

Note that the quantum error process $e_t = y_t$ does not commute with the perturbing quantum process f_t in Fock space due to the multiplication table

$$\mathrm{d}f\mathrm{d}w = i\hbar \mathrm{d}t, \quad \mathrm{d}w\mathrm{d}f = -i\hbar \mathrm{d}t.$$

This corresponds to the canonical commutation relations for the normalized derivatives $e(t)$ and $f(t)$, so that the true Heisenberg principle is fulfilled at the boundary $\sigma\tau = \hbar/2$. Thus our quantum stochastic model of nondemolition observation is the minimal perturbation model for the given accuracy λ of the continual indirect measurement of the position $X(t)$.

The stochastic decoherence equation for this model with $y_t = w_t$,

$$L = \left(\frac{\lambda}{2}\right)^{1/2} \hat{x}, \quad H = \frac{\hat{p}^2}{2m} + \phi(\hat{x})$$

was derived and solved in [24, 30, 27, 31] for the case of linear and quadratic potential ϕ. Here $\hat{x}$ is multiplication by x, and $\hat{p} = \frac{\hbar}{i}\frac{\mathrm{d}}{\mathrm{d}x}$.

The stochastic posterior nonlinear equation in this case is

$$\mathrm{d}\psi(t) + \left(\frac{i}{\hbar}H + \frac{\lambda}{4}\tilde{x}(t)^2\right)\psi(t)\,\mathrm{d}t = \left(\frac{\lambda}{2}\right)^{1/2} \tilde{x}(t)\,\psi(t)\,\mathrm{d}\tilde{y}_t,$$

where $\tilde{x}(t) = \hat{x} - q(t)\,\hat{1}$, $\quad q(t,\omega) = \mathrm{Tr}\left(\hat{x}P_{\psi_w(t)}\right)$ is the posterior expectation (statistical prediction) of $\hat{x}$, and $\mathrm{d}\tilde{y}_t = \mathrm{d}y_t - (2\lambda)^{1/2} q(t)\,\mathrm{d}t$ defines the innovating process $\tilde{y}_t$ which is equivalent to the standard Brownian motion $\tilde{w}_t$ with respect to the output (not input!) probability measure $\Pr(\infty, \mathrm{d}\omega)$.

Let us give the explicit solution of this stochastic wave equation for the free particle ($\phi = 0$) and the stationary Gaussian initial wave packet. One can show [24, 31] that the nondemolition observation of such particle is described by filtering of quantum noise which results in the continual collapse of any wave packet to the Gaussian stationary one centered at the posterior expectation $q(t) = \langle\psi(t)\,|\hat{x}\psi(t)\rangle$ with finite dispersion $\|(q(t) - \hat{x})\,\psi(t)\|^2 \to (\hbar/2\lambda m)^{1/2}$. The center can be found from the linear Newton equation

$$\frac{\mathrm{d}^2}{\mathrm{d}t^2}z(t) + 2\kappa\frac{\mathrm{d}}{\mathrm{d}t}z(t) + 2\kappa^2 z(t) = -g(t),$$

for the deviation process $z(t) = q(t) - y(t)$, with $z(0) = q_0 - y(0)$, $z'(0) = v_0 - y'(0)$. Here $\kappa = (\lambda\hbar/2m)^{1/2}$ is the decay rate which is also the frequency of effective oscillations, $q_0 = \langle\hat{x}\rangle$, $v_0 = \langle\hat{p}/m\rangle$ are the initial expectations, $y(t)$ is a (generalized) trajectory of the observable process $Y(t)$, and $g(t) = y''(t)$ is the effective gravitation for the particle in the moving framework of $y(t)$. The continuous collapse $z(t) \to 0$ of the posterior expectation $q(t)$ towards a linear trajectory $y(t)$ is illustrated in the Appendix.

4. Conclusion: A quantum message from the future

Although Schrödinger didn't derive the stochastic filtering equation for the continuously decohering wave function $\chi(t)$, describing the state of the semiclassical system including the observable nondemolition process y_t in continuous time in the same way as we did it for his cat just in one step, he did envisage a possibility of how to get it 'if one introduces two symmetric systems of waves, which are traveling in opposite directions; one of them presumably has something to do with the known (or supposed to be known) state of the system at a later point in time' [32]. This desire coincides with the "transactional" attempt of interpretation of quantum mechanics suggested in [33] on the basis that the relativistic wave equation yields in the nonrelativistic limit two Schrödinger type equations, one of which is the time reversed version of the usual equation: 'The state vector ψ of the quantum mechanical formalism is a real physical wave with spatial extend and it is identical with the initial "offer wave" of the transaction. The particle (photon, electron, etc.) and the collapsed state vector are identical with the completed transaction.' There was no mathematical proof of this statement in [33], and it is obviously not true for the deterministic state vector $\psi(t)$ satisfying the conventional Schrödinger equation, but we are going to show that this interpretation is true for the stochastic wave $\chi(t)$ satisfying our decoherence equation.

First let us note that the stochastic equation for the offer wave $\chi(t)$ for the standard input probability measure μ can be represented in Fock space as

$$\mathrm{d}\chi(t) + K\chi(t)\,\mathrm{d}t = L\mathrm{d}y_t\chi(t), \quad \chi(0) = \psi \otimes \delta_0,$$

where $y = \Lambda^+ + \Lambda_- + \varepsilon\Lambda$. It coincides on the noise vacuum state δ_0 with the stochastic Schrödinger equation

$$\mathrm{d}\varphi(t) + K\varphi(t)\,\mathrm{d}t = \frac{i}{\hbar}L\mathrm{d}f_t\varphi(t), \quad \varphi(0) = \psi \otimes \delta_0$$

corresponding to the generalized Heisenberg equation with the Langevin force because $if_t = \hbar(\Lambda^+ - \Lambda_-)_t$ has the same action as $\hbar y_t$ on δ_0 since $\Lambda_-\delta_0 = 0$, $\Lambda\delta_0 = 0$ (the annihilation process Λ_- is zero on the vacuum δ_0, as well as the number process Λ). Thus when extended from δ_0 to any initial Fock vector φ_0 quantum stochastic evolution is the HP unitary propagation [21] which is a unitary cocycle on Fock space over $L^2(\mathbb{R}_+)$ with respect to the free time-shift evolution $\varphi(t,s) = \varphi(0, s+t)$ in the Fock space. This free plain wave evolution in the half of space $s > 0$ in the extra dimension is the input, or offer wave evolution for our three dimensional (or more?) world located at the boundary of $\mathbb{R}_+$. The single offer waves do not interact in the Fock space until they reach the boundary $s = 0$ where they produce the quantum jumps described by the stochastic differential equation.

As has been recently shown in [34], by doubling the Fock space it is possible to extend the cocycle to a unitary group evolution which will also include the free propagation of the output waves in the opposite direction. The conservative boundary condition corresponding to the interaction with our world at the boundary, is including the creation, annihilation and exchange of the input-output waves. The corresponding "Schrödinger" boundary value problem is the second quantization of the Dirac wave equation on the half line, with a boundary condition in Fock space which is responsible for the stochastic interaction of quantum noise with our world in the course of the transaction of the input-output waves. The nondemolition continual observations are represented in this picture by the measuring at the boundary of the arrival times and positions of the particles corresponding to the quantized waves in Fock space with respect to an "offer state" the input vacuum, dressed into the output wave. The continual reduction process for our world wave function then is simply represented as the decohering input wave function in the extended space, which is filtered from the corresponding mixture of pure states by the process of innovation of the initial knowledge during the continual measurement. The result of this filtering gives the best possible prediction of future states which is allowed by the quantum causality. As was shown on the example of a free quantum particle under observation, the filtering appear as a dissipation, oscillation and gravitation as a result of nondemolition observation.

Acknowledgement:

I would like to acknowledge the help of Professor R L Hudson and some of my students attending the lecture course on Modern Quantum Theory who were the first who read and commented on these notes containing the answers on some of their questions. The best source on history and drama of quantum theory and its interpretation is in the biographies of the great inventors, Schrödinger, Bohr and Heisenberg [3, 36, 37], and on the conceptual development of this theory before the rise of quantum probability – in [35].

APPENDIX: **The continuous collapse of a free quantum particle under the stationary nondemolition observation.**

The posterior position expectation $q(t)$ in the absence of effective gravitation, $y''(t) = 0$, collapses to the registered linear trajectory $y(t) = ut - q$ with the rate $\kappa = (\lambda\hbar/2m)^{1/2}$, remaining not collapsed, $q_0(t) = v_0 t$ in the framework where $q_0 = 0$, only in the classical limit $\hbar/m \to 0$ or absence of observation $\lambda = 0$:

$$q_0(t) = v_0 t, \quad q(t) = ut + e^{-\kappa t}\left(q\cos\kappa t + \left(q + \kappa^{-1}(v_0 - u)\right)\sin\kappa t\right) - q$$

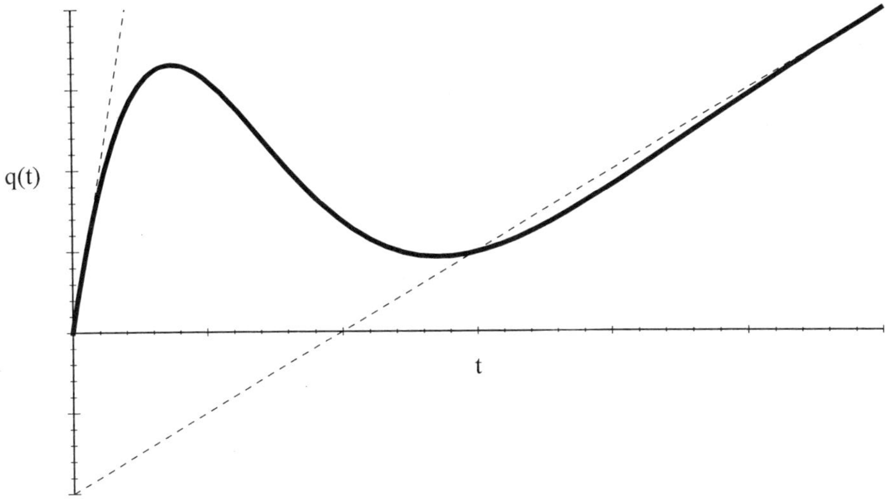

References

[1] E. Schrödinger, *Quantization as an Eigenvalue Problem.* Ann. Phys. **79**, 361–76 (1926).

[2] E. Schrödinger, *Abhandlundgen zur Wellenmechanik.* Leipzig: J.A. Barth (1926).

[3] W. Moore, *Schrödinger life and thought.* Cambridge University Press (1989).

[4] W. Heisenberg, *On the Perceptual Content of Quantum Theoretical Kinematics and Mechanics.* Z. Phys. **43**, 172–198 (1925). English translation in: J. A. Wheeler and Wojciech Zurek, eds. *Quantum Theory and Measurement.* (Princeton University Press, 1983), pp. 62–84.

[5] J. von Neumann, *Mathematische Grundlagen der Quantummechanik.* Springer , Berlin, 1932.

[6] A. Einstein, B. Podolski & N. Rosen, *Can Quantum-Mechanical Description of Physical Reality be Considered Complete?* Phys.Rev. **47**, 777–800 (1935).

[7] N. Bohr, Phys. Rev. **48**, 696–702 (1935).

[8] E. Schrödinger, Naturwis. **23**, 807–12, 823–8, 844–9 (1935).

[9] G. Birkgoff & J. von Neumann, *The Logic of Quantum Mechanics.* Annals of Mathematics **37**, 823–843 (1936).

[10] G. C. Wick, A. S. Wightman & E. P. Wigner, *The Intrinsic Parity of Elementary Particles.* Phys. Rev. **88**, 101–105 (1952).

[11] J. P. Jauch & G. Piron, *Can Hidden Variables be Excluded in Quantum mechanics? (1963).*

[12] V.P. Belavkin, *Nondemolition Principle of Quantum Measurement Theory.* Foundations of Physics, **24**, No. 5, 685–714 (1994).

[13] R. L. Stratonovich & V. P. Belavkin, *Dynamical Interpretation for the Quantum Measurement Projection Postulate.* Int. J. of Theor. Phys., **35**, No. 11, 2215–2228 (1996).

[14] G. Ludwig, Math. Physics, **4**, 331 (1967), **9**, 1 (1968).

[15] E. B. Davies & J. Lewis, Commun Math. Phys., **17**, 239–260 (1970).

[16] E. B. Ozawa, J. Math. Phys., **25**, 79–87 (1984).

[17] V. P. Belavkin, *Quantum Stochastic Calculus and Quantum Nonlinear Filtering.* Journal of Multivariate Analysis,**42**, No. 2, 171–201 (1992).

[18] L. Accardi, A. Frigerio & J. Lewis, Publ. RIMS Kyoto Univ., **18**, 97 (1982)

[19] K. Itô, *On a Formula Concerning Stochastic Differentials.* Nagoya Math . J., **3**, 55-65, (1951).

[20] V. P. Belavkin, *On Quantum Itô Algebras and Their Decompositions.* Lett. in Math. Phys. **45**, 131–145 (1998).

[21] R. L. Hudson & K. R. Parthasarathy, *Quantum Itô's Formula and Stochastic Evolution.* Comm. Math. Phys., **93**, 301–323 (1984).

[22] R. L. Stratonovich, *Conditional Markov Processes and Their Applications to Optimal Control.* Moscow State University, (1966).

[23] V. P. Belavkin, *Quantum Filtering of Markov Signals with Wight Quantum Noise.* Radiotechnika and Electronika, **25**, 1445–1453 (1980).

[24] V. P. Belavkin, In Modelling and Control of Systems, ed A. Blaquière, Lecture Notes in Control and Information Sciences, **121**, 245–265, Springer 1988.

[25] V. P. Belavkin, *A Continuous Counting Observation and Posterior Quantum Dynamics.* J. Phys. A: Math. Gen. **22**, L1109–L1114 (1989).

[26] V. P. Belavkin, *A Posterior Schrödinger Equation for Continuous Nondemolition Measurement.* J. Math. Phys. **31**, 2930–2934 (1990).

[27] V. P. Belavkin, *A New Wave Equation for a Continuous Nondemolition Measurement.* Phys. Lett.A, **140**, 355–362 (1989).

[28] V. P. Belavkin, *A Stochastic Posterior Schrödinger Equation for Counting Nondemolition Measurement.* Letters in Math. Phys. **20,** 85–89 (1990).

[29] A. Barchielli & V. P. Belavkin, *Measurements Continuous inTtime and a posteriori States in Quantum Mechanics.* J. Phys. A: Math. Gen. **24**, 1495–1514 (1991).

[30] L. Diosi, Phys. Rev. A **40**, 1165–74 (1988).

[31] V. P. Belavkin & P. Staszewski, *Nondemolition Observation of a Free Quantum Noise.* Phys. Rev. **45**, N0. 3 1347–1356 (1992).

[32] E. Schrödinger, Sitzberg Preus Akad. Wiss. Phys.–Math. Kl. 144-53 (1931).

[33] J. G. Cramer, Rev. Mod. Phys., **58**, 647–87.

[34] V. P. Belavkin, *On the Equivalence of Quantum Stochastics and a Dirac Boundary Value Problem, and an Inductive Stochastic Limit.* Submitted for publication, January 2000.

[35] M. Jammer, *The Conceptual Development of Quantum Mechanics.* McGraw-Hill, 1966.

[36] A. Pais, *Niels Bohr's Times*, Clarendon Press - Oxford 1991.

[37] D. C. Cassidy, *Uncertainty. Werner Heisenberg.* W.H.Freeman, New-York, 1992.

Decoherence and its Implications in Quantum Computation and Information Transfer
T. Gonis and P.E.A. Turchi (Eds.)
IOS Press, 2001

Hardy's Paradox Revisited

Yakir Aharonov[a,b,c], Alonso Botero[c,d], Sandu Popescu[e,f], Benni Reznik[a], Jeff Tollaksen[g]

[a] *School of Physics and Astronomy, Tel Aviv University, Tel Aviv 69978, Israel.*
[b] *Department of Physics, University of South Carolina, Columbia, SC 29208.*
[c] *Department of Physics, Texas A & M University, College Station, TX 7784-4242, USA.*
[d] *Centro Internacional de Física, Ciudad Universitaria, Bogotá, Colombia.*
[e] *H.H. Wills Physics Laboratory, University of Bristol, Tyndall Avenue, Bristol BS8 1TL, UK*
[f] *BRIMS, Hewlett-Packard Laboratories, Stoke Gifford, Bristol BS12 6QZ, UK*
[g] *Department of Physics, Boston University, Boston, MA 02215.*

Abstract. Using as an illustration a gedanken-experiment due to Hardy[1], (a variation on the concept of "interaction-free measurements" invented by Elitzur and Vaidman[2]), we present a novel approach to quantum mechanical paradoxes.

Quantum mechanics is in a very strange situation. On one hand, since its discovery more that seven decades ago, it has been most successful in explaining an enormous range of physical phenomena. On the other hand, it is safe to say that an intuitive understanding of quantum mechanics is still missing. New, unexpected, and surprising phenomena are very frequently discovered, and almost every attempt to go beyond Schrodinger's equation and explain the behavior of quantum particles in terms related to our everyday macroscopic-world concepts leads to absurd conclusions - to "paradoxes".

A beautiful illustration of quantum mechanical paradoxes is due to Hardy[1]. He shows that in a two-particle Mach-Zehnder interferometer, realistic trajectories inferred from one particle's detection are in direct contradiction with the trajectories inferred from the other particle's detection. In effect, Hardy derives a paradoxical inference in which an electron and a positron in some way manage to "be" and "not to be" at the same time and at the same location.

The general way of "resolving" quantum paradoxes such as Hardy's has been to point out that they rely on counter-factual reasoning. That is, the contradictions arise only when we make inferences about experiments which did not actually take place. Had we actually performed the relevant measurements, we are told, then standard measurement theory predicts that the system would have been disrupted in such a way that no paradoxical implications would arise. [3].

We will show however that the paradoxical reality implied counter-factually has in fact new, *experimentally accessible* consequences. These observable consequences are the result of *weak measurements*, which allow us to test - to some extent - assertions that have been otherwise regarded as counter-factual.

The problem with the counter-factual statements is that if we actually perform measurements to test them, we disturb the system significantly, and in such disturbed conditions no paradoxes arise. Our main point is that if one doesn't perform absolutely precise measurements but is willing to accept some finite accuracy, then one can limit the disturbance on

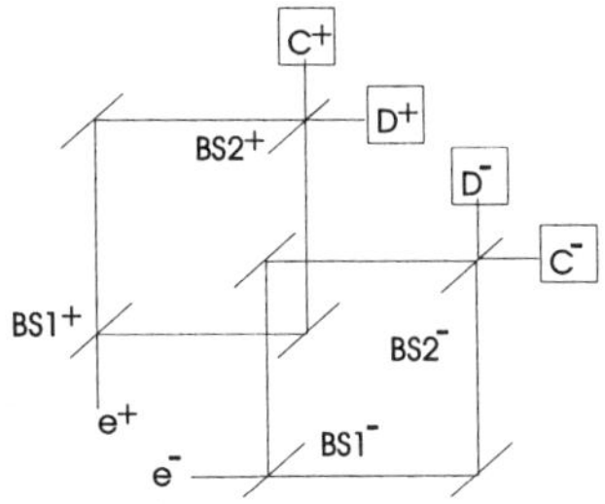

Figure 1: Hardy's gedanken-experiment

the system. For example, according to Heisenberg's uncertainty relations, an absolutely precise measurement of position reduces the uncertainty in position to zero but produces an infinite uncertainty in momentum. On the other hand, if we measure the position only up to some finite precision $\Delta x = \Delta$ we can limit the disturbance of momentum to a finite amount $\Delta p \geq \hbar/\Delta$. We use such limited disturbance measurements to experimentally test the paradoxes implied by the counter-factual statements. What we find is that the paradox in no way disappears - the results of our measurements turn out to be most surprising and to show a strange, but very consistent structure.

Hardy's gedanken-experiment is a variation on the interaction-free measurements (IFM) invented by Elitzur and Vaidman [2]. The set-up consists of two "superposed" Mach-Zehnder interferometers (MZI)(see Fig 1), one with a positron and one with an electron.

Consider first a single interferometer, for instance that of the positron (labeled by +). By adjusting the arm lengths, it is possible to arrange specific relative phases in the propagation amplitudes for paths between the beam-splitters $BS1^+$ and $BS2^+$ so that the positron, entering the interferometer as described in fig.1, can only emerge towards the detector C^+. However, the phase difference can be altered by the presence of an object, for instance in the lower arm, in which case detector D^+ may be triggered. In the usual IFM setup, this is illustrated by the dramatic example of a sensitive bomb that absorbs the particle with unit probability and subsequently explodes. In this way, if D^+ is triggered, it is then possible to infer the presence of the bomb without "touching" it, i.e., to know both that there was a bomb and that the particle went through the path where there was no bomb.

In the double MZI setup, things are arranged so that if each MZI is considered separately, the electron can only be detected at C^- and the positron only at C^+. However, because there is now a region where the two particles overlap, there is also the possibility that they will annihilate each other. We assume that this occurs with unit probability if both particles happen to be in this region. According to quantum mechanical rules, the presence of this interference-destroying alternative allows for a situation similar to that of the IFM in which detectors D^- and D^+ may click in coincidence (in which case, obviously, there is no annihilation).

Indeed, let us label the arms of the interferometers as "overlapping", O, and "non-overlapping", NO. The state of the electron and positron, after passing through $BS1^-$ and $BS1^+$ is

$$\frac{1}{2}\left(|\mathrm{O}\rangle_{\mathrm{p}} + |\mathrm{NO}\rangle_{\mathrm{p}}\right)\left(|\mathrm{O}\rangle_{\mathrm{e}} + |\mathrm{NO}\rangle_{\mathrm{e}}\right). \tag{1}$$

The detectors C^+ and D^+ measure the projectors on the states $\frac{1}{\sqrt{2}}\left(|\mathrm{O}\rangle_{\mathrm{p}} + |\mathrm{NO}\rangle_{\mathrm{p}}\right)$ and $\frac{1}{\sqrt{2}}\left(|\mathrm{O}\rangle_{\mathrm{p}} - |\mathrm{NO}\rangle_{\mathrm{p}}\right)$ respectively, and similarly for the detectors C^- and D^-. Each interferometer is so arranged that the free propagation doesn't add any supplementary phase difference between the arms; if it were not for the electron-positron interaction, the detectors D^- and D^+ would never click.

We are interested in measuring the electron and positron when they traveled in the interferometers beyond the moment when they could have annihilated; we are interested in the cases when annihilation didn't occur. The state at this moment becomes

$$|\Psi\rangle = \frac{1}{\sqrt{3}}|\mathrm{NO}\rangle_\mathrm{p}|\mathrm{O}\rangle_\mathrm{e} + \frac{1}{\sqrt{3}}|\mathrm{O}\rangle_\mathrm{p}|\mathrm{NO}\rangle_\mathrm{e} + \frac{1}{\sqrt{3}}|\mathrm{NO}\rangle_\mathrm{p}|\mathrm{NO}\rangle_\mathrm{e}, \tag{2}$$

which is obtained from (1) by projecting out the term $|\mathrm{O}\rangle_\mathrm{p}|\mathrm{O}\rangle_\mathrm{e}$ corresponding to annihilation.

The simultaneous clicking of D^- and D^+ implies finding the electron-positron pair in the state

$$|\Phi\rangle = \frac{1}{2}\left(|\mathrm{NO}\rangle_\mathrm{p} - |\mathrm{O}\rangle_\mathrm{p}\right)\left(|\mathrm{NO}\rangle_\mathrm{e} - |\mathrm{O}\rangle_\mathrm{e}\right). \tag{3}$$

From the overlap of $|\Psi\rangle$ and $|\Phi\rangle$ we see that this possibility is indeed allowed, and occurs with probability $1/12$.

Suppose now that D^- and D^+ do click. Trying to "intuitively" understand this situation leads to paradox. Based on the interferometers setup, we should infer from the clicking of D^- that the positron must have gone through the overlapping arm; otherwise nothing would have disturbed the electron, and the electron couldn't have ended in D^-. On the other hand, a similar argument starting from the clicking of D^+, leads to the conclusion that the electron must have also gone through the overlapping arm. But then if both the electron and the positron went through the overlapping arms, they should have annihilated, and couldn't have reached the detectors. Hence the paradox.

Other contradictions can be obtained as well. From the clicking of D^- we infer that the positron must have gone through the overlapping arm; otherwise nothing would have disturbed the electron, and the electron couldn't have ended up in D^-. Furthermore, from the fact that there was no annihilation we also deduce that the electron must have gone through the non-overlapping arm. However, from the clicking of D^+ we deduce that the electron is the one which went through the overlapping arm and the positron went through the non-overlapping arm!

The above statements about the positions of the electron and positron are counter-factual, i.e. we haven't actually measured the positions. Suppose however that we try to measure, say, the position of the electron, for example by inserting a detector $\mathcal{D}_O^-$ in the overlapping arm of the electron MZI. We find that the electron is always in the overlapping arm - the detector $\mathcal{D}_O^-$ always clicks - in accordance with our previous counterfactual statements. Indeed, before measuring the position of the electron by the detector $\mathcal{D}_O^-$, (and given that there was no annihilation), the electron-positron pair is in the state (2). In this state the measurement of the position of the electron can yield both answers, namely "electron not in the overlapping arm" (no click at $\mathcal{D}_O^-$) and "electron in the overlapping arm" (click at $\mathcal{D}_O^-$). The first answer occurs with probability 2/3. However, after the measurement the state of the electron-positron pair becomes

$$\frac{1}{\sqrt{2}}(|\mathrm{O}\rangle_\mathrm{p} + |\mathrm{NO}\rangle_\mathrm{p})|\mathrm{NO}\rangle_\mathrm{e}, \tag{4}$$

the (normalized) projection of (2) on the Hilbert subspace corresponding to the electron in the non-overlapping arm. But in this state the positron has zero probability of arriving at D^+. Thus in all cases in which the positron arrives at D^+, the detector $\mathcal{D}_O^-$ also clicks, indicating that the electron did indeed go through the overlapping arm.

On the other hand, $\mathcal{D}_O^-$ disturbs the electron and the electron could end up in the D^- detector even if no positron were present! Hence, when we actually measure the position of the electron, we can no longer infer from a click at D^- that a positron should have traveled through the overlapping arm of the positron MZI in order to disturb the electron. The paradox disappears.

Let us now however measure the positions of the electron and positron in a more "gentle" way, such that we do not totally disturb the physical observables which do not commute with position. A way of doing this is to follow von Neumann's measuring procedure. That is, to couple a measuring device to the observed system via the interaction Hamiltonian

$$H_{int} = g(t)\hat{P}\hat{A}, \tag{5}$$

where $g(t)$ is the coupling constant describing the turning "on" and "off" of the measuring interaction, $\hat{A}$ is the measured observable and $\hat{P}$ is the canonical conjugate momentum of $\hat{Q}$ which describes the position of the pointer of the measuring device. The measurement is made "gentle" by letting the coupling constant to be small and the time of interaction small. Let $\Psi_{MD}(Q)$ be the initial state of the measuring device.

As shown by Aharonov et al. [4] in the weak coupling regime the final state of the measuring device is

$$\Psi_{MD}(Q - A_w), \tag{6}$$

which describes the pointer shifted to a surprising value, A_w, called the "weak value" of the observable $\hat{A}$ and given by

$$A_w = \frac{\langle\Phi|\hat{A}|\Psi\rangle}{\langle\Phi|\Psi\rangle}. \tag{7}$$

Here $|\Psi\rangle$ is the "pre-selected state", i.e. the initial state of the system - in our case the state (2) - while $|\Phi\rangle$ is the "post-selected state" - in our case the state (3) which describes finding the electron at D^- and the positron at D^+.

Note that in contrast to ordinary expectation values, weak values can lie outside the range of eigenvalues of $\hat{A}$ and are generally complex! Their real and imaginary parts are given by the corresponding effects on the pointer $\hat{Q}$ and its conjugate $\hat{P}$ respectively.

The above behaviour of the measuring device may look strange indeed; we want to emphasize however that there is nothing strange about the measurement itself - it is an ordinary, standard measurement of $\hat{A}$, only that the coupling g with the measured system is made weaker. In fact, it can be shown that *any* external system, that interacts linearly with an observable $\hat{A}$ of a pre- and post-selected system, will react, in the limit that the coupling is sufficiently small, as if the value of $\hat{A}$ is A_w. (For a detailed description of how weak values arise, and their significance see [5]).

Finally and most importantly, in the weak regime different measurements do not disturb each other so non-commuting variables $\hat{A}$ and $\hat{B}$ can be measured simultaneously and they yield the same weak values A_w and B_w as when measured separately.

Let us now investigate Hardy's paradox by using weak measurements. What we would like to test are question such as "Which way does the electron go?", "Which way does the positron go?", "Which way does the positron go when the electron goes through the overlapping arm?" etc.. In other words, we would like to measure the single-particle "occupation" operators

$$\hat{N}^{+}_{\mathrm{NO}} = |\mathrm{NO}\rangle_{\mathrm{p}}\langle\mathrm{NO}|_{\mathrm{p}} \qquad \hat{N}^{+}_{\mathrm{O}} = |\mathrm{O}\rangle_{\mathrm{p}}\langle\mathrm{O}|_{\mathrm{p}}$$
$$\hat{N}^{-}_{\mathrm{NO}} = |\mathrm{NO}\rangle_{\mathrm{e}}\langle\mathrm{NO}|_{\mathrm{e}} \qquad \hat{N}^{-}_{\mathrm{O}} = |\mathrm{O}\rangle_{\mathrm{e}}\langle\mathrm{O}|_{\mathrm{e}} \tag{8}$$

which tell us separately about the electron and the positron and also the *pair* occupation operators

$$\hat{N}^{+,-}_{\mathrm{NO},\mathrm{O}} = \hat{N}^{+}_{\mathrm{NO}}\hat{N}^{-}_{\mathrm{O}} \qquad \hat{N}^{+,-}_{\mathrm{O},\mathrm{NO}} = \hat{N}^{+}_{\mathrm{O}}\hat{N}^{-}_{\mathrm{NO}}$$
$$\hat{N}^{+,-}_{\mathrm{O},\mathrm{O}} = \hat{N}^{+}_{\mathrm{O}}\hat{N}^{-}_{\mathrm{O}} \qquad \hat{N}^{+,-}_{\mathrm{NO},\mathrm{NO}} = \hat{N}^{+}_{\mathrm{NO}}\hat{N}^{-}_{\mathrm{NO}} \tag{9}$$

which tell us about the simultaneous locations of the electron and positron. We note a most important fact, which is essential in what follows: the weak value of a product of observables is *not* equal to the product of their weak values. Hence, we have to measure the pair occupation operators independently from the single-particle occupation numbers.

Since we will be performing weak measurements of these observables, i.e. using probes which interact weakly with the electron-positron system and produce only limited disturbance, we can perform all these tests simultaneously. We will show that the results of our measurements echo, to some extent, the counter-factual statements, but go far beyond that. They are now true observational statements and, if anything, they are even more paradoxical. Indeed, using the definition of the weak value (7) and the pre- and post-selected states (2), (3) we obtain

$$N^{-}_{Ow} = 1, \qquad N^{+}_{Ow} = 1 \tag{10}$$

$$N^{-}_{NOw} = 0, \qquad N^{+}_{NOw} = 0 \tag{11}$$

$$N^{+,-}_{O,Ow} = 0 \tag{12}$$

$$N^{+,-}_{O,NOw} = 1, \qquad N^{+,-}_{NO,Ow} = 1 \tag{13}$$

$$N^{+,-}_{NO,NOw} = -1. \tag{14}$$

What do all these results tell us? The single-particle occupation numbers (10) are consistent with the intuitive statements that "the positron must have been in the overlapping arm otherwise the electron couldn't have ended at D^{-}" and also that "the electron must have been in the overlapping arm otherwise the positron couldn't have ended at D^+". But then what happened to the fact that they could not be both in the overlapping arms since this will lead to annihilation? Quantum mechanics is consistent with this too - the pair occupation number $N^{+,-}_{\mathrm{O},\mathrm{Ow}} = 0$ shows that there are zero electron-positron pairs in the overlapping arms!

We also feel intuitively that "the positron must have been in the overlapping arm otherwise the electron couldn't have ended at D^-, and furthermore, the electron must have gone through the non-overlapping arm since there was no annihilation". This is confirmed by $N^{+,-}_{\mathrm{O},\mathrm{NO}} = 1$. But we also have the statement "the electron must have been in the overlapping arm otherwise the positron couldn't have ended at D^- and furthermore the positron must have gone through the non-overlapping arm since there was no annihilation". This is confirmed too, $N^{+,-}_{\mathrm{NO},\mathrm{Ow}} = 1$. But these two statements together are at odds with the fact that there is in fact just one electron-positron pair in the interferometer. Quantum mechanics solves the paradox in a remarkable way - it tells us that $N^{+,-}_{\mathrm{NO},\mathrm{NOw}} = -1$, i.e. that there is also *minus* one electron-positron pair in the non-overlapping arms which brings the total down to a single pair!

Finally, the intuitive statement that "The electron did not go through the non-overlapping arm since it went through the overlapping arm" is also confirmed - a weak measurement finds no electrons in the non-overlapping arm, $N^-_{NOw} = 0$. But we know that there *is* one electron in the non-overlapping arm as part of a pair in which the positron is in the overlapping arm, $N^{+,-}_{O,NO} = 1$; how is it then possible to find no electrons in the non-overlapping arm? The answer is given by the existence of the *minus* one electron-positron pair, the one with the electron and positron in the non-overlapping arms, which contributes a further *minus* one electron in the non-overlapping arm, bringing the total number of electrons in the non-overlapping arm to zero:

$$N^-_{NOw} = N^{+,-}_{O,NOw} + N^{+,-}_{NO,NOw} = 1 - 1 = 0. \tag{15}$$

Let us now turn to the question of how to perform the weak measurements described above. First of all, we note that, as discussed earlier, an important property of weak measurements, is that what are usually mutually disturbing measurements, "commute" in this limit, i.e., they no longer disturb each other and can be performed simultaneously. Hence, in principle, the whole set of predictions (10)-(14) for the single and pair occupation numbers can be experimentally verified simultaneously. But as we have also mentioned before, this comes for a price - the measurements are necessarily imprecise. How imprecise? It can be easily seen that for the measurements considered here (where the measured operators have only two distinct eigenstates), the weak regime is obtained when the shift of the pointer is smaller than the uncertainty ΔQ [6]. Thus in a single experiment we obtain little information about the value of the weak values. That is, every single measurement may yield an outcome which may be quite far from the weak value (the spread of the outcomes around the weak value is large). Nevertheless, by repeating the measurements (i.e. performing a large number of independent measurements on identically prepared systems), A_w can be determined to any desired accuracy [7]. (A different, improved version of the weak measurements will be discussed later in the paper.)

The single particle occupation can be inferred by a weak measurement of the charge along each arm. For example by sending a massive charged test particle close enough to the relevant path (but sufficiently distant from others) and then using the induced transverse momentum transfer as a pointer variable. The weakness condition is met by preparing the test particle to be in a localized state in the transverse direction, and hence ensuring that momentum transfer is small enough. The measurement must be repeated many times. Finally, after measuring the momentum transfer in each experiment, one evaluates the mean of the result of the separate trials, which is taken to stand for the weak value [7].

In each experiment one can simultaneously also measure the pair occupation operator by introducing a weak interaction between the electron and the positron. For instance to observe $\hat{N}^{+,-}_{\mathrm{NO,NO}}$, we let the non-overlapping trajectories pass through two boxes, just before they arrive to the final two beam splitters. The electron and positron are temporarily captured in the boxes and then released. This will not modify the experiment, provided that no extra phases are generated while the particles cross the boxes. Now suppose that the boxes are connected by a very rigid spring of natural length l. While the electron and positron pass through the boxes the relative deviation in the equilibrium length of the spring produced by the electrostatic force between the two boxes will be

$$\frac{\delta l}{l} = \frac{F_{e,p}}{Kl} \simeq -\frac{e^2}{Kl^3} N^{+,-}_{\mathrm{NO,NO}} \,. \tag{16}$$

where K is the spring constant. The relative shift in the equilibrium position plays the role of the pointer variable with the ratio $g = e^2/Kl^3$ as a dimensionless coupling constant. In other words, when an electron-positron pair is present in the boxes, due to their electrostatic attraction the spring will be compressed. On the other hand, if only the electron, or only the positron, or neither of them is present in the boxes, then here is no electrostatic force and the spring is left undisturbed. In the weak regime however, we will observe a systematic stretching of the spring! This is indicative of a negative pair occupation $N^{+,-}_{\mathrm{NO,NO}}$ which implies an electrostatic *repulsion* between the two boxes[1].

In the above set-up, the measuring devices have to be quite imprecise in order to ensure that they do not disturb each other, and therefore the experiment has to be repeated many times to learn the weak values. A different version of the experiment allows us however to measure all weak values with great precision in one single experiment [9].

To summarize, we have suggested a new set of gedanken-experiments in connection with Hardy's set-up. We find that these experiments yield strange and surprising outcomes. However, since they are experimental results they cannot be dismissed as the original counter-factual statements can. Whatever one's ultimate view about quantum mechanics, one has to understand and explain the significance of these outcomes. The results of the weak measurements appear to be even stranger than Hardy's original paradox. We would like to argue however that the situation is in fact far better. The weak values obey a simple, intuitive, and, most important, *self-consistent* logic. This is in stark contrast with the logic of the original counter-factual statements which is not internally self-consistent. Strangeness by itself is not a problem; self-consistency is the real issue. Indeed, the fact that light has the same velocity in all reference frames is certainly highly unusual, but everything works in a self consistent way, and because of this special relativity is rather easy to understand. We are convinced that, due to its self-consistency, the logic of weak measurements will lead to a deeper understanding of the nature of quantum mechanics.

---

[1]Note that since the electrostatic energy is invariant under a reversal of signs in the charges, this "negativeness" is not the same thing as charge conjugation.

Acknowledgements We thank A. C. Elitzur, S. Dolev and L. Vaidman for discussions. Y. A. and B.R. acknowledge the support from grant 471/98 of the Israel Science Foundation, established by the Israel Academy of Sciences and Humanities, and NSF grant PHY-9971005.

References

[1] L. Hardy, Phys. Rev. Lett. **68** (1992) 2981.

[2] A. C. Elitzur and L. Vaidman, Found. of Phys. **23**(1993) 987.

[3] B. D'Espagnat, Veiled Reality, chapt. 11, Addison-Wesley, (1995).

[4] Y. Aharonov, L. Vaidman, Phys. Rev. **A 41** (1990) 11; J. Phys. **A 24** (1991) 2315.

[5] Y. Aharonov, S. Popescu, D. Rohrlich and L. Vaidman, Phys. Rev. **A 48** (1993) 4084.

[6] A first order approximation in the shift of the pointer of the final state of the measuring device (7) yields immediately the weak value.

[7] A. Botero and B. Reznik, *Phys. Rev.* **A 61** (2000) 050301(R).

[8] L. Vaidman, Phys. Rev. Lett. **70** (1993) 3369; Found. of Phys. 26 (1996) 895.

[9] Y. Aharonov, A. Botero, S. Popescu, B. Reznik and J. Tollaksen, in preparation.

Decoherence and its Implications in Quantum Computation and Information Transfer
T. Gonis and P.E.A. Turchi (Eds.)
IOS Press, 2001

Approximation of single-photon states by postselection from correlated pairs

Ondřej Haderka, Jan Peřina jr.
Joint Laboratory of Optics of Palacký University and Inst. of Physics AS CR, Av. 17. listopadu 50A, 772 00 Olomouc, Czech Republic

Abstract. The possibility to produce single-photon states by selection from entangled pairs obtained by parametric frequency down-conversion is suggested and the efficiency of the procedure is analyzed. A realistic model of such device is developed. The principle of the selection mechanism is clarified. It is shown that, taking into account unavoidable physical noises, it is very difficult to use this device for the fabrication of photon-number squeezed states. To achieve considerable squeezing, high degree of correlation and high mean photon numbers of the correlated photon pairs are necessary.

1 Introduction

Correlated photon pairs produced by spontaneous parametric frequency down-conversion [1, 2] have recently been widely used in experimental quantum physics. They have been successfully applied for the research of fundamental questions of quantum theory [3] as well as their use for direct applications of quantum mechanical laws in optical communications has been suggested [4, 5].

Correlated photon pairs are generated by splitting of pump photons into signal and idler photons in a $\chi^{(2)}$ medium. The energies (wavelengths) of the signal and idler photons obey energy conservation law and the down-converted beams emerge in well defined directions satisfying the momentum conservation law. The energy conservation law obeyed at every elementary nonlinear interaction causes strong quantum correlations between the signal and idler beams. These correlations then manifest themselves in the fact that both photons occur in a sharp time window (of the order of hundreds fs in typical nonlinear crystals). The state describing such correlated fields cannot be factorized into a product of states of signal and idler fields. This property, also called entanglement, leads to a number of interesting consequences, the most notable of which is probably nonlocal projection of the result of a measurement performed on one of the beams on the other beam.

In this work we make use of another nonclassical property of correlated photon pairs. There is no classical field for which the joint probability of photodetection at two different points in space where the mean intensities are equal is much larger than that for two detections at the same point. However, exactly this is the case of down-converted light [1]. This is a direct manifestation of the fact that down-converted photons are always produced in pairs. This suggests the use of down-converted photon pairs as possible sources capable of providing close approximations of single-photon states because when a photon is detected in the signal beam we know that its twin must be present in the idler beam.

Single-photon states represent an eagerly needed source in many areas of optics. E.g., it is possible to generate strong photon-number squeezed states [3] but it is not possible to weaken them to a single-photon level while preserving the squeezing. In secure quantum communications, the practical existence of such a source would help formulating general security proofs [6] of quantum key distribution protocols and it would also make practical schemes more efficient [5].

In general, by post selection from correlated photon pairs we can prepare a variety of interesting states of light applicable in experimental quantum optics. In this work we focus on producing a source of single photon states and we develop a realistic model which tells us how good this source can be.

2 Model of a source of single-photon states

We assume that the postselecting device is placed in the signal beam. This device yields a simple yes-no result (a trigger) and, based on this result, the state in the idler beam is either processed or rejected. Clearly the idea to produce single photon states requires photon number measurement in the signal beam. If the result of the measurement gives single-photon state in the signal, single photon is expected to be also in the idler beam. It is, however, not an easy task to construct a practical photon-number measuring device. Generally used photon counting detectors - avalanche photodiodes or photomultipliers - use many order noisy amplification processes that smear out resolution of small photon numbers. An often suggested quantum non-demolition measurement based on Kerr nonlinearity [1] needs a medium with strong nonlinearity that is not currently available. In our work we use a model of a photon number measuring device based on $1 \times N$ coupler [7].

In our model [8] we suppose that the down-conversion process is pumped by either cw or pulsed laser beam. The signal and idler beams are selected by filters and pinholes (geometrically and spectrally filtered) but they are still considered multimode as it is usually the case in the laboratory (inclusion of the single-mode case in the model is straightforward and does not change anything except for some quantitative measures). The filtering is in general imperfect, i.e. sometimes only one of the members of the pair is led to the detector. Detectors have limited quantum efficiency and they are not capable of resolving the photon number, they just click in the presence of the signal containing one or more photons. In addition, there are noise detections coming both from dark counts of the detectors and from the stray light in the laboratory. The scheme simulating all these imperfections is given in Fig. 2.

The detector with quantum efficiency η non-sensitive to the impinging number of photons $|n\rangle$ can be represented with the following 'clicking' operator

$$E^{click} = \sum_{n=1}^{\infty} [1 - (1-\eta)^n] |n\rangle \langle n| + d \sum_{n=0}^{\infty} (1-\eta)^n |n\rangle \langle n| , \qquad (1)$$

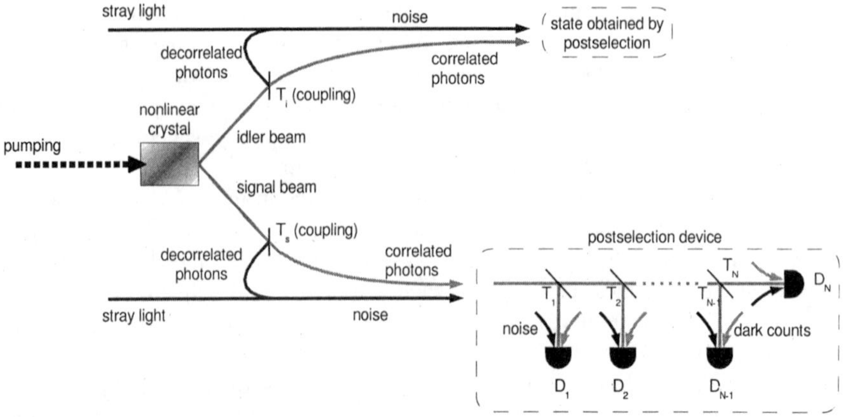

Figure 1: Scheme of the model. Photons of the pumping beam may split in the nonlinear crystal to pairs of correlated frequency-downconverted photons. Signal beam is then coupled to a postselection device consisting of an $1 \times N$ coupler and N detectors. Coupling is imperfect and decorrelated photons contribute to the noise impinging on the detectors that exhibit also internal dark count noise. Idler beam is undergoes similar coupling problems and after postselection it contains desired state whenever triggered.

where d is a total noise count rate that is computed as

$$d = d_{dark} + (1 - d_{dark})d_{noise} \tag{2}$$

when both dark counts and noise coming from stray light are taken into account. Detailed derivation of operator (1) is given elsewhere [8]. Similarly, the operator expressing the fact that the detector does not react to the incident light can be formulated as

$$E^{noclick} = (1 - d)\sum_{n=0}^{\infty} (1 - \eta)^n |n\rangle \langle n| \, . \tag{3}$$

The light field emerging from the output of the nonlinear crystal can be expressed as an entangled state

$$|\psi\rangle = \sum_{n=0}^{\infty} c_n |n, n\rangle_{s,i}, \tag{4}$$

where the indices s,i refer to the signal and idler beams, respectively. Here we describe the multimode field[1] in terms of one effective mode and the coefficients c_n therefore reflect the Poissonian statistics of a multimode field

$$|c_n|^2 = \frac{\mu^n}{n!} e^{-\mu}, \tag{5}$$

μ being the mean photon number per detection interval. Both signal and idler beams experience losses when coupled to the detectors and filtered spectrally by interference filters and

[1]The single-mode case would easily be included in the model by replacing (5) with an analogous formula for the statistics of thermal field.

geometrically by pinholes and by the setup of the experiment. We represent all these coupling losses by transmission coefficients T_s, T_i. After this coupling the state (now partially decorrelated) is described by the statistical operator

$$\rho_{s,i} = \sum_{l_i=0}^{\infty} |l_i\rangle_{i\,i}\langle l_i| \sum_{n=l_i}^{\infty} |c_n|^2 \begin{pmatrix} n \\ l_i \end{pmatrix} T_i^{l_i} R_i^{n-l_i} \sum_{l_s=0}^{n} \begin{pmatrix} n \\ l_s \end{pmatrix} T_s^{l_s} R_s^{n-l_s} |l_s\rangle_{s\,s}\langle l_s|, \tag{6}$$

where we consider only diagonal elements of the statistical operator in the Fock basis (off-diagonal elements do not contribute to detection of photon numbers $|n\rangle\langle n|$). When a selection measurement represented by the operator E^{sel} (see below) is performed in the signal beam, the state obtained in the idler beam can be computed by tracing over signal:

$$\rho_i^{post} = \mathrm{Tr}_s \left\{ E^{sel} \rho_{s,i} \right\}. \tag{7}$$

One can imagine a variety of selection measurements. In this work, having possible quantum cryptographic applications on mind, we are interested in producing single-photon states, or, more realistically, to approach them as close as possible. For that purpose it would be optimal to trigger on single-photon states in the signal only. Unfortunately, existing detectors prevent us from doing that in any simple way. Nevertheless, we can approximate such selection measurement using a $1 \times N$ coupler and N detectors. Provided that mean photon number of the signal $\mu \ll N$ and the detectors exhibit moderate quantum efficiencies and dark count rates, only one detector clicks on single-photon signal while multiple detectors click on multi-photon signals with high probability. We choose the selection operator in the form:

$$E^{sel} = \sum_{k=1}^{N} E_k^{click} \prod_{j=1, j\neq k}^{N} E_j^{noclick}. \tag{8}$$

Substituting (8) to (7) we get

$$\rho_i^{post} = \frac{1}{p^{post}} \sum_{l_i=0}^{\infty} |l_i\rangle_{i\,i}\langle l_i| \sum_{n=l_i}^{\infty} |c_n|^2 \begin{pmatrix} n \\ l_i \end{pmatrix} T_i^{l_i} R_i^{n-l_i} \sum_{k=1}^{N} Q_k(n), \tag{9}$$

$$Q_k(n) = \left[\prod_{j=1}^{N} (1-d_j)\right] \left\{ \frac{1}{1-d_k} A_k^n(T_s) - B^n(T_s) \right\}, \tag{10}$$

$$B(T_s) = T_s \left[\sum_{j=1}^{N} T_j (1-\eta_j)\right] + R_s, \tag{11}$$

$$A_k(T_s) = B(T_s) + T_s T_k \eta_k, \tag{12}$$

where $R_s = 1 - T_s$ and $R_i = 1 - T_i$. The quantity

$$p^{post} = \mathrm{Tr}_i \left\{ \rho_i^{post} \right\} \tag{13}$$

serves as a normalization constant for (9) and has the meaning of selection probability, i.e., the rate of triggering. Substituting (5) to (9) we arrive at

$$\rho_i^{post} = \frac{1}{p^{post}} \left[\prod_{j=1}^{N} (1-d_j)\right] \sum_{n=0}^{\infty} |n\rangle_{i\,i}\langle n| \frac{1}{n!} \times \tag{14}$$

$$\times \left[\sum_{k=1}^{N} \frac{1}{1-d_k} \left[\mu T_i A_k(T_s)\right]^n e^{-\mu[1-R_i A_k(T_s)]} - N \left[\mu T_i B(T_s)\right]^n e^{-\mu[1-R_i B(T_s)]} \right]$$

with

$$p^{post} = \left[\prod_{j=1}^{N}(1-d_j)\right]\sum_{k=1}^{N}\frac{1}{1-d_k}e^{-\mu[1-A_k(T_s)]} - N\,e^{-\mu[1-B(T_s)]}. \tag{15}$$

Since we are not interested in the statistics of light in the signal beam, we can account for the stray light and decorrelated photons in the model using the term d_{noise} in (2). It can be shown [8] that under the condition that the photon number in the signal beam is much lower than the number of modes, the decorrelated photons exhibit Bose-Einstein statistics. The relevant noise term for the k-*th* detector then reads

$$d_k^{noise} = 1 - \frac{1}{1+\frac{\eta_k \mu_k^{noise}}{1-\mu_k^{noise}}}, \qquad \mu_k^{noise} = \frac{\alpha_s T_k R_s \mu}{1+\alpha_s T_k R_s \mu}, \tag{16}$$

where the term α_s is inserted to account for the intensity of noise. $\alpha_s < 1$ attenuates the decorrelated photons contribution, while $\alpha_s > 1$ serves to include stray light with thermal statistics.

On the other hand, photon statistics in the idler beam is important for the purpose of further computations. We therefore mix the noise field [9]

$$\rho_i^{noise} = \sum_{n=0}^{\infty}|n\rangle_{i\,i}\langle n|\,\frac{(\alpha_i R_i \mu)^n}{(1+\alpha_i R_i \mu)^{n+1}} \tag{17}$$

with the idler and obtain the statistical operator in the form

$$\rho_i^{mix} = \sum_{n=0}^{\infty}|n\rangle_{i\,i}\langle n|\sum_{m=0}^{n}\rho_i^{post}(m)\,\rho_i^{noise}(n-m), \tag{18}$$

where $\rho(n) = \langle n|\rho|n\rangle$. The factor α_i has similar meaning as α_s in (16).

It is easy to see that for the purpose of selecting single-photon states, optimum performance is obtained using a symmetric $1\times N$ coupler. Therefore, we discuss somewhat simplified symmetrized model where the following conditions are applied:

$$d_j \to d, \quad \eta_j \to \eta, \quad T_j \to \frac{1}{N} \quad \Rightarrow \quad A_j \to A. \tag{19}$$

Eq. (18) can now be replaced by

$$\begin{aligned}\rho_i^{mix} &= \frac{N}{p^{post}}\sum_{n=0}^{\infty}|n\rangle_{i\,i}\langle n|\sum_{m=0}^{n}\frac{1}{m!}\times \\ &\times \left[\frac{[\mu T_i A(T_s)]^m}{1-d}e^{-\mu[1-R_i A(T_s)]} - N\,[\mu T_i B(T_s)]^m\,e^{-\mu[1-R_i B(T_s)]}\right]\frac{(\alpha_i R_i \mu)^{n-m}}{(1+\alpha_i R_i \mu)^{n-m+1}},\end{aligned} \tag{20}$$

with

$$p^{post} = (1-d)^N\left(\frac{N}{1-d}e^{-\mu[1-A(T_s)]} - N\,e^{-\mu[1-B(T_s)]}\right). \tag{21}$$

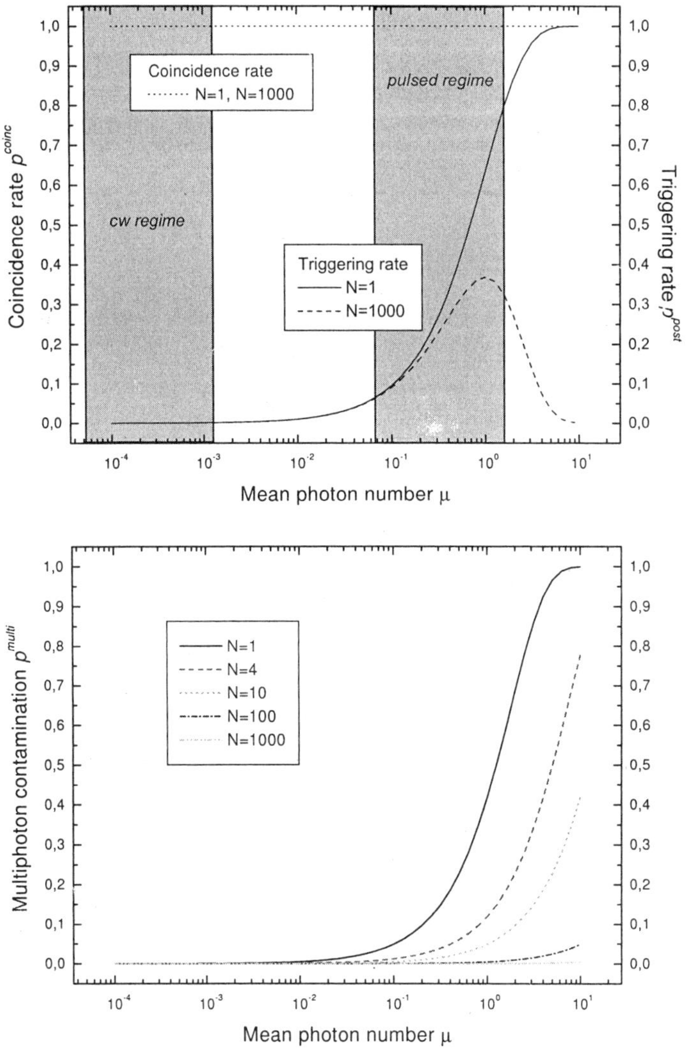

a)

b)

Figure 2: The idealistic model with noiseless ($d = 0$) and 100% efficient detectors ($\eta = 1$), and optimum coupling ($T_s = T_i = 1$). a) Triggering and coincidence rates. b) Multiphoton contamination. Detection interval $\tau = 1$ ns is considered.

3 Results

We will characterize the device using a number of quantities that show the properties of the idler beam with respect to its expected single-photon character:

- Triggering rate
$$p^{post}.$$
- The coincidence rate (conditional probability that the idler beam contains one or more photons provided that it was triggered)
$$p^{coinc} = \sum_{m=1}^{\infty} \rho_i^{mix}(m) .$$
- Vacuum probability (probability of finding zero photons in the triggered idler state)
$$p^{vac} = \rho_i^{mix}(0) .$$
- Multiphoton contamination (probability of finding more than one photon in a nonempty triggered idler state)
$$p^{multi} = \frac{1 - [\rho_i^{mix}(0) + \rho_i^{mix}(1)]}{1 - \rho_i^{mix}(0)} .$$
- Fano factor (standard quantity to describe photon number squeezing)
$$F = \frac{\sum_{m=1}^{\infty} m^2 \rho_i^{mix}(m) - \left[\sum_{m=1}^{\infty} m \, \rho_i^{mix}(m)\right]^2}{\sum_{m=1}^{\infty} m \, \rho_i^{mix}(m)} .$$

We start with the simplest case, namely the optimum (idealistic) performance of the device (see Fig. 2). In Fig. 2a we can see triggering and coincidence rates in the cases of single or thousand detectors in the postselection device, the latter case representing the many-detectors limit. We can see that the triggering rate grows up to unity with increasing mean photon number μ for $N = 1$, while it shows a maximum close to $\mu = 1$ for large N and then falls down to zero. This behavior is caused by the fact that for strong intensities there is high probability of multi-photon states that are recognized by the many-detector device and the trigger is not issued, contrary to the single-detector case. The coincidence rate is unity regardless of mean photon number, as in this oversimplified model there are no noises preventing the occurrence of photons in triggered idler state. We mark two shaded regions in Fig. 2a denoting typical experimental ranges for cw or pulsed pumping of the down-conversion process. We refer to them below.

In Fig. 2b we show the multi-photon contamination of triggered idler states for several values of N. Clearly, the more detectors are used in the device, the better exclusion of multi-photon states is achieved, as expected. Vacuum probability is always zero in this case as follows from the fact that coincidence rate is unity. Therefore Fano factor dependence (not plotted) copies the multi-photon contamination curves as this is the only cause of broadening of the photon number distribution.

The model illustrated in Fig. 2 does not represent a realistic case. We show it in order to give a closer idea of how the photon-number measuring postselection device may contribute to the selection of single-photon states. Vacuum states are eliminated using the correlated character of the signal and idler beams (vacui do not trigger) while multi-photon contributions are eliminated by approximate measurement of photon number in the signal beam. In reality, however, such idealized device is impossible to construct. Real detectors exhibit non-negligible noise and limited quantum efficiency and the signal and idler beam are imperfectly coupled to the detectors. The performance of simulated experiment with realistic silicon avalanche detectors is shown in Fig. 3.

Triggering rate (Fig. 3a) is now lower than in the previous case due to losses in the device and the maximum in the many-detectors limit is shifted to higher mean photon numbers (out of bounds in Fig. 3a) for the same reason. Coincidence rate now approaches unity only in the high intensity limit. It shows growth with intensity in the pulsed regime while it is almost constant in the cw regime when one detector device is used. The more detectors we apply, the lower is the coincidence rate due to the increase of 'vacuum' triggers. This is especially critical in the cw regime.

This fact also reflects in the plot of vacuum probability in Fig. 3b. The application of multiple detectors in the device now brings only slight improvement in the limitation of multiphoton states (Fig. 3b). The dependence of the Fano factor is a combination of the vacuum and multi-photon contributions. The use of many-detector device now gives only slight advantage in the high energy limit and big disadvantage in the low energy limit in the effort for the generation of a photon-number-squeezed state.

The model in Fig. 3 is still unrealistic in the following sense. The decorrelated photons (cf. Fig. 1) are simply removed from the signal and idler beams. We include them into the detector noise in the signal beam and mix them with the idler beam by setting the α_s and α_i coefficients to unity. Inclusion of these noise photons makes the device even less efficient, as one would expect. The coincidence rate is further lowered and almost all the advantages given by the application of many-detector devices are canceled for low coupling efficiencies T_s, T_i as in Fig. 3.

The role of coupling efficiencies is well illustrated in Fig. 4. Fig. 4a plots the dependence of vacuum probability and multi-photon contamination in pulsed regime. The lower the coupling, the higher is the vacuum contribution because the statistic of the idler mode resembles more and more that of a thermal state. The limit $(1+\mu)^{-1}$ for $T_s, T_i \rightarrow 0$ stems from the properties of thermal light as can be shown analytically by limiting the r.h.s. of eq. (20). When more detectors are used, vacuum contribution grows a little more due to the fact that more detectors bring more 'vacuum' triggers through dark counts and shift the statistic even more to the thermal state. As for the multiphoton contamination, it works the better the more detectors are used. It is, however, efficient only for good coupling as it works for correlated states only. This can also be seen in Fig. 4b where the dependence of Fano factor on coupling is depicted. Clearly, for the generation of photon-number-squeezed states, high coupling is necessary.

Similar treatment may also be performed for the cw configuration. Here the multiphoton contribution is low and therefore also the improvement stemming from exclusion of multiphoton states by postselection is negligible. Fano factor dependencies are given mainly by the vacuum contribution.

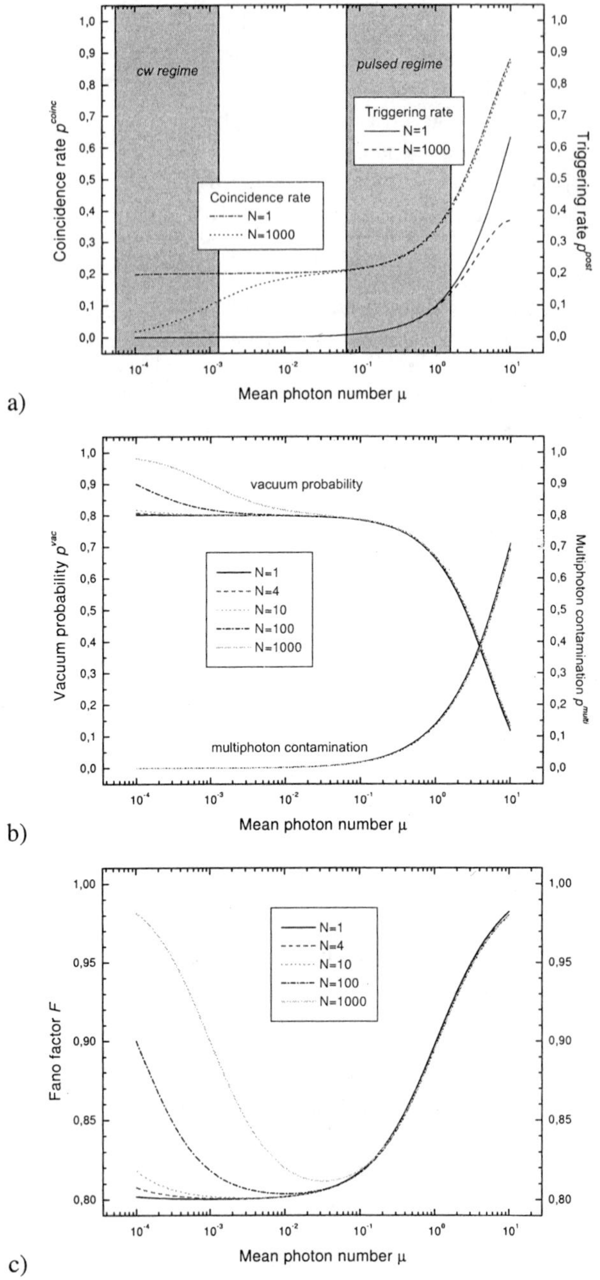

Figure 3: Model with realistic detectors ($d^{dark} = 100\ \mathrm{s}^{-1}$, $\eta = 50\%$) and realistic coupling ($T_s = T_i = 0.2$). a) Triggering and coincidence rates. b) Vacuum probability and multiphoton contamination. c) Fano factor.

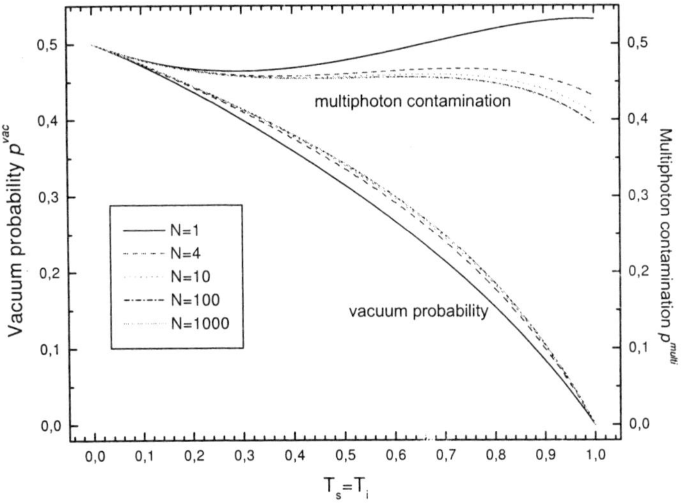

a)

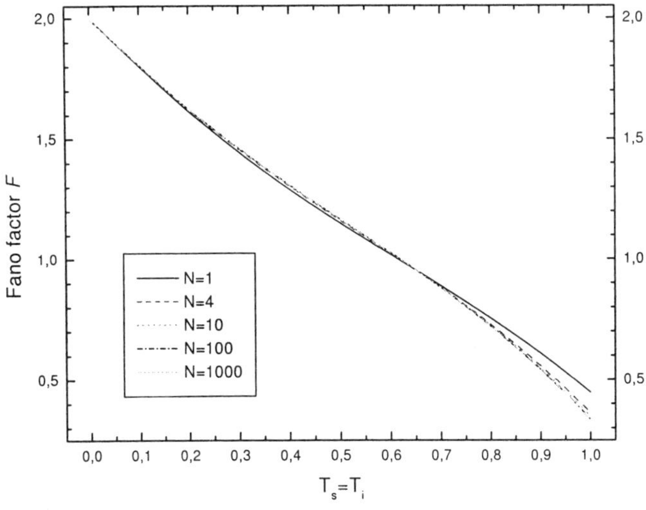

b)

Figure 4: Realistic model ($d^{dark} = 100\ s^{-1}$, $\eta = 50\%$) for pulsed pumping ($\mu = 1$) that also includes decorrelated photons in the noise ($\alpha_s = \alpha_i = 1$). a) Vacuum probability and multiphoton contamination. b) Fano factor.

4 Conclusion

We have analyzed the possibility to use correlated photon pairs for the generation of single photon states. Based on an approximate photon-number measurement performed in the signal beam, some realizations of the state in the idler beam are postselected. The model includes physical imperfections caused by alignment of the laboratory setup and accounts for detection noises and limited quantum efficiency. It is shown, that while ideal device would produce true single-photon states, in real devices physical imperfections deteriorate the statistics of the generated states. Dark counts of the detectors and noise coming from stray light and decorrelated photons increase vacuum contribution in the postselected state. Limited detector quantum efficiency and losses in the apparatus prevent from sufficient exclusion of multiphoton states. This makes the production of a state close to a single-photon state difficult. With pulsed pumping and optimized coupling, however, considerable photon-number-squeezing might be achieved.

Acknowledgement

This work has been supported by the Czech Ministry of Education (project no. VS96028) and the Czech National Security Authority (project no. 19982003012).

References

[1] L. Mandel, E. Wolf, Optical Coherence and Quantum Optics, chap. 22.4, Cambridge University Press, Cambridge (1995).

[2] D. F. Walls, G. J. Milburn, Quantum Optics, chap. 5, Springer, Berlin (1995).

[3] J. Peřina, Z. Hradil, B. Jurčo, Quantum Optics and Fundamentals of Physics, chap. 8, Kluwer, Dordrecht (1994).

[4] A. K. Ekert, Quantum cryptography based on Bell's theorem, Phys. Rev. Lett. 67, 661-663 (1991).

[5] N. Lütkenhaus, Security against individual attacks for realistic quantum key distribution, Phys. Rev. A 61, 052304 (2000).

[6] D. Mayers, Unconditional security in quantum cryptography, quant-ph/9802024v4.

[7] P. Törmä, T. Kiss, I. Jex, H. Paul, Fock state detection and preparation using multiports, Jemná mechanika a optika No. 11-12 (1996).

[8] J. Peřina jr., O. Haderka, in preparation.

[9] J. Peřina, Quantum Statistics of Linear and Nonlinear Optical Phenomena, Kluwer, Dordrecht (1991).

Decoherence and Quantum Computing

Decoherence and its Implications in Quantum Computation
and Information Transfer
T. Gonis and P.E.A. Turchi (Eds.)
IOS Press, 2001

Quantum Communication Complexity
(a survey)

Gilles Brassard *

Département IRO, Université de Montréal
C.P. 6128, succursale centre-ville
Montréal (Québec), Canada H3C 3J7

brassard@iro.umontreal.ca

Abstract

Can quantum communication be more efficient than its classical counterpart? Holevo's theorem rules out the possibility of communicating more than n bits of classical information by the transmission of n quantum bits—unless the two parties are entangled, in which case twice as many classical bits can be communicated but no more. In apparent contradiction, there are distributed computational tasks for which quantum communication cannot be simulated efficiently by classical means. In extreme cases, the effect of transmitting quantum bits cannot be achieved classically short of transmitting an exponentially larger number of bits.

In a similar vein, can entanglement be used to save on classical communication? It is well known that entanglement on its own is useless for the transmission of information. Yet, there are distributed tasks that cannot be accomplished at all in a classical world when communication is not allowed, but that become possible if the non-communicating parties share prior entanglement. This leads to the question of how expensive it is, in terms of classical communication, to provide an exact simulation of the spooky power of entanglement.

1 Introduction

It is well known that the use of quantum information allows for tasks that would be provably impossible in a classical world, such as the transmission of unconditionally confidential information between parties that share only a short secret key [2, 4]. Apart from this *quantum cryptography*, are there advantages to be gained in setting up an infrastructure that would facilitate the transmission of quantum information? In particular, are there advantages to be gained in terms of communication *efficiency*? A different but related question concerns quantum entanglement: can entangled parties make better use of a classical communication channel than their non-entangled counterparts? Better still: can entangled parties benefit from their entanglement if they are not allowed any form of direct communication?

* Supported in part by Canada's NSERC and Québec's FCAR.

There are good reasons to believe at first that the answer to all the above questions is negative. In particular, Holevo's theorem [18] states that no more than n bits of expected classical information can be communicated between unentangled parties by the transmission of n quantum bits—henceforth called *qubits*—regardless of the coding scheme that could be used. If the communicating parties share prior entanglement, twice as much classical information can be transmitted [5], but no more. This applies even if the communication is not restricted to be one-way [14]. It is thus reasonable to expect that no significant savings in communication can be achieved by the transmission of qubits, and possibly no savings at all if the communicating parties do not share prior entanglement. As for the last question, it is well known that entanglement alone cannot be used to signal information—otherwise faster-than-light communication would be possible and causality would be violated—and thus it would seem that entanglement is useless if it is not supplemented by direct forms of communication. Here we survey striking results to the effect that all the intuition in this paragraph is wrong.

After a review of classical communication complexity in Section 2, we consider in Section 3 the situation in which quantum communication is allowed. In Section 4, we revert to classical communication but allow unlimited prior entanglement between the communicating parties. Section 5 investigates in more detail the power of prior entanglement when no direct communication is allowed to take place, which we call spooky communication complexity. In Section 6, we determine how expensive it is to simulate the effect of entanglement in a purely classical world. Finally, we conclude with open problems in Section 7. Although we do not cover the important topic of lower bounds for quantum communication complexity, we encourage the reader to consult [20, 14] for early results and [11] for powerful new techniques.

2 Classical Communication Complexity

Let Alice and Bob be two distant parties who wish to collaborate on a common task that depends on distributed inputs. More precisely, let X, Y and Z be sets and consider a function $f : X \times Y \to Z$. Assume Alice and Bob are given some $x \in X$ and $y \in Y$, respectively, and their goal is to compute $z = f(x, y)$. Sometimes, we add a *promise* $P(x, y)$ for some Boolean function P, in which case Alice and Bob are required to compute the correct answer $f(x, y)$ only when $P(x, y)$ holds. Whether or not there is a promise, the obvious recipe is for Alice to communicate x to Bob, which allows him to compute z. Once obtained, Bob can then communicate z back to Alice if both parties need to know the answer. If we are concerned with the amount of communication required to achieve this task—paying no attention to the computing effort involved in the process—could there be more efficient solutions for some functions f? For *all* functions?

The answer is obviously positive for the first question. For example, if $X = Y = \{0,1\}^n$ for some integer n, $Z = \{0,1\}$ and $f(x, y)$ is defined to be 0 if and only if x and y have the same Hamming weight (the same number of 1s), it suffices for Alice to communicate the Hamming weight of x to Bob for him to verify the condition. Thus, about $\lg n$ bits [1] of communication are sufficient for this task, which is much more economical than if Alice had transmitted all n bits of her input to Bob. The answer

[1] The symbol "lg" is used to denote the base-two logarithm.

to the second question, however, is negative: There are functions for which the obvious solution *is* optimal. For instance, n bits of communication are necessary and sufficient in the worst case for Bob to decide whether or not Alice's input is the same as his. This unsurprising statement is not easy to prove, but a host of techniques have been developed to handle that kind of questions. See [21] for a survey.

A more interesting scenario takes place when we do not insist on the correct answer to be obtained with certainty. If we accept a small error probability, we can do significantly better on the above-mentioned equality-testing problem. Let $\varepsilon > 0$ be the tolerated error probability and let p be the smallest prime number larger than n/ε. Let $\mathbb{F}$ be the finite field with p elements. Upon receiving their inputs x and y, Alice forms polynomial $P(z) = x_1 + x_2 z + x_3 z^2 + \cdots + x_n z^{n-1}$ over $\mathbb{F}$ and Bob forms $Q(z) = y_1 + y_2 z + y_3 z^2 + \cdots + y_n z^{n-1}$. Then, Alice chooses a random element $w \in \mathbb{F}$. She computes $v = P(w)$ and transmits both w and v to Bob, who computes $Q(w)$ and compares the answer with v. If $Q(w) \neq v$, it has been established that $x \neq y$. Otherwise, Bob can claim with confidence that $x = y$ because two distinct polynomials of degree smaller than n cannot agree on more than n distinct points and therefore the proportion of points in $\mathbb{F}$ on which $P(z)$ and $Q(z)$ agree must be less than $n/\#\mathbb{F} = n/p < \varepsilon$. Note that $p \leq 2n/\varepsilon$ since there is always a prime number between any number and its double, and hence each of w and v can be written with no more than $2 + \lg n + \lg \frac{1}{\varepsilon}$ bits. The communication complexity of this protocol is therefore at most twice this many bits, which is much less than n for any fixed error probability ε when n is large enough.

In the classical model of communication complexity, it is often allowed for Alice and Bob to *share random variables* even though one may argue that this does not make much sense from a mathematical point of view. In this scenario, we assume that Alice draws a random bit string (or integer) according to some specific distribution—or sometimes even a random real number—and she tells Bob the outcome of this draw in an *initialization phase.* This communication is not accounted for in the complexity of the protocol because it takes place *before* Alice and Bob are given their respective inputs. When correctness of the protocol is analysed for a given input, probabilities are taken over the possible choices of that random variable, as if it had been chosen *after* the inputs were determined [2]. In this model, the equality-testing problem can be solved with error probability ε with only $m = \lceil \lg 1/\varepsilon \rceil$ bits of communication, regardless of the value of n. In the initialization phase, Alice and Bob share m random bit strings $a_1, a_2, \ldots, a_m$, each of length n. Once they receive their inputs x and y, Alice computes $b_i = x \cdot a_i$ for each i, where $x \cdot a$ is the *inner product* [3] between bit strings x and a. Alice transmits $b_1, b_2, \ldots, b_m$ to Bob, who verifies whether or not $b_i = y \cdot a_i$ for each i. If not, it has been established that $x \neq y$. Otherwise, Bob can claim with confidence that $x = y$ because the probability of error of this strategy is $2^{-m} \leq \varepsilon$ since it is $^1/_2$ independently for each i.

---

[2] Note that drawing shared random bits once Alice and Bob are separated makes perfect sense in a quantum world if they share entanglement: they simply have to measure corresponding qubits from $|\Phi^+\rangle = \frac{1}{\sqrt{2}}|00\rangle + \frac{1}{\sqrt{2}}|11\rangle$ pairs in the computational basis.

[3] To compute the inner product between two bit strings of equal length, line them up one under the other and count the number of positions in which they both have a 1. If this number is even, the inner product is 0; otherwise it is 1. Mathematically, it is the exclusive or (sum modulo 2) of the bitwise AND of the two strings.

3 Quantum Communication Complexity

The topic of classical communication complexity was introduced and first studied by Andrew Yao in 1979 [26]. Almost 15 years elapsed before the same pioneer thought of asking how the situation might change if Alice and Bob were allowed to exchange quantum rather than classical bits [27]. It seems at first that no savings in communication are to be expected at all because of Holevo's theorem [18], which states that no more than n bits of expected classical information can be communicated between unentangled parties by the transmission of n qubits. (It was implicit in Yao's original model that Alice and Bob were not allowed to share prior entanglement in the initialization phase.)

The first hint that quantum communication could be more efficient than classical communication was given in August 1997 by Richard Cleve, Wim van Dam, Michael Nielsen and Alain Tapp [14] in a probabilistic setting[4]. Alice and Bob are given two-bits vectors x_1x_2 and y_1y_2, respectively. They must both decide if $x_1y_1 + x_2y_2$ is even or odd, and they are restricted to two bits of communication. Shared random variables are allowed. It is proven in [14] that no classical protocol for this task can give the correct answer with a probability better than $^7/_9$. Yet, if two quantum bits of communication are allowed, instead of two classical bits, the correct answer can be obtained with an improved probability of $^4/_5$.

A more convincing case for the superiority of quantum communication came the following year when Harry Buhrman, Richard Cleve and Avi Wigderson [9] proved that quantum communication can be *exponentially* better than classical communication in the error-free model, provided the inputs respect a given promise, and almost quadratically better in the bounded-error promise-free model. Subsequently, Ran Raz proved that an exponential advantage exists to quantum communication even in the bounded-error promise-problem model [23].

The first exponential separation [9] was inspired by the famous Deutsch–Jozsa problem [15], which was used in 1992 to show for the first time that quantum computers could be exponentially faster than classical computers in an oracle-based [6] error-free setting. More specifically, Buhrman, Cleve and Wigderson defined the following scenario, where $\Delta(x,y)$ denotes the *Hamming distance* between bit strings x and y, which is the number of bit positions on which they differ. Let k be an integer, $n = 2^k$, $X = Y = \{0,1\}^n$ and $Z = \{0,1\}$. Function $f : X \times Y \to Z$ is the equality function: $f(x,y) = 1$ if and only if $x = y$. We have seen already that this communication complexity problem requires n bits of classical communication if errors are not tolerated. Now, we introduce the promise $P(x,y)$, which is defined to be true if and only if $\Delta(x,y) \in \{0, n/2\}$. In other words, $P(x,y)$ holds if and only if either $x = y$ or x and y differ on exactly half their positions. It is proven in [9] that the error-free equality-testing problem requires at least cn classical bits of communication, for some real positive constant c and all sufficiently large n, even when the correct answer is required only when the promise holds. (The hard part of that proof is taken from [16].) Even though quantum communication cannot be significantly more efficient than classical communication for the straight equality-testing problem [11], it is shown in [9]

[4] For the sake of historical completeness, it is easy to modify a protocol given three months earlier by Buhrman, Cleve and van Dam [8] (original `quant-ph` version) to achieve a similar goal, and indeed this is done in the final version of that paper, which is to appear in *SIAM Journal on Computing*.

that it can be solved with certainty using as few as k quantum bits of communication whenever the promise holds, which is exponentially better than the cn bits that would be required in a classical scenario. A quantum protocol for this problem is easily derived from the first example of "spooky communication" given in Section 5.

This exponential "superiority" of quantum over classical communication is not entirely convincing because it vanishes as soon as we tolerate an arbitrarily small probability of error. Indeed, we have seen that the equality-testing problem can be solved with a *constant* number of bits of classical communication, for any fixed error probability, when shared random variables are allowed[5]. The other problem featured in [9] is more interesting, even though the quantum superiority is merely almost quadratic, because it applies in the more realistic bounded-error model. Consider the following scenario. Alice and Bob are very busy and they would like to find a time when they are simultaneously free for lunch. They each have an engagement calendar, which we think of as an n-bit string x (resp y), where $x_i = 1$ (resp. $y_i = 1$) means that Alice (resp. Bob) is free for lunch on day i. Mathematically, they want to find an index i such that $x_i = y_i = 1$ or establish that such an index does not exist. Balasubramanian Kalyanasundaram and Georg Schnitger [19] proved in 1987 that this task requires at least cn classical bits of expected communication in the worst case, for some real positive constant c and all sufficiently large n, even when the answer is only required to be correct with probability at least $^2/_3$. Intuitively, this means that lunch cannot be scheduled short of exchanging a constant fraction of the appointment calendar. In sharp contrast, it is shown in [9] that this problem can be solved with the exchange of at most $d\sqrt{n}\lg n$ quantum bits for some constant d and all sufficiently large n. This is accomplished by implementing a distributed version of Grover's quantum search algorithm [17] in which we search for a 1 in the bitwise AND of x and y. A $\lceil \lg n \rceil$-qubit quantum register is shuttled back and forth between Alice and Bob for each of the approximately $\sqrt{n}$ iterations of Grover's algorithm.

4 Substituting Entanglement for Communication

A slightly different model was introduced by Richard Cleve and Harry Buhrman [13]. Assume Alice and Bob are restricted to communicating classical information. Are there tasks for which they could save on the required amount of communication if they share prior entanglement? Again, it is tempting to think that this is not possible because entanglement cannot be used to increase the capacity of a classical channel: Alice cannot communicate more than n expected bits of classical information to Bob if less than n bits are actually transmitted between them—even if they are allowed two-way communication and unlimited use of entanglement. And again, this intuition is wrong.

In their original paper [13], Cleve and Buhrman were able to show that entanglement can be used to save one bit of classical communication, but only in a three-party scenario. Still, this was the very first example of a distributed task that could be solved more efficiently (in terms of communication) in our quantum world than would be possible in a sad classical world, because it predates [14] by four months. Subsequently, Buhrman, Cleve and van Dam [8] found a *two-party* distributed problem that can be

[5] Even if we do not allow shared random variables, the problem can be solved with $2k + c$ bits of classical communication, where $k = \lg n$ and c is a constant that depends only on the error probability.

solved with a probability of success exceeding 85% if prior shared entanglement is available, whereas the probability of success in a classical world could not exceed 75% with the same amount of communication, even if shared random variables are allowed.

The first problem for which communication complexity could be reduced by more than a constant additive amount was also discovered in this shared-entanglement scenario, rather than in Yao's original qubit-transmission scenario described in the previous section: Harry Buhrman, Wim van Dam, Peter Høyer and Alain Tapp [10] gave a k–party distributed task that requires roughly $k \lg k$ bits of communication in a classical world, yet it can be carried out with exactly k bits of classical communication if the parties are allowed to share prior entanglement.

The exponential and almost-quadratic improvements mentioned in the previous section [9, 23] apply just as well in the shared-entanglement scenario. This is obvious since the effect of any protocol that requires the communication of ℓ quantum bits can be achieved by the transmission of 2ℓ classical bits—through *quantum teleportation* [3]—provided ℓ bits of shared entanglement are available.

There is yet another natural communication complexity scenario, in which the parties are allowed to share prior entanglement *and* to communicate quantum bits. For the sake of brevity, we shall not elaborate on this approach here.

5 Spooky Communication Complexity

An even more intriguing question is to determine if entanglement can be used *instead of* communication. Are there tasks that would be impossible to achieve in a classical world if Alice and Bob were not allowed to communicate, yet those tasks can be performed without *any* form of communication provided the participants share prior entanglement? In the words of Alain Tapp, this would provide a form of *pseudo telepathy* because it would give the *illusion* of communication between Alice and Bob when in fact no such communication takes place. And indeed there would be no communication because entanglement cannot be used to signal information: nothing Alice can do locally on her quantum system can cause a measurable change in Bob's, no matter how they are entangled.

A moment's thought suffices to realize that pseudo telepathy is possible if we are content with probabilistic tasks. Define the *EPR task* as follows. Once separated, Alice and Bob are given each a real number x and y, respectively, between 0 and π. They are to produce each a single bit: a for Alice and b for Bob. Alice's output must be equally likely to be 0 or 1, and so must Bob's output, but the required correlation is that $a = b$ with probability $\cos^2(x - y)$. It is precisely the essence of Bell's theorem [1] that such correlations cannot be established in a classical world if communication between Alice and Bob is not allowed, even if the inputs are restricted to binary choices $x \in \{0, \pi/6\}$ and $y \in \{0, 5\pi/6\}$. Yet, it is easy for participants who share a $|\Phi^+\rangle = \frac{1}{\sqrt{2}}|00\rangle + \frac{1}{\sqrt{2}}|11\rangle$ state to realize this task. If we think of this state as a pair of entangled polarized photons, it suffices for Alice and Bob to measure their photons at polarization angles x and y, respectively, and the outcomes of the measurements provide the required outputs a and b.

But is pseudo telepathy possible when there is a deterministic criterion to decide if the goal has been achieved and when errors are not tolerated? This brings us to our last form of communication complexity, which we call *spooky communication complexity*. As usual, let X, Y and Z be sets and consider a function $f : X \times Y \to Z$ such that it is not possible to compute the value of $f(x, y)$ with certainty from knowledge of either x or y alone. It follows that if Alice and Bob are given $x \in X$ and $y \in Y$, respectively, they cannot compute $f(x, y)$ without communication. Can such a function f exist so that Alice and Bob—or at least one of them—could compute $f(x, y)$ nevertheless provided the participants share prior entanglement? Of course not, since this would allow for faster-than-light communication! Thus, we have to define spooky communication complexity in a more subtle way, through a *relation* rather than a function.

Let X, Y, A and B be sets and consider a relation $R \subseteq X \times Y \times A \times B$. In an *initialization phase*, Alice and Bob are allowed to discuss strategy and share random variables. They are also allowed to share entanglement. After Alice and Bob are physically separated, they are given $x \in X$ and $y \in Y$, respectively. Without being allowed any forms of communication, their goal is to produce $a \in A$ and $b \in B$, respectively, such that $(x, y, a, b) \in R$. We say that *spooky communication*, or *pseudo telepathy*, takes place if this task could not be fulfilled with certainty in a classical world, whereas it can provided Alice and Bob share prior entanglement. The amount of spooky communication *complexity* is measured in the number of bits of entanglement that are required to succeed in the worst case. The *spooky advantage* is defined as the function that relates the spooky complexity to the number of classical bits of communication that would be needed in the worst case to succeed in the classical setting.

The first example of spooky communication was provided by Gilles Brassard, Richard Cleve and Alain Tapp [7] as yet another variation on the Deutsch–Jozsa problem [15]. Let k be an integer, $n = 2^k$, $X = Y = \{0,1\}^n$ and $A = B = \{0,1\}^k$. The *Deutsch–Jozsa relation* R is defined as follows, where $\Delta(x, y)$ denotes again the Hamming distance between x and y.

$$(x, y, a, b) \in R \iff \begin{cases} x = y \text{ and } a = b, \text{ or} \\ \Delta(x, y) = n/2 \text{ and } a \neq b, \text{ or} \\ \Delta(x, y) \notin \{0, n/2\} \end{cases}$$

In other words, Alice and Bob are promised that either their inputs are the same, or that they differ on exactly half the bits. They must produce identical outputs if and only if their inputs are the same. But if the promise is not fulfilled, there are no conditions on what Alice and Bob produce. The challenge comes from the fact that the outputs a and b must be exponentially shorter than the inputs x and y.

It is not immediate that the Deutsch–Jozsa relation requires communication to be established in the classical setting. After all, it *can* be established when $n = 2$ (easily) and $n = 4$ (think about it!). But it is proven in [7] that there exists a positive constant c such that the Deutsch–Jozsa relation cannot be established classically with fewer than cn bits of communication provided n is sufficiently large, based on the similar lower bound from [9] that we had mentioned in Section 3. On the other hand, we show below that the Deutsch–Jozsa relation *can* be established in the spooky setting with as few as k bits of entanglement. It follows that the spooky advantage of this problem is exponential because n is exponential in k.

To establish the Deutsch–Jozsa relation, Alice creates a $2k$-qubit register in state

$$\sum_{z \in \{0,1\}^k} 2^{-k/2} \, |z, z\rangle \,,$$

which is the same as k pairs in state $|\Phi^+\rangle$ up to ordering of the qubits. She keeps the first k qubits of that register and gives the other k qubits to Bob. After Alice and Bob are separated, they receive their respective inputs x and y. To each integer i, $1 \leq i \leq n$, associate the bit string $z_i \in \{0,1\}^k$ that represents number $i-1$ in binary. Now, Alice applies to her register the unitary transformation that maps $|z_i\rangle$ to $(-1)^{x_i}\,|z_i\rangle$ for each i, and Bob does the same to his register, but with y_i instead of x_i. This produces joint state

$$\sum_{i=1}^{n} 2^{-k/2}\,(-1)^{x_i}(-1)^{y_i}|z_i, z_i\rangle \;=\; \sum_{i=1}^{n} 2^{-k/2}\,(-1)^{x_i \oplus y_i}|z_i, z_i\rangle\,.$$

Next, Alice applies the Walsh–Hadamard transform, which sends $|0\rangle$ to $\frac{1}{\sqrt{2}}|0\rangle + \frac{1}{\sqrt{2}}|1\rangle$ and $|1\rangle$ to $\frac{1}{\sqrt{2}}|0\rangle - \frac{1}{\sqrt{2}}|1\rangle$, to each of the k qubits of her register, and Bob does the same on his register. Finally, Alice and Bob measure their registers in the computational basis. The resulting classical strings, a and b, are their final output. It is proven in [7] that this process accomplishes the required job.

6 Classical Simulation of Entanglement

Once we have established that pseudo telepathy is possible, the next natural question is to determine how much classical communication is necessary and sufficient to simulate the effect of k bits of entanglement. It follows from the previous section (and Bell's theorem [1] for small values of k) that at least $c2^k$ bits are required, for some constant $c > 0$ and all $k \geq 1$. But are these many bits of classical communication *sufficient* to simulate *everything* that can be accomplished with k bits of entanglement?

In fact, can the effect of entanglement be simulated *at all* with a finite amount of classical communication? As the simplest possible example, can the EPR task, as defined in Section 5, be simulated by classical communication? In particular, we must have $a = b$ if $x = y$ and $a \neq b$ if $|x - y| = \pi/2$. Surely, it is not possible for Alice to communicate her input x to Bob, for this would require an infinite amount of communication. Yet, it is shown in [7] that *four* bits of classical communication are sufficient in the worst case for an exact simulation of the EPR task, provided Alice and Bob are allowed to share a continuous real random variable in the initialization phase—admittedly an unreasonable proposition. The essence of the idea is best explained if we further restrict the inputs x and y to be between 0 and 1, rather than between 0 and π. In this case, *a single bit* of classical communication suffices to simulate the EPR task exactly. This restriction is somewhat bizarre since it translates to requiring the angles to be between 0 and approximately 57.3 degrees. Without this restriction, a rather painful piecewise construction has to be implemented, as explained in [7], and we need four bits of classical communication to take care of the various possible cases.

In the initialization phase, Alice and Bob share a boolean variable c, which is equally likely to be 0 or 1, and a continuous real variable r chosen uniformly in the interval $(0,1)$. After they are separated, they receive their angles x and y, respectively. Alice outputs $a = c$, a random bit as required. Then, she tells Bob if $a < x$ with a single bit of classical communication. This allows Bob to determine whether or not r lies in between x and y. In case it does not, Bob outputs the same bit $b = c$ as Alice. Note in particular that if $x = y$, then we get $a = b$ with certainty, as required. On the other hand, if r does lie between x and y, then Bob outputs $b = 1 - c$, a bit complementary to Alice's, with probability $\sin(2|y - r|)$, otherwise he outputs $b = c$ just like Alice.

The probability that $a = b$ is calculated as an integral over the various possibilities for r, as if it had been chosen after x and y are fixed. For simplicity, assume that $x \leq y$. The probability that $a = b$ is 1 if $0 \leq r \leq x$ or $y \leq r \leq 1$ since, in that case, r does not lie between x and y. Otherwise, if $x < r < y$, the probability that $a = b$ is $1 - \sin(2(y - r))$. Therefore, the global probability that $a = b$ is

$$\int_{r=0}^{x} 1 \, \mathrm{d}r + \int_{r=x}^{y} [1 - \sin(2(y-r))] \, \mathrm{d}r + \int_{r=y}^{1} 1 \, \mathrm{d}r$$

$$= \tfrac{1}{2} + \tfrac{1}{2}\cos(2(y-x)) = \cos^2(x-y)\,,$$

as required. It is tempting to "improve" on this approach and make it work for all angles between 0 and π, simply with an appropriate change in the probability function $\sin(2|y-r|)$ that determines whether or not Bob will output the same bit as Alice when r lies between x and y. Unfortunately, any such attempt will result in "probabilities" that are either negative or greater than 1!

It is shown in [7] how to simulate an arbitrary von Neumann measurement with only *eight* classical bits of communication in the worst case, but it is left as an open question to determine whether or not the effect of an arbitrary positive-operator-valued measurement (POVM) can be simulated with a bounded amount of classical communication in the worst case.

A different approach to the classical simulation of entanglement was taken independently by Michael Steiner [25], who showed that the EPR relation can be simulated exactly with significantly fewer *expected* bits of classical communication, provided we accept that there be no upper limit on the required amount of communication in the case of bad luck. Steiner's technique was subsequently refined by Nicolas Cerf, Nicolas Gisin and Serge Massar [12], who showed that as few as 1.19 expected bits of classical communication suffice to simulate exactly an arbitrary von Neumann measurement. Even an arbitrary POVM can be simulated by this technique, at the expected cost of 6.38 bits of classical communication.

Building on [25, 12], Serge Massar, Dave Bacon, Nicolas Cerf and Richard Cleve [22] discovered that the exact classical simulation of quantum entanglement can be achieved without any need for shared random variables, provided we are satisfied with an *expected* bounded amount of classical communication. In particular, they show how to simulate the effect of an arbitrary POVM on one bit of entanglement with less than 20 bits of expected classical communication. Conversely, they also show that the exact simulation of quantum entanglement with a worst-case bounded amount of classical communication (as in the scenario of [7] described earlier in this section) is *not* possible without an infinite amount of shared randomness.

Finally, the question asked at the beginning of this section is almost resolved in [7]. It is still unknown if there exists a constant c such that $c2^k$ bits of classical communication are sufficient to simulate exactly the effect of k bits of entanglement for all values of k. However, it is shown in [7] that as few as $(3k+6)2^k$ expected bits of classical communication suffice to simulate the outcome of any POVM that Alice and Bob could perform on their respective shares of k bits of entanglement. This simulation protocol does not require Alice and Bob to share random variables in the initialization phase. No similar results are known for worst-case bounded communication even if we allow the sharing of continuous random variables.

7 Conclusions and Open Problems

We have seen a variety of scenarios according to which quantum mechanics allows for a significant improvement in the efficiency of communication, compared to what would be possible in a classical world. This is surprising because the transmission of n quantum bits cannot serve to communicate more than n classical bits of information, and because quantum entanglement on its own cannot be used to communicate at all. Perhaps the most interesting aspect of quantum communication complexity is that the advantage provided by quantum mechanics has been established rigourously. This is in sharp contrast with the field of quantum *computing*, in which it is merely *believed* that quantum mechanics allows for an exponential speedup in some computational tasks, such as the factorization of large numbers [24]. Indeed, it has not yet been ruled out that there might exist an efficient factorization algorithm for the classical computer.

Several interesting questions are still open. The exponential advantage of quantum communication over classical communication has been established only in the case of promise problems, in both the error-free [9] and bounded-error [23] scenarios. Could there be a (total) function $f : X \times Y \to Z$, where $X = Y = \{0,1\}^n$, for which the distributed computation of f would be exponentially more efficient with quantum communication compared to classical communication?

We have seen at the end of Section 4 that the amount of classical communication required for the accomplishment of a distributed task in the presence of unlimited entanglement cannot be more than twice the amount of quantum communication that would suffice for the same task, because quantum teleportation can be used to transmit quantum bits through a classical channel. How about the other direction? Could there be a task that can be accomplished with a small amount of classical communication in the presence of unlimited entanglement, but that would require a much larger amount of quantum communication if prior entanglement were not available?

We have seen in Section 5 that the Deutsch-Jozsa relation can be established classically without communication when $n = 2$ or $n = 4$, but not when n is arbitrarily large. But how large must "large" be? In particular, can it be established for $n = 8$? It is interesting to note that the Deutsch-Jozsa relation becomes easier and easier to *fake* when n becomes larger. Indeed, if Alice and Bob share k random variables $t_1, t_2, \ldots, t_k \in \{0,1\}^n$ in the initialization phase, and if they output $a_i = x \cdot t_i$, and $b_i = y \cdot t_i$, respectively, their probability of being caught with $a = b$ if in fact $\Delta(x, y) = n/2$ goes down as 2^{-k}. A nice open question is to determine the task on n input and k output bits that can be handled with certainty given sufficient entanglement and no communication, but for which the probability of success would be as small as possible in a classical world.

Finally, several open questions are given in Section 6 concerning the classical simulation of quantum entanglement. Is it possible to achieve the EPR task with fewer than four bits of classical communication in the worst case? Is it possible to simulate an arbitrary POVM with a *worst-case* bounded amount of classical communication? How much classical communication is sufficient in the worst case to simulate the effect of k bits of entanglement? In the expected case? We have seen how classical communication can be used to simulate entanglement for tasks that did not involve classical communication in the quantum setting. How about the classical simulation of tasks that use not only quantum entanglement but also classical (or perhaps quantum) communication?

Acknowledgements

I am grateful to Richard Cleve for introducing me to the fascinating field of quantum communication complexity and to Alain Tapp for sharing with me some of his best thoughts on the topic. In addition to discussions with Richard and Alain, this paper has benefited from insightful comments from Harry Buhrman and Ronald de Wolf.

References

[1] J. S. Bell, "On the Einstein–Podolsky–Rosen paradox", *Physics*, Vol. 1, 1964, pp. 195 – 200.

[2] C. H. Bennett and G. Brassard, "Quantum cryptography: Public key distribution and coin tossing", *Proceedings of IEEE International Conference on Computers, Systems and Signal Processing*, Bangalore, India, December 1984, pp. 175 – 179.

[3] C. H. Bennett, G. Brassard, C. Crépeau, R. Jozsa, A. Peres and W. K. Wootters, "Teleporting an unknown quantum state via dual classical and Einstein–Podolsky–Rosen channels", *Physical Review Letters*, Vol. 70, no. 13, 29 March 1993, pp. 1895 – 1899.

[4] C. H. Bennett, G. Brassard and A. K. Ekert, "Quantum Cryptography", *Scientific American*, October 1992, pp. 50 – 57.

[5] C. H. Bennett and S. J. Wiesner, "Communication via one- and two-particle operators on Einstein–Podolsky–Rosen states", *Physical Review Letters*, Vol. 69, no. 20, 16 November 1992, pp. 2881 – 2884.

[6] A. Berthiaume and G. Brassard, "Oracle quantum computing", *Journal of Modern Optics*, Vol. 41, no. 12, December 1994, pp. 2521 – 2535.

[7] G. Brassard, R. Cleve and A. Tapp, "Cost of exactly simulating quantum entanglement with classical communication", *Physical Review Letters*, Vol. 83, no. 9, 30 August 1999, pp. 1874 – 1877.

[8] H. Buhrman, R. Cleve and W. van Dam, "Quantum entanglement and communication complexity", *SIAM Journal on Computing*, to appear, 2000. Originally posted in May 1997 as `arXiv:quant-ph/9705033`.

[9] H. Buhrman, R. Cleve and A. Wigderson, "Quantum vs. classical communication and computation", *Proceedings of the 30th Annual ACM Symposium on Theory of Computing*, May 1998, pp. 63 – 68.

[10] H. Buhrman, W. van Dam, P. Høyer and A. Tapp, "Multiparty quantum communication complexity", *Physical Review A*, Vol. 60, no. 4, October 1999, pp. 2737 – 2741. Originally posted in October 1997 as `arXiv:quant-ph/9710054`.

[11] H. Buhrman and R. de Wolf, "Communication complexity lower bounds by polynomials", manuscript, 1999. Available as `arXiv:cs:CC/9910010`.

[12] N. Cerf, N. Gisin and S. Massar, "Classical teleportation of a quantum bit", *Physical Review Letters*, Vol. 84, no. 11, 13 March 2000, pp. 2521 – 2524.

[13] R. Cleve and H. Buhrman, "Substituting quantum entanglement for communication", *Physical Review A*, Vol. 56, no. 2, August 1997, pp. 1201 – 1204. Originally posted in April 1997 as `arXiv:quant-ph/9704026`.

[14] R. Cleve, W. van Dam, M. Nielsen and A. Tapp, "Quantum entanglement and the communication complexity of the inner product function", *Quantum Computing and Quantum Communication: Proceedings of First NASA International Conference (QCQC'98)*, Lecture Notes in Computer Science, Vol. 1509, Springer–Verlag, 1999, pp. 61 – 74. Originally posted in August 1997 as `arXiv:quant-ph/9708019`.

[15] D. Deutsch and R. Jozsa, "Rapid solution of problems by quantum computation", *Proceedings of the Royal Society of London*, Series A, Vol. 439, 1992, pp. 553 – 558.

[16] P. Frankl and V. Rödl, "Forbidden intersections", *Transactions of the American Mathematical Society*, Vol. 300, no. 1, 1987, pp. 259 – 286.

[17] L. K. Grover, "Quantum mechanics helps in searching for a needle in a haystack", *Physical Review Letters*, Vol. 79, no. 2, 14 July 1997, pp. 325 – 328.

[18] A. Holevo, "Bounds for the quantity of information transmitted by a quantum communication channel", *Problemy Peredachi Informatsii*, Vol. 9, no. 3, 1973, pp. 3 – 11. English translation in *Problems of Information Transmission*, Vol. 9, 1973, pp. 177 – 183.

[19] B. Kalyanasundaram and G. Schnitger, "The probabilistic communication complexity of set intersection", *Proceedings of 2nd Annual IEEE Conference on Structure in Complexity Theory*, June 1987, pp. 41 – 47.

[20] I. Kremer, *Quantum Communication*, Master's thesis, Hebrew University, Computer Science Department, 1995.

[21] E. Kushilevitz and N. Nisan, *Communication Complexity*, Cambridge University Press, 1997.

[22] S. Massar, D. Bacon, N. Cerf and R. Cleve, "Classical simulation of quantum entanglement without local hidden variables", manuscript, September 2000. Available as `arXiv:quant-ph/0009088`.

[23] R. Raz, "Exponential separation of quantum and classical communication complexity", *Proceedings of the 31st Annual ACM Symposium on Theory of Computing*, May 1999, pp. 358 – 367.

[24] P. W. Shor, "Polynomial-time algorithms for prime factorization and discrete logarithms on a quantum computer", *SIAM Journal on Computing*, Vol. 26, 1997, pp. 1484 – 1509.

[25] M. Steiner, "Towards quantifying non-local information transfer: finite-bit non-locality", *Proceedings of 29th Winter Colloquium on the Physics of Quantum Electronics*, Snowbird, Utah, 1999. Also available as `arXiv:quant-ph/9902014`.

[26] A. C.–C. Yao, "Some complexity questions related to distributed computing", *Proceedings of the 11th Annual ACM Symposium on Theory of Computing*, 1979, pp. 209 – 213.

[27] A. C.–C. Yao, "Quantum circuit complexity", *Proceedings of the 34th Annual IEEE Symposium on Foundations of Computer Science*, November 1993, pp. 352 – 361.

Decoherence and its Implications in Quantum Computation and Information Transfer
T. Gonis and P.E.A. Turchi (Eds.)
IOS Press, 2001

Entanglement and Decoherence in Cavity Quantum Electrodynamics Experiments

Serge HAROCHE
Laboratoire Kastler Brossel, Département de Physique de l'Ecole Normale Supérieure, Paris, France

Abstract. Cavity quantum electrodynamics experiments performed with Rydberg atoms and microwave photons interacting in a cavity provide a testing ground for the physics of quantum information. We have realized quantum phase gates using atoms and photons as « qubits », various engineered entanglement experiments involving up to three particles at a time and ideal non-destructive measurements of single photons. We have also generated and studied quantum superpositions of coherent fields with different phases, the so called « Schrödinger cat states » of the field, and used them to perform complementarity and decoherence studies.

1. Introduction

Fundamental quantum concepts such as state superpositions and entanglement can be put to use to process and transfer information in non-classical ways[1,2]. Teleportation of quantum states [3-6], distribution of cryptographic keys for secure communication ([7-9]), operation of simple quantum logic gates [10-11] have been recently realized. What makes this new physics so fascinating is the fact that the logic at work in it is quite counter-intuitive, completely at odds with our common-sense experience based on the observation of every day events which present a perfectly classical appearance. The fundamental problem we have to face in order to understand the possibilities and assess the limitations of this new physics is related to the deep understanding of the so-called *quantum-classical boundary*.

Why do large systems, made of atoms and molecules obeying quantum mechanics, do generally behave classically? The widely admitted answer now – at least among the tenants of the orthodox interpretation of quantum mechanics – is that *decoherence phenomena*, related to the coupling of large systems with their natural *environment*, play an essential role in the emergence of classicality [12-14]. Decoherence has the effect to degrade the degree of coherence and entanglement in large systems, before destroying it altogether, leading eventually to full classical behavior.

The success of quantum information applications will thus rely *in fine* on our ability to keep the effects of the coupling to the environment minimal, and in our skills for exploiting imperfect entanglement, corrupted to some unavoidable degree by the decoherence phenomenon. A wide range of theoretical and experimental studies are now addressing these issues, constituting a new domain at the frontier between physics and information science which can be called "*the physics of quantum information*"[1].

If information theory can largely rely on "gedanken experiments», in the end the success of this new science will be measured by its actual experimental realizations. Information is indeed at its heart a concrete science, the logical operations being implemented by interacting physical systems, the "bit" carriers. Experimentalists in quantum optics, in nuclear magnetic resonance [15] and in solid state physics [16,17] are thus competing to demonstrate operations of increasing complexity on real physical systems.

Among these fields, quantum optics is the only one so far in which entanglement between quantum bits (or "qubits") has been demonstrated without ambiguity. In this domain, the information is encoded in qubits made of atoms or photons. Three systems have been so far investigated. One deals with twin beams of photons produced by down-conversion processes, the entanglement taking place between the polarizations, momenta or energies of these photons [4,5,7-9]. Another system operates on trapped ions, the information being encoded either in internal or in external degrees of freedom of the ions [10].

The third system is a mixed one, where the qubits are atoms and photons. It is realized in cavity quantum electrodynamics (CQED) experiments, with very excited Rydberg atoms crossing one at a time a high Q superconducting cavity [18,19]. This chapter describes recent CQED experiments realized at Ecole Normale Supérieure in Paris, in which various ways to build and manipulate entanglement in systems made of a few particles have been demonstrated. These experiments also provide a very simple testing ground for decoherence theory. They illustrate some fascinating aspects of quantum logic at work, and also point to the formidable difficulties lying ahead of us for the practical exploitation of entanglement in complex systems made of many particles.

We start (section 2) by describing briefly our experimental system. We then discuss two kinds of entanglement experiments, based respectively on resonant and non-resonant atom-photon processes. Resonant interactions (section 3) have allowed us to operate quantum logic gates [11], to implement a novel quantum non-demolition procedure detecting single photons in a cavity [20] and to demonstrate a general method for entangling several particles in a deterministic way [21]. Non-resonant interactions (section 4) have lead us to generate mesoscopic superpositions of fields states made of several photons (so called "Schrödinger cat states")[22]. We have used these superpositions to perform complementarity and decoherence experiments which constitute a first step in the exploration of the quantum-classical boundary.

2. The Cavity QED entangling machine

2.1. The set-up

Our set-up, whose scheme is shown in Fig.1, has been described in [20-21]. We recall only briefly here its main features. A thermal beam of Rubidium atoms is velocity selected by laser optical pumping and prepared in the circular Rydberg state with principal quantum number 51 (level e in the following) or 50 (level g) by a succession of laser and radiofrequency induced transitions in box B. Levels e and g are shown in Fig.1 inset. In these circular states, the electron orbit is a large circle around the atom's core of 2500 A diameter, resulting in a very large electric dipole matrix elements for transitions between adjacent Rydberg levels. This ensures extremely strong atom-field coupling, an essential condition for atom-field entanglement to occur before field or atomic relaxation processes can set in. A third level (the circular state with principal quantum number 49, called i in the following) plays also a role in the experiments. The atoms cross one at a time the cavity C sustaining a Gaussian field mode (waist = 6 mm), resonant or nearly resonant with the e $\Rightarrow$ g transition at 51.1 GHz. The cavity is made of two superconducting spherical niobium mirrors enclosed by an aluminum ring with small holes for atom access (a cross section of the ring with enlarged holes for clarity is shown in Fig.1). The cavity mode is represented by the vertical pencil-shaped shade along cavity axis in Fig.1. The atom and field relaxation times, 30 and 1 ms respectively, are much longer than the atom cavity crossing time (of the order of 20 μs). This ensures that quantum coherences survive much longer than the atom-cavity coupling time.

The atomic preparation is pulsed. Combined with the atomic velocity selection, it ensures that the position of each atom along the beam downstream from box B is known with $\pm$ 1 mm. The atoms are detected, after C, by a state selective field-ionization detector D, which discriminates between e, g and i (detection efficiency 40%). Sequences of atoms can be sent across the cavity with a delay between atoms of the order of 100 μs, exceeding the atom-cavity crossing time. As a result, one atom at most interacts with the cavity field at any time. In order to avoid the presence of more than one atom in each pulse of each sequence, the lasers preparing the Rydberg states have a low intensity, exciting on average 0.1 to 0.2 atom per pulse. The preparation process being Poissonian, each pulse contains randomly either zero or one atom. The probability to prepare a sequence of two or three atom is accordingly low and we must use the detection to post-select the useful events out of a large number of runs in which at least one of the atoms of the sequence is missing.

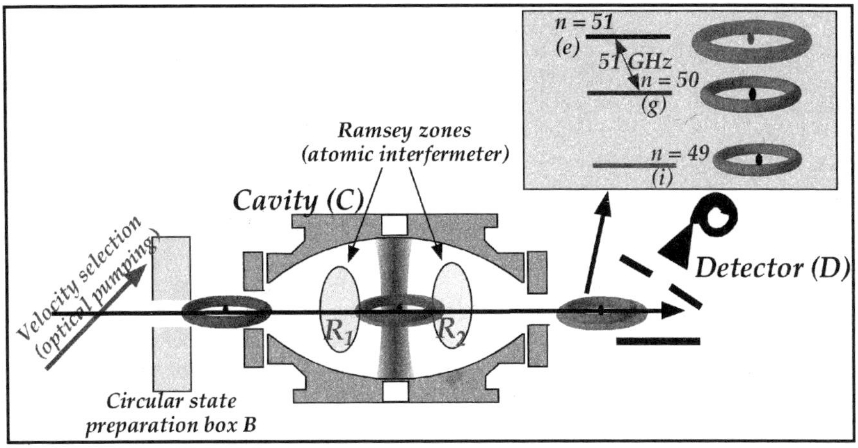

Figure 1. Scheme of the cavity QED apparatus. In the inset: energy diagram of the relevant Rydberg states e (principal quantum number n =51), g (n= 50) and i (n = 49).

A field can be prepared in C either by the emission of one of the atoms of the sequence, called *source* atom, or by an external coherent solid state microwave source S, coupled into C by a waveguide (not shown on Figure). In this latter case a coherent field with a Poissonian photon number distribution is prepared in C. The whole set-up is cooled to 1.3 K. The mean number of thermal photons in C, dominated by microwave leaks, is 0.7. Radiation cooling is provided, before each experimental sequence, by pulses of atoms crossing C in level g. They absorb the residual thermal field and reduce the field background to 0.12 photon on average. The whole experimental sequence then occurs before the thermal background has had time to relax back to its equilibrium value.

2.2. Qubit superpositions and Ramsey interferences

The atoms can be prepared in a linear superposition of energy states, before the interaction with C, and mixed again after this interaction, making it possible to study qubit coherences. For this purpose we use two pulses of classical microwave, applied in the zones called R_1 and R_2 in Fig.1, which are generated by a second microwave source S' and injected in the cavity structure through a small hole in the ring (S' and this hole are not shown in Fig.1.) This auxiliary microwave field produces inside the ring-mirror structure a standing wave pattern. Antinodes of this pattern, sandwiching the central cavity mode, are used to perform the pulses at the time the atom is crossing them. Different sequences of pulses can be applied on successive atoms by commuting the field source S'. Applying on an atom two successive $\pi/2$ R_1 and R_2 pulses, resonant on the g $\Rightarrow$ i transition, amounts to operating a Ramsey separated field interferometer. The pulses mixing Rydberg states act as atomic internal state beam splitters. The probability to detect (after R_2) in level i the atom initially prepared in level g (before R_1) is the squared sum of two amplitudes, corresponding to the atom crossing the R_1-R_2 interval in level g or i. The relative phase of these two amplitudes is swept by tuning the frequency ν of S'. By repeating the experiment many time, we reconstruct the g $\Rightarrow$ i transition probability, which oscillates when the frequency ν is swept, exhibiting the well known Ramsey fringe pattern [23]. Typically, the fringes have a 72% contrast instead of the ideal 100%. The interaction of the atom with the quantum field in C between R_1 and R_2 modifies the phase and the amplitude of the fringes, revealing useful information about the atom-quantum field coupling. Microwave R_1 and R_2 pulses mixing levels e and g can also be applied on the atoms. In this case, the classical transition occurs at the frequency of the quantum mode C. As described in [21], special caution has then to be exercised to prevent the R_1 and R_2 pulses to leak photons into C. Note finally that the fields used to mix the atomic levels in zones R_1 and R_2 have a very short relaxation time, much smaller than the decay time of the photons in the cavity mode C. This explains why the R_1

and R_2 pulses can be described classicaly and do not produce any entanglement between atom and radiation. They can be considered as mere classical tools to manipulate the atomic state superpositions[24].

3. Resonant entanglement experiments

3.1. Quantum Rabi oscillation and entanglement procedure

The manipulations on the state of the combined photon-atom system rely on the resonant quantum Rabi rotation experienced by each atom in C [25]. When C is exactly resonant with the $e \Rightarrow g$ transition, the atom-cavity system undergoes oscillations between the states $|e, 0>$ and $|g, 1>$ (atom in e or g with either 0 or 1 photon in C). The frequency of this oscillation at cavity center is $\Omega/2\pi = 47$ kHz. The Rabi frequency varies while the atom is crossing the Gaussian field mode and the Rabi rotation integrates this variation. It is convenient to define the effective atom-C interaction time t_i which would produce, if the coupling Ω were constant, the same final Rabi angle on the system. After the atom-cavity interaction, the combined system ends up in the state :

$$|\psi> = \cos(\Omega t_i/2)\ |e, 0> + \sin(\Omega t_i/2)\ |g, 1> \quad (1)$$

which is in general an entangled atom-field state. It is essential, in order to adjust the atom-field entanglement to be able to control t_i . This is achieved in two steps. First we fix the atomic velocity to the value v =503 m/s such that $t_i = \sqrt{\pi}\, w/v$ corresponds to $\Omega t_i = 2\pi$ (2π Rabi pulse). Then, we apply, when each atom is inside C, a voltage across the mirrors which produces an electric field F on the atom, Stark switching in this way the atomic transition out of resonance with C. The atom-field evolution is then frozen at a well defined time, thus reducing the effective interaction time. Any time between $\sqrt{\pi}\, w/v$ and 0 can be realized in this way. Three Rabi pulses are useful. When the condition $\Omega t_i = \pi/2$ is realized, the atom and the field end up in the maximally entangled state:

$$|\psi> = (1/\sqrt{2})\ (|e, 0> + |g, 1>) \quad (2)$$

This entanglement survives after the atom has left the cavity. A measurement on the atom has then an immediate effect on the cavity field, which is prepared either in a Fock state (0 or 1) by detecting the atom in level e or g, or in a superposition of Fock states if the atomic levels are mixed by R_2 prior to detection. We can perform in this way Eintein-Podolsky-Rosen (EPR, [26]) experiments with a mixed pair, half matter, half radiation.

When the $\Omega t_i = \pi$ condition is satisfied, we realize an exchange of excitation between field and matter:

$$|e,0> \Rightarrow |g, 1> \ ; \ |g, 1> \Rightarrow -|e, 0> \quad (3)$$

This pulse can be used to transfer information between the atomic and field qubits, to copy an atomic coherence $c_e|e> + c_g|g>$ into a field one and back. The storage of the information into the cavity constitutes a quantum memory for the atomic qubit [27].

Finally, the $\Omega\, t_i = 2\pi$ condition produces, when there is 1 photon in C, a global dephasing of the combined system's wave function, without changing the photon number. No change is obtained if C is empty because $|g, 0>$ is not coupled to any other state by the atom-field interaction :

$$|g, 0> \Rightarrow |g, 0>; \ |g, 1> \Rightarrow -|g, 1> \quad (4)$$

This effect is analogous to the well-known phase shift experienced by a spin 1/2 undergoing a 2π rotation. The fact that it is conditioned to the value of the photon qubit (0 or 1) is essential for the operation of quantum logic gates. Similar phase shift effects are exploited in the ion trap gates [10]. Note also that multiple of 2π Rabi pulses leaving the photon number unchanged are essential to generate "trapping states" of the Rydberg atom micromaser [28].

By combining Rabi pulses of various duration on successive atoms, we engineer entanglement on a train of particles crossing the cavity one at a time. We now present in more details the entanglement experiments we have realized in this way.

3.2. Two atom entanglement experiment

We have combined a $\pi/2$ and a π Quantum Rabi pulse acting on two atoms crossing successively the cavity in order to entangle two atoms. The principle of the experiment is sketched in Fig2. The first atom, initially prepared in level e has a 50% probability to emit a photon in C. The second atom, initially in g, interacts with the field the exact time required in order to absorb with unit probability the photon, if there is one. The experiment appears thus as an energy exchange between the two atoms, mediated by the cavity field. After the two atoms have emerged form the cavity, the system is in a symmetric superposition of a state in which the two atoms have retained their initial energies and a state in which they have exchanged them. This is an EPR kind of state, engineered in this way for the first time with massive particles [29]. Similar EPR pairs of atoms have been realized since with ions in traps [30]. We have checked the quantum correlations between the two atoms by measuring atomic energies and atomic state superpositions on both atoms, using the Ramsey pulse R_2 to detect atomic coherences. The results of these experiments are presented in [29].

3.3. Quantum phase gate

We can prepare in C a field containing 0 or 1 photon, or a linear combination of 0 and 1 photon states, by sending an atom in a superposition of e and g levels and having it undergo a π pulse in C (initially in the vacuum)[27]. We can also inject a small coherent field in C, produced by S, with an average photon number much smaller than 1. This field is then a superposition of 0 and 1 states, with a negligible probability to contain more than 1 photon. Such a 0/1 photon field will be called the *field qubit*. We can also send across C an atom prepared in level g or i, or in any linear superposition of g and i. We just have to excite the atom in g, then subject it to a pulse R_1 resonant on the g $\Rightarrow$ i transition. We will call it the *atom qubit*. Assume now that this qubit, when in g, undergoes a 2π Rabi pulse in C when it contains 1 photon. In the 2 x 2 basis of the field and atom qubits, the evolution of the system is described by the following truth table, which completes Eq.(4):

$$|i,0> \Rightarrow |i,0>;\ |i,1> \Rightarrow |i,1>;\ |g,0> \Rightarrow |g,0>;\ |g,1> \Rightarrow -|g,1> \qquad (5)$$

The first two relations express the fact that the photon mode is not resonant with transitions originating from level i, which plays the role of a *reference* level. Equations (5) define the operation of a quantum phase gate (QPG). When the initial field and atom qubits are sent

Figure 2. Scheme of experiment generating pairs of entangled atoms: the first atom in level e, the second in level g are sent across an initially empty cavity (at left). If the interaction times are properly set, the two atoms exit the cavity in an entangled superposition of states (the two possible classical outcomes are represented at right; the + sign symbolizes the existence of a coherent superposition)

through C in state superpositions, the operation of the gate results in an entanglement of the two qubits. For example:

$$(1/\sqrt{2})(|i>+|g>)(c_0|0>+c_1|1>)\Rightarrow\{c_0|0>(|i>+|g>)+c_1|1>(|i>-|g>)\}/\sqrt{2} \quad (6)$$

$$=\{|i>(c_0|0>+c_1|1>)+|g>(c_0|0>-c_1|1>)\}/\sqrt{2} \quad (7)$$

where c_0 and c_1 are the probability amplitudes corresponding to 0 or 1 photon in the cavity. Eqs (6) and (7) exhibit the symmetry of the QPG. Either the field in state 1 is the *control* qubit which produces a phase shift on the atom in g (Eq.6), or conversely it is the atom in g which is the *control* qubit dephasing the field in state 1 (Eq.7).

The correlations implied by Eqs(6) and (7) and the coherent operation of the QPG have been checked by experiments which have shown that the gate does operates as an entangling machine for the field and atom qubits [11]. We have also shown that, combined with classical microwave pulses operating on the atom qubit, this gate can be turned into a XOR gate realizing a QND measurement of a single photon.

3.4. QND Measurement of a Single Photon

Let us subject the atomic qubit, initially in level g, to two classical $\pi/2$ microwave pulses R_1 and R_2 whose frequency ν is swept across the $g \Rightarrow i$ transition frequency ν_{gi} and let us measure the probability that the atom finally undergoes this transition (experimental scheme shown in Fig.3a). This probability, recorded versus ν, exhibits a Ramsey fringe pattern, which can be used to measure the phase difference between the "paths" corresponding to the atom crossing C in level g or i. We can for example adjust the frequency of the Ramsey pulses so that, if there is 0 photon in C, the atom ends up in the initial state g with, in principle, unit probability. If there is 1 photon in C instead, the atomic coherence produced by R_1 has its phase shifted by π when the atom crosses C. The second pulse in R_2 now completes the transition between the two atomic qubit states, exchanging their energies and the atom ends up in level i. In short, the atomic qubit flips if and only if there is one photon in C. The field qubit remains in any case unaltered. This defines a XOR gate.

Figures 3b,c show the Ramsey fringes expected when the cavity contains 0 or 1 photon respectively. The XOR gate is obtained when the frequency of the Ramsey fields is set at a fringe extremum, shown by the vertical line on the right side of figure 3b,c. We see that the final atomic state is an indicator of the photon number, which remains unaltered at the end of the process. Detection of the atomic state in D after R_2 amounts to an ideal quantum non-demolition (QND) measurement [31] of a single photon field.

To demonstrate the QND operation, we first generate in C a 0 or 1 photon field. This is achieved by sending first a *source* atom undergoing a $\pi/2$ Rabi pulse. According to Equ.2, the detection of this atom in e or g prepares with equal probabilities either 0 or 1 photon. We then record the Ramsey fringes produced on a second *meter* atom undergoing a 2π Rabi pulse. We detect separately the fringes corresponding to the e and g outcomes of the *source* atom measurement. We obtain (see Fig.3d) two fringe patterns with opposite phases. This demonstrates that, as expected, a single photon in C dephases by π the atomic coherence. We note that at the fringe extremum marked by the vertical line, the *meter* atom has a large probability to exit in level g if there is 0 photon in C and in level i if there is 1 photon.

In order to check the non-destructive character of this procedure, we have also performed the repeated measurement of a small thermal field. We have left a blackbody field with an average number of 0.2 photon build up in C. We have then sent a first *meter* atom to perform a QND measurement of this field, then a second *probe* atom, initially in g and undergoing a π Rabi pulse in C, thus absorbing with unit probability the field left in C after the *meter* atom. By analyzing the correlations between the final states of the *meter* and *probe*, we have demonstrated the QND nature of the *meter* atom measuring process. We have found that the *probe* has a much larger probability to be excited if the *meter* has indicated the presence of a photon.

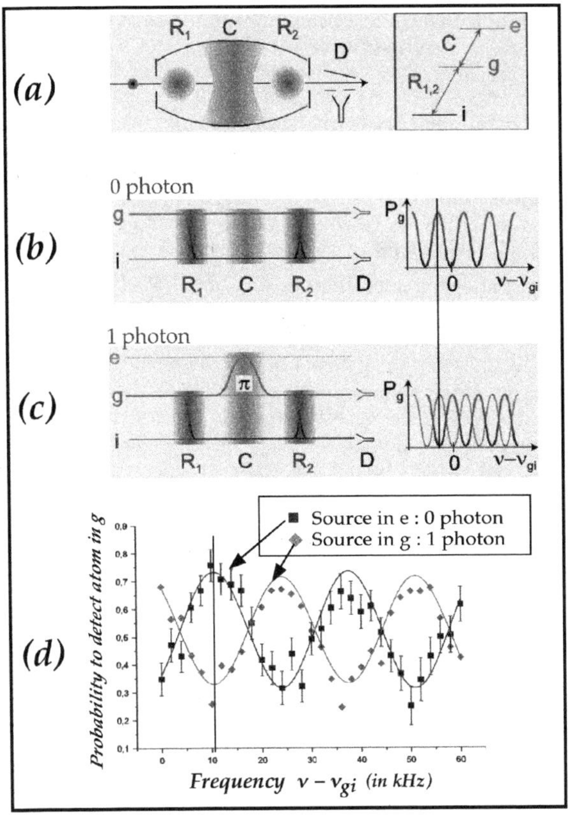

Figure 3. Quantum non-demolition measurement of a single photon: (a) Scheme of the Ramsey interferometer (made of the two Ramsey zones R_1 and R_2), with the cavity mode C in between. At right, energy levels relevant to the experiment. The cavity mode is resonant with the e-g transition and the Ramsey interferometer works on the g-i transition. (b) and (c) Principle of experiment: At left, representation of the two interfering « paths » followed by an atom in the interferometer when there is respectively 0 and 1 photon in C : the interference between these paths produces a fringe pattern in the probability to finally detect the atom in g, when the frequency of the Ramsey pulses is tuned (shown at right). The fringes are phase shifted by π when the photon number changes from 0 to 1. (d) Experimental fringes recorded on a "meter" atom versus the Ramsey pulse frequency ν, after a first source atom has prepared either 0 (squares) or 1 (diamonds) photon in C. (From ref [20], Reprinted by permission from Nature (Vol 400, p 239 (1999)).

We have also performed a double QND experiment, sending two successive *meter* atoms across C and checking that their indications are correlated. With these experiments, we have analyzed the limitations of our QND scheme. This Ramsey interferometric method allows us to tell, with an 80% success rate, whether there is 0 or 1 photon in C immediately after the passage of the *meter* atom. The probability that this atom absorbs the photon, due to imperfections of the 2π Rabi pulse, is about 10%.

This QND measurement must be contrasted with usual photodetection procedures based on the destruction of the photon. Here, for the first time, a single photon can be detected and kept stored in a box for further measurements. This opens, as we show below, interesting perspectives for processing quantum information.

It is also interesting to compare our scheme with previous QND measurements in quantum optics, based on the interferometric measurement of the dispersive phase shift induced by a *signal* light field on a *meter* beam [32]. This shift is sensitive to the intensity of the *signal*, via its coupling with the *meter* beam in a non-linear medium. The *signal* field must be intense enough to produce an observable non-linear effect in this medium so that QND detection is generally sensitive only to macroscopic photon fluxes. Formally, our QND scheme looks quite similar. The 0/1 photon signal induces a phase shift on an atom *meter*, which is also detected by an interferometric method. Here however, the non-linearity of the Cavity QED system is extremely large, making it possible to detect non-destructively single light quanta.

Our QND method does not apply to fields containing more than one photon. The Rabi frequency in a field of n photons is $\Omega\sqrt{n}$, so that an interaction time corresponding to a 2π Rabi pulse for $n = 1$ performs with n photons a $2\pi\sqrt{n}$ rotation. The probability that the photon number changes after the interaction is $\sin^2(\pi\sqrt{n}) \neq 0$ for $n > 1$. This violates the QND requirement as soon as the field state has a component outside the 0/1 photon subspace. We have proposed a dispersive phase shift method that overcomes this limitation [33]. If the cavity mode and the $e \Rightarrow g$ transition are detuned, the *meter* atom can no longer absorb the field, whatever the photon number is. The atomic levels undergo an energy light shift, linear function of n, which can be measured by Ramsey interferometry [34]. We have described detection procedures allowing us to pin-down the photon number [35]. The detection of several atoms is then required to read out n, because we need then at least $\mathrm{Log}_2(n)$ bits of information. Experimental realization of this dispersive QND method is in progress.

3.5. Three particle entanglement

The atom-cavity QPG can be combined with a $\pi/2$ atom-photon Rabi pulse to prepare a three-particle-entangled state, according to a procedure proposed independently in [36] and [37]. We describe now the principle of this experiment [21] which illustrates how our apparatus can be used to *knit* entanglement in a complex system by a *step-by-step* succession of operations. A first atom A_1 undergoes a $\pi/2$ Rabi pulse in the initially empty cavity mode, resulting in the entangled state of Eq.2. We then use the 0/1 photon field produced in C as a qubit for the operation of a QPG entangling it with a second atom A_2. This atom, first prepared by R_1 into the linear superposition $(1/\sqrt{2})(|i\rangle + |g\rangle)$, undergoes a 2π Rabi pulse in C. When it exits C, the A_1-A_2-C system is prepared in the entangled state:

$$|\Psi_{triplet}\rangle = (1/2)\left[\,|e_1\rangle(|i_2\rangle + |g_2\rangle)|0\rangle + |g_1\rangle(|i_2\rangle - |g_2\rangle)|1\rangle\right] \quad (8)$$

where the indices correspond to the atom numbers. This state can be rewritten as:

$$|\Psi_{triplet}\rangle = (1/2)\left[\,|i_2\rangle(|e_1,0\rangle + |g_1,1\rangle) + |g_2\rangle(|e_1,0\rangle - |g_1,1\rangle)\right] \quad (9)$$

describing an A_1-C entangled pair whose phase is conditioned to the state of A_2.

Because $|\Psi_{triplet}\rangle$ involves two levels for each subsystem, it is equivalent to an entangled state of three spins 1/2. Let us define the states $|+_j\rangle$ ($|-_j\rangle$) (with $j = 1,2$) as $|+_1\rangle = |e_1\rangle$ ($|-_1\rangle = |g_1\rangle$), $|-_2\rangle = (|g_2\rangle \pm i_2\rangle)/\sqrt{2}$ and $|+_C\rangle = |0\rangle$ ($|-_C\rangle = |1\rangle$). With these notations, $|\Psi_{triplet}\rangle$ takes the form of the GHZ three-spin state [38]:

$$|\Psi_{triplet}\rangle = (1/\sqrt{2})\left(\,|+_1,+_2,+_C\rangle - |-_1,-_2,-_C\rangle\right) \quad (10)$$

In order to analyze the entanglement, we must detect the two atoms and the field in C. We cannot however measure this field directly. We must first copy it on a third atom A_3 crossing C after A_2 [27]. The A_3-C interaction is set so that A_3, initially in g, is not affected if C is empty but undergoes a π Rabi pulse in a single photon field (Eq.3). Within a phase, A_3 maps the state of C. According to Eq.(3), the cavity state $|+_C\rangle$ is mapped into $|g_3\rangle$ and $|-_C\rangle$ into $|e_3\rangle$. Thus by detecting A_1, A_2 and A_3, we measure a set of observables belonging to the three parts of the entangled triplet. We have checked in details this entanglement by performing measurements on the three atoms and analyzing the correlations between them

[21]. Note that three- [39] and four-particle [40] entanglement have been recently studied in other quantum optics experiments.

4. Non resonant entanglement

4.1. Single atom index effect

Let us now consider the situation where the cavity mode is slightly detuned from the $e \Rightarrow g$ transition frequency (level i does not play any role here). Since energy can no longer be conserved during elementary photon absorption and emission processes, the atom and the field cannot exchange energy and cannot get entangled in a resonant way. This does not mean however that the two subsystems do not interact. The mere presence of an atom in the cavity slightly shifts its frequency. This is a refractive index effect, produced by a single atom [19]. This effect depends upon the atom's energy state. When the cavity mode frequency is larger than the atomic transition one, the field frequency is increased or decreased, depending upon whether the atom is in level g or e. What happens if the atom crosses the cavity in a linear superposition of these levels? The cavity mode is then in a superposition of states corresponding to two different frequencies at once. This peculiar situation leads to the possibility of studying new kinds of entanglement.

4.2. Schrödinger cat states

Let us start by injecting a small field in C. We use for this the classical microwave source S coupled to C by a waveguide. The connection is opened for a short time, then interrupted, resulting in the trapping in C of a coherent field with a photon number of the order of 10. This field decays within a time T_{cav} of a few hundred microseconds. If we neglect first relaxation, we can describe the electric field stored in the cavity mode as a periodic function of time which can be associated with a complex number α whose modulus and argument represent the field amplitude and phase. This c-number can be in turned viewed as a vector in "phase plane" (Fresnel representation). The tip of this vector lies in a small uncertainty circle describing the conjugated amplitude and phase field fluctuations.
Assume now that we send an atom across C (fig.4a). The period of the field oscillations is slightly modified as the atom interacts with the field mode. This results in a kick of the field phase, after the atom has left C. Equivalently, the vector representing the field rotates in phase space by an angle $+\phi$ or $-\phi$, depending on the state (e or g) of the atom (fig 4b). If the atom is sent in the linear superposition $|e\rangle + |g\rangle$, the combined atom field system evolves into an entangled state which writes:

$$|\Psi\rangle = |e, \alpha \exp(i\phi)\rangle + |g, \alpha \exp(-i\phi)\rangle \qquad (11)$$

The situation is reminiscent of the famous " Schrödinger's cat " [41] entangled with a single atom in a superposition of states corresponding to its " live " and " dead " states. We note that the field vector acts as a kind of meter pointing in two different directions correlated to the atom's energy (fig 4c). In other words, the field plays the role of an apparatus measuring the atom's energy in a quantum non-demolition way. The situation is symmetric of the one described above (atom used to measure non-destructively the field's energy).

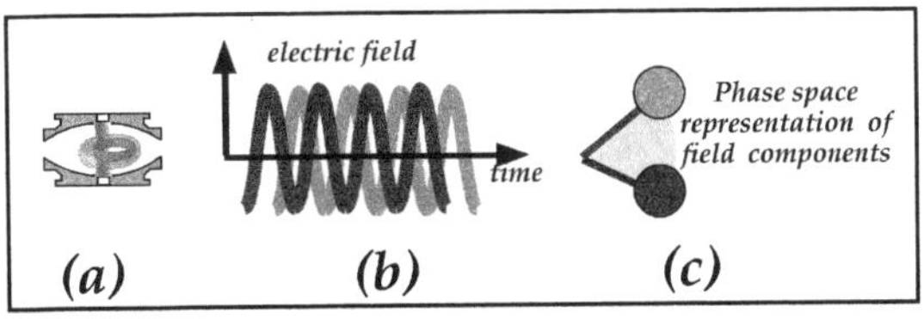

Figure 4. How to prepare a Schrödinger cat state of the field in the cavity: (a): an atom in a superposition of states crosses C. (b): it imparts to the field two phases at once. (c): Each phase component is represented by a vector pointing in a definite direction in phase space. The tip of each vector lies in a small uncertainty circle. The "atom + field" system ends up in an entangled state, the atomic energy state being correlated to the field phase.

4.3. A complementarity experiment

This leads us to the description of an experiment demonstrating in a simple way the complementarity principle. Ramsey interferences can be obtained in the e $\Rightarrow$ g atomic transition probability by subjecting atoms to two $\pi/2$ microwave pulses in R_1 and R_2, before and after their interaction with C. The frequency of these pulses is now swept across the frequency ν_{eg} of the e $\Rightarrow$ g transition. Interferences fringes are however observed *only* if nothing in the apparatus allows us to distinguish the two excitation channels (atom excited from g to to e in R_1 and thus crossing the cavity in level e and atom excited in R_2, after having crossed C in g). What happens then if C contains a small coherent field, non-resonant with the e $\Rightarrow$ g transition? Since the phase of this field is kicked by an angle depending upon the atom's state in C, this field acts as a "which path" detector [42,43] able to reveal the atomic " path " through the interferometer. According to the complementarity principle, the fringes then tend to vanish. This is what we observe (Fig. 5). When the phase rotation is too small to tell us unambiguously the path, the fringe contrast is merely reduced. The fringes disappear altogether when the phase rotation is large, making certain the atom's path. The field rotation is adjusted by changing the detuning between the atom transition and the cavity mode (the smaller the detuning, the larger the single atom index effect and the field phase kick). We have described here this " which path " experiment in qualitative terms. A quantitative description is very easy to derive. In fact, the fringe contrast is merely proportional to the scalar product of the two field components. Since in this experiment we observe the atom

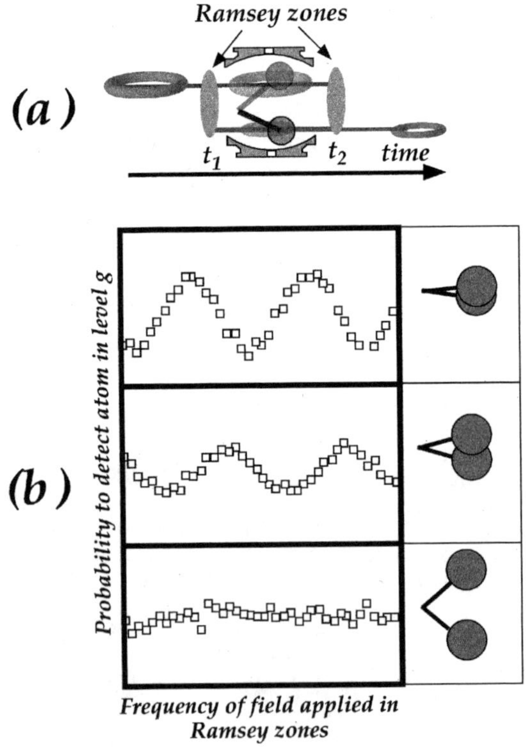

Figure 5. Complementarity experiment: (a) Scheme of principle: an atom follows two interfering paths between the two Ramsey zones and the phase of the field stored in the cavity provides a « which path » information. (b): the probability to detect downstream the atom in level g is recorded, as a function of the frequency of the field applied in the Ramsey zones, for three different values of the field dephasing angle. The fringe contrast decreases when the separation of the field components (represented in the insets at right) is increased.

and not the field, the signal is obtained by tracing the atom-field density operator over the field variables. This procedure introduces directly the overlap integral between the two field components in the calculation of the interference term. Similar complementarity experiments have been performed with other experimental systems [44].

4.4. Decoherence caught in the act

In the complementarity experiment, we have measured an interference effect characterizing the atom state superposition, which is influenced by the presence of the field in C. What can we say about the quantum coherence of this field? How long does the superposition of the two field states survive in C? To answer this question, we must analyze in more details the nature of the field's " environment ". Since the field losses are mainly due to photon scattering on mirrors surface imperfections, we can describe this environment as being made of the free space around the cavity, which can be filled by the scattered photons. The mean field energy damping time in this experiment is T_{cav} = 160 microseconds. If the cavity contains a field with, in average, n photons, a small field with about one photon escapes in the environment within a characteristic time T_{cav}/n. This microscopic field is entangled through its phase with the field left in C.

This entanglement with the environment provides a way to determine, at least in principle the phase of the field in C. By measuring the phase of the tiny component escaping in the outside world, one would immediately reduce the field left in C to one or the other term, thereby destroying the quantum coherence. This is the very essence of the decoherence phenomenon as analyzed by W. Zurek [12]. Thus, after a time of the order of T_{cav}/n, the quantum coherence between the two field components in C has vanished. This explains why macroscopic fields, corresponding to huge n values, behave classicaly since they experience a quasi-instantaneous decoherence process. In our experiment, however, n is of the order of 3 to 10 only. The decoherence time is then long enough to allow for the observation of transient interference signals associated to the two components of our " Schrödinger cat " state in C.

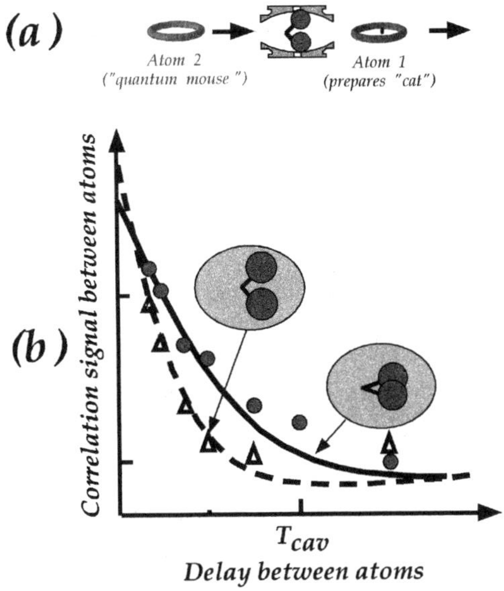

Figure 6. Decoherence experiment: (a) Scheme of principle: Atom 1 prepares in C a superposition of field states with different phases and atom 2 (the « quantum mouse ») probes it after a variable delay. (b): A correlation signal between energy measurements on the two atoms is recorded, by averaging many experimental runs. This signal displays an interference effect which decays as a function of the delay between the atoms. Experimental results for two different values of the field component separation are shown here (cercles and triangles). The curves result from numerical calculations based on decoherence theory.

To perform this observation, we send in C, after the first atom who prepares the " cat state " a second atom, playing so to speak the role of a " *quantum mouse* " used to probe in C the quantum coherence of the field. This second atom kicks in turn the phase of the field components in C, recombining them partially. This recombination produces, in a correlation signal between the two atoms, an interference term sensitive to the quantum coherence between the cat-state components left by the first atom in C. This interference signal, shown in Fig.6, decreases when the delay between the two atoms is increased. This phenomenon occurs faster and faster when the two components of the cat are separated more and more. This provides a direct illustration of the main features of environment induced decoherence, which acts faster and faster as the size of the systems become more and more macroscopic. Note the good agreement between the experimental points and the theoretical curves, deduced from a very simple calculation based on decoherence theory [45]. In some way, this experiment, which verifies the main aspect of the decoherence theory, constitutes a step in the exploration of the quantum classical boundary. Similar "Schrödinger cat" experiments [46] and decoherence studies [47] have been performed with an ion in a trap. (Note however that in this case, the loss of coherence was due to the perturbing effect of a classical noise acting on the system and not to quantum entanglement with the environment).

5. Conclusion

We have described a very versatile system allowing us to entangle particles according to a preset deterministic program, to generate Schrödinger cat state superpositions of fields and to study their decoherence. The experiments will be extended in the future to study larger systems involving more atoms, more photons and more cavities. For such experiments, we need an improved version of our apparatus, with better cavities keeping the photons stored for longer times and better atomic sources allowing us to produce velocity selected Rydberg atoms on demand, and not statistically, as it is presently the case. We also need to improve on our atom detection efficiency. We are working on these improvements and hope to be able soon to explore further the quantum classical boundary with this very promising system.

Acknowledgments

Laboratoire Kastler Brossel is a Unité Mixte du Centre National de la Recherche Scientifique (CNRS) et de l'Université Pierre et Marie Curie. The ENS Cavity QED group is supported in part by the European Community and Japan Science and Technology Corporation (International Cooperation Research Project, Quantum Entanglement Project).

References

[1] D.Bouwmeester, A.Ekert and A.Zeilinger (editors), *The Physics of Quantum Information*, Springer Verlag, Berlin, Heidelberg, 2000.
[2] C.H.Bennett and D.P. DiVicenzo, *Nature*, **404**, (2000) 247-255.
[3] C.H.Bennett, G.Brassard, C.Crepeau, R.Jozsa, A.Peres and W.K.Wooters, *Phys.Rev.Lett.* **70**, (1993) 1895- 1898.
[4] D.Bouwmeester, J.W. Pan, K.Mattle, M.Eibl, H.Weinfurter and A.Zeilinger, *Nature*, **390**, (1997) 575-579.
[5] D.Boschi, S.Branca, F.De Martini, L.Hardy and S.Popescu, *Phys.Rev.Lett.*, **80**, (1998) 1121-1124.
[6] A. Furusawa, J.L. Sorensen, S.L.Braunstein, C.A. Fuchs, H.J.Kimble and E.S. Polzik, *Science*, **282**, (1998) 706-709.
[7] T.Jennewein, C.Simon, G.Weihs, H.Weinfurter and A.Zeilinger *Phys.Rev.Lett.* **84** (2000) 4729-4732
[8] D.S.Naik, C.G.Peterson, A.G.White, A.J.Berglund and P.G.Kwiat, *Phys.Rev.Lett.* **84** (2000) 4733-4736
[9] W.Tittel, J.Brendel, H.Zbinden and N.Gisin, *Phys.Rev.Lett.* **84** (2000) 4737-4740
[10] C. Monroe, D.M. Meekhof, B.E.King, W.M.Itano and D.J.Wineland, *Phys.Rev.Lett.* , **75**, (1995) 4714-4717.
[11] A.Rauschenbeutel, G.Nogues, S.Osnaghi, P.Bertet, M.Brune, J.M.Raimond and S.Haroche, *Phys.Rev.Lett.*, **83** (1999) , 5166-51669.
[12] W.Zurek, « *Decoherence and the transition from quantum to classical* », *Physics Today*, Vol **44**, No 10, (1991) p 36-44; W.Zurek, *Physics World*, January 1997, p25
[13] D.Giulini, E,Joos, C.Kiefer, J.Kupsch, I.O.Stamatescu and D.Zeh, « *Decoherence and the appearance of a classical world in quantum theory* », edited by D.Giulini et al, Springer Verlag, Berlin Heidelberg 1996.
[14] R.Omnes, « *The Interpretation of Quantum Mechanics* », Princeton University Press (1994).

[15] N.A. Gershenfeld and I.L.Chuang, *Science,* **275**, (1997) 350-356.
[16] J.R.Friedman, V.Patel, W.Chen, S.K.Tolpygo and J.E.Lukens, *Nature* **406** (2000) 43-45
[17] C.H.van der Wal, A.C.J.ter Haar, F.K.Wilhelm, R.N.Schouten, C.J.P.M.Harmans, T.P.Orlando, S.Lloyd and J.E.Mooij, *Science* **290** (2000) 773-777
[18] S. Haroche, « *Cavity Quantum Electrodynamics* » in *Fundamental Systems in Quantum Optics,* edited by Dalibard, J. et al, North Holland, Amsterdam, 1992, pp. 767- 940.
[19] S.Haroche and J.M.Raimond in *« Cavity Quantum Electrodynamics »*, P.Berman (editor), Academic Press, Boston (1994) p 123-170.
[20] G.Nogues, A.Rauschenbeutel, S.Osnaghi, M.Brune, J.M.Raimond and S.Haroche, *Nature*, **400**, (1999) 239-242.
[21] A.Rauschenbeutel, G.Nogues, S.Osnaghi, P.Bertet, M.Brune, J.M.Raimond and S.Haroche, *Science*, **288**, (2000) 2024- 2028.
[22] M.Brune, E.Hagley, J.Dreyer, X.Maître, A.Maali, C.Wunderlich, J.M.Raimond and S.Haroche, Phys.Rev.Lett. **77**, (1996) 4887-4890.
[23] N.Ramsey, *Molecular Beams,* Oxford University Press, New York (1985)
[24] J.I.Kim, K.M.Fonseca Romero, A.M.Horiguti, L.Davidovich, M.C.Nemes and A.F.R. de Toledo Piza, *Phys.Rev.Lett,* **82** (1999) 4737-4740.
[25] M.Brune, F.Schmidt-Kaler, A.Maali, J.Dreyer, E.Hagley, J.M.Raimond and S.Haroche, *Phys.Rev.Lett*, 76, (1996) 1800-1803.
[26] A.Einstein, B.Podolsky and N.Rosen, *Phys.Rev.*, 47, (1935) 777-780.
[27] X.Maître, E.Hagley, G.Nogues, C.Wunderlich, P.Goy, M.Brune, J.M.Raimond and S.Haroche, *Phys.Rev.Lett.* **79**, (1997) 769-772.
[28] M.Weidinger, B.Varcoe, R.Heerlein and H.Walther, *Phys.Rev.Lett.*, **82**, (1999) 3795-3798.
[29] E.Hagley, X. Maître, G.Nogues, C.Wunderlich, M.Brune, J.M.Raimond and S.Haroche *Phys.Rev.Lett.*, **79**, (1997) 1-5.
[30] Q.A.Turchette, C.S.Wood, B.E.King, C.J.Myatt, D.Leibfried, W.M.Itano, C.Monroe and D.J.Wineland, *Phys.Rev.Lett.* , **81**, (1998) 3631-3634.
[31] V.B.Braginski and F.Y.Khalili, *Rev.Mod.Phys.*, **68**, (1996) 1-11.
[32] P.Grangier, A.L.Levenson and J.P.Poizat, *Nature*, **396**, (1998) 537-542.
[33] M.Brune, S.Haroche, V.Lefevre, J.M.Raimond and N.Zagury, *Phys.Rev.Lett*, **65**, (1990) 976-979.
[34] M.Brune, P.Nussenzveig, F.Schmidt-Kaler, F.Bernardot, A.Maali, J.M.Raimond and S.Haroche *Phys.Rev.Lett.*, **72**, (1994) 3339-3342.
[35] S.Haroche, M.Brune and J.M.Raimond, *J.Phys.II France*, **2**, (1992) 659-670.
[36] S.Haroche, G.Nogues, A.Rauschenbeutel, S.Osnaghi, M.Brune and J.M.Raimond, « Quantum Knitting in Cavity QED » in *Laser Spectroscopy, XIV International Conference*, edited by Blatt, R., et al, World Scientific, Singapore, 1999, pp 140-149.
[37] S.B.Zheng, *J.Opt.B.*, **1**, (1999) 534-538.
[38] D.M.Greenberger, M.A. Horne and A.Zeilinger, *Am.J.Phys.* **58**, (1990) 1131-1143.
[39] J.W.Pan, D.Bouwmeester, M.Daniell, H.Weinfurter and A.Zeilinger, *Nature*, **403**, (2000) 515-519.
[40] C.A.Sackett, D.Kielpinski, B.E.King, C.Langer, V.Meyer, C.J.Myatt, M.Rowe, Q.A. Turchette, W.M.Itano and D.J.Wineland *Nature*, **404**, (2000) 256-259.
[41] E.Schrödinger, *Naturwissenschaften* **23** (1935), p 807-849, translated in english in « Quantum Theory and Measurement », J.A.Wheeler and W.Zurek, editors, Princeton University Press, (1983) p. 152-167.
[42] S.Haroche, M.Brune and J.M.Raimond, Applied.Physics **B 54**, (1992) 355
[43] M.O.Scully, B.E.Englert and H.Walther, *Nature*, **351**, (1991) 111-116
[44] T.Pfau, S.Spälter, C.Kurtsiefer, C.R. Ekstrom and J.Mlynek, *Phys.Rev.Lett.* **73** (1994) 1223-1227; M.S.Chapman, T.D.Hammond, A.Lenef, J.Schmiedmeyer, R.A.Rubinstein, E.Smith and D.E.Pritchard, *Phys.Rev.Lett.* **75** (1995) 3783
[45] J.M.Raimond, M.Brune and S.Haroche, *Phys.Rev.Lett.* **79** (1997) 1964-67; X. Maître, E. Hagley, J. Dreyer, A. Maali, C. Wunderlich, M. Brune, J.M. Raimond, S. Haroche, *Journ. Mod. Optics*. **44**, (1997) 2023
[46] C.Monroe, D.Meekhof, B.E.King and D.J.Wineland, *Science,* **272** (1996) 1131-1135
[47] C.J.Myatt, B.E.King, Q.A.Turchette, C.A.Sackett, D.Kielpinski, W.M.Itano, C.Monroe and D.Wineland *Nature,* **403** (2000) 269-273.

Decoherence and its Implications in Quantum Computation and Information Transfer
T. Gonis and P.E.A. Turchi (Eds.)
IOS Press, 2001

Decoherence and long-lived Schrödinger cats in BEC

Diego A. R. Dalvit[1], Jacek Dziarmaga[1,2], and Wojciech H. Zurek[1]

1) Los Alamos National Laboratory, T-6, Theoretical Division, MS-B288, Los Alamos, NM 87545, USA

2) Institute of Physics, Jagiellonian University, Kraków, Poland

Abstract

We consider quantum superposition states in Bose-Einstein condensates. A decoherence rate for the Schrödinger cat is calculated and shown to be a significant threat to this macroscopic quantum superposition of BEC's. An experimental scenario is outlined where the decoherence rate due to the thermal cloud is dramatically reduced thanks to trap engineering and "symmetrization" of the environment. We show that under the proposed scenario the Schrödinger cat belongs to an approximate decoherence-free pointer subspace.

1 Introduction

Microscopic quantum superpositions are an everyday physicist's experience. Macroscopic quantum superpositions, despite nearly a century of experimentation with quantum mechanics, are still encountered only very rarely. Fast decoherence of macroscopically distinct states is to be blamed [1]. In spite of that, recent years were a witness to an interesting quantum optics experiment [2] on decoherence of a few photon superpositions. Moreover, matter-wave interference in fullerene C_{60} has been observed [3]. Another experiment has succeeded in "engineering" the environment in the context of trapped ions [4]. Recently the first detection of a macroscopic Schrödinger cat state in an rf-SQUID was reported [5]. All these successes tempt one to push similar investigations of basic quantum mechanics into the rapidly progressing field of Bose-Einstein condensation (BEC) of alkali metal atomic vapors [6] The condensates can contain up to 10^7 atoms in the same quantum state. What is more, it is possible to prepare condensates in two different internal states of the atoms. Some of these pairs of internal levels are immiscible, and their condensates tend to phase separate into distinct domains with definite internal states [7]. The immiscibility seems to be a prerequisite to prepare a quantum superposition in which all atoms are in one or the other internal state, $|\psi\rangle = (|N,0\rangle + |0,N\rangle)/\sqrt{2}$, where N is the total number of condensed atoms. There are at least two theoretical proposals how to prepare a macroscopic quantum superposition in this framework [8, 9].

Other proposals involve non-unitary evolution of the BEC towards the Schrödinger cat state by means of continuous quantum measurements [10, 11]. Neither of these proposals addresses the crucial question of decoherence.

This paper is a simplified and more pedagogical version of our previous work [12]. All technical details of the calculations, in particular the computation of the decoherence rate for the BEC superposition state, can be found in that reference.

2 The condensate

Let us first make a short summary of some of the ideas for creating quantum superposition states of Bose-Eintein condensates. The methods that have been thus far proposed start with two weakly interacting dilute Bose condensates of atoms in different internal states. Let us call these internal states A and B. Atoms interact through s-wave collisions, characterized by a single parameter a which is known as the scattering length. We shall consider the case of repulsive interaction $a > 0$; it is known that for the opposite case the condensate is unstable above a critical total number of particles. It is assumed that the inter-scattering lengths for collisions $A - A$ and $B - B$ are the same, i.e. $a_{AA} = a_{BB}$, which is in general different from the intra-scattering length a_{AB}. When the self-energy of atom-atom interactions of the BEC is much less than the mode energy spacing, then one can treat the condensate in the two-mode approximation. Moreover, the condensate is shined with a appropriately chosen laser that introduces a Josephson-like coupling of strength λ that interchanges internal atomic states in a coherent manner. The condensate two-mode Hamiltonian is

$$H_{\rm C} = \epsilon_g(a^\dagger a + b^\dagger b) - \lambda(a^\dagger b + b^\dagger a) + \frac{u_{\rm c}}{2}(a^\dagger a^\dagger aa + b^\dagger b^\dagger bb) + v_{\rm c}(a^\dagger b^\dagger ab) \ , \qquad (1)$$

Here ϵ_g is the energy of the lowest single-particle state in the trap (assumed to be the same for A and B), $u_{\rm c} = 4\pi\hbar^2 a_{AA}/m$ and $v_{\rm c} = 4\pi\hbar^2 a_{AB}/m$.

This hamiltonian was studied in great detail in [8, 9, 13]. The precise quantum state of the condensate for all values of $u_{\rm c}$ and $v_{\rm c}$ was studied by means of a Wigner-like distribution on the two-mode Bloch sphere. It was found that when the repulsion between atoms in different states is bigger that the repulsion between atoms in the same state, that is when $v_{\rm c} > u_{\rm c}$, then the condensate tends to phase separate in space into distinct domains with definite internal states. This condition is known as the immiscibility condition. More importantly, the ground state of the system is a quantum superposition state. When the "purity" factor defined as

$\epsilon \equiv (\lambda/(v_c - u_c)N)^N$ is much less than one, then the lowest energy subspace contains two macroscopic superpositions

$$|+\rangle = \frac{1}{\sqrt{2\,N!}}[(a^\dagger)^N + (b^\dagger)^N]\,|0,0\rangle \equiv \frac{1}{\sqrt{2}}(|N,0\rangle + |0,N\rangle) \ ,$$
$$|-\rangle = \frac{1}{\sqrt{2\,N!}}[(a^\dagger)^N - (b^\dagger)^N]\,|0,0\rangle \equiv \frac{1}{\sqrt{2}}(|N,0\rangle - |0,N\rangle) \ . \qquad (2)$$

The lower $|+\rangle$ and the higher $|-\rangle$ states are separated by a small energy gap of $E_- - E_+ = N(u_c - v_c)\epsilon \ln \epsilon$. In the extreme case of $\lambda = 0$ (no laser applied) the two states are completely degenerate. If we just have $\epsilon < 1$, the $|\pm\rangle$ states in fact contain an admixture of intermediate states $|N-1,1\rangle, \ldots, |1, N-1\rangle$ such that their overlap is $\langle +|-\rangle = \epsilon$. For $\epsilon \gg 1$ they shrink to

$$(a^\dagger \pm b^\dagger)^N\,|0,0\rangle \ . \qquad (3)$$

which does not correspond to a macroscopic superposition. From now on we assume the pure case, $\epsilon \ll 1$.

The proposal of Gordon and Savage [9] for obtaining these superposition states is similar to a typical experiment in quantum optics with non-linear systems. In the BEC case, the non-linearity is provided by the collisional interactions and the Josephson coupling. The idea is to prepare an initial state with all the atoms in the same internal state and then turn on the Josephson coupling for some amount of time adequately chosen. After the Josephson coupling is turned off, the quantum state of the system has been modified, and a Schrödinger cat state has been formed. Other proposals for obtaining basically the same kind of superposition use a continuous quantum measurement process [11].

3 The thermal cloud

As we have already mentioned, neither of the proposals for generating quantum superpositions in BEC addresses the crucial question of decoherence. In fact, the condensate is an open quantum system which is in contact with an environment of non-condensed thermal particles. This interaction may be responsible for the loss of coherence between the components of the quantum superposition states of the previous section. If the decoherence time is very small, then the existence of these states in BEC would be merely of academic interest, since the expectations of observing them in the lab would vanish. Therefore it is important to understand how the thermal cloud affects the longevity of the BEC cats.

The Hamiltonian for the dilute environment formed by non-condensed particles is

$$H_{\rm E} = \sum_s \left[\epsilon_s (a_s^\dagger a_s + b_s^\dagger b_s) - \lambda (a_s^\dagger b_s + b_s^\dagger a_s)\right] . \quad (4)$$

ϵ_s is the single particle energy of level s. After the transformation

$$S_s = \frac{a_s + b_s}{\sqrt{2}} \; , \; O_s = \frac{a_s - b_s}{\sqrt{2}} \quad (5)$$

$H_{\rm E}$ adopts a diagonal form

$$H_{\rm E} = \sum_s \left[(\epsilon_s - \lambda) S_s^\dagger S_s + (\epsilon_s + \lambda) O_s^\dagger O_s\right] . \quad (6)$$

In Figure 1 we make a schematic plot of the energy levels of the non-condensed system before ($\lambda = 0$) and after ($\lambda \neq 0$) applying the Josephson coupling. For simplicity we consider an isotropic harmonic trap with a dip in its center, where the condensate particles will be localized (we shall return to this point later). We see that when the laser coupling is applied, the states S_s (which are symmetric under the exchange $a \leftrightarrow b$) and the states O_s (which are antisymmetric under the same exchange) feel the same, but shifted, trapping potentials. The two ladders of energy eigenstates of the environment are separated by a gap equal to 2λ. Each of the levels is occupied according to the thermal Bose distribution. When the gap is much bigger that the typical thermal energy $2\lambda \gg k_{\rm B}T$, then the antisymmetric states are unoccupied. We will see in the next section that this is a prerequisite for reducing decoherence.

4 Condensate-Non condensate interactions

Condensed atoms interact with non-condensed atoms via two-body collisions. The exact interaction hamiltonian V can be easily obtained from the total (Gross-Pitaevskii like) hamiltonian

$$\begin{aligned} H \; = \; & \int d^3x \left[v(\phi_A^\dagger \phi_A)(\phi_B^\dagger \phi_B) + \left\{ \frac{u}{2} \phi_A^\dagger \phi_A^\dagger \phi_A \phi_A + \nabla \phi_A^\dagger \nabla \phi_A + U(r) \phi_A^\dagger \phi_A - \lambda \phi_A^\dagger \phi_B \right\} \right. \\ & \left. + \{A \leftrightarrow B\} \right] , \end{aligned} \quad (7)$$

after splitting the quantum fields ϕ_A and ϕ_B into condensate and bath parts, i.e.

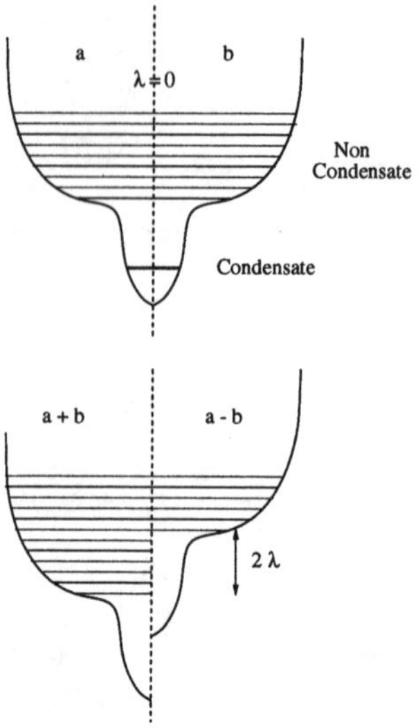

Figure 1: Environmental single-particle energy levels before and after turning on the Josephson-like coupling.

$$\begin{aligned} \phi_A(\vec{x}) &= a\, g(\vec{x}) + \sum_s a_s\, u_s(\vec{x})\ , \\ \phi_B(\vec{x}) &= b\, g(\vec{x}) + \sum_s b_s\, u_s(\vec{x}), \end{aligned} \tag{8}$$

where $g(\vec{x})$ is the wave function of the condensate localized in the dip. After introducing the symmetric and antisymmetric operators S_s and O_s for the thermal bath, the interaction hamiltonian can be splitted into two terms, $V = V_S + V_O$, which are respectively symmetric and antisymmetric under the interchange $a \leftrightarrow b$ of system operators. For the sake of conciseness, we shall refrain from writing down the whole expressions for these two interaction hamiltonians. Suffice it to say that each of them has two possible energy-conserving interaction verteces, both proportional to the interaction couplings u or v:

- There are inelastic two-body collisions that do not conserve the number of condensed particles (condensate feeding and depletion processes). There are in turn two possible diagrams for these processes, and they are shown in Figure 2.

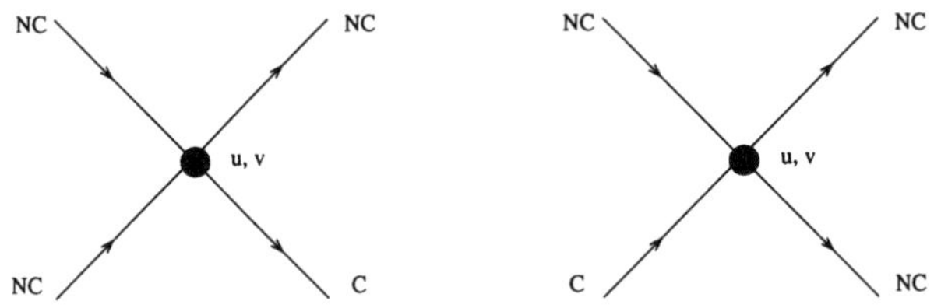

Figure 2: Inelastic two-body collisions between condensate and non-condensate particles

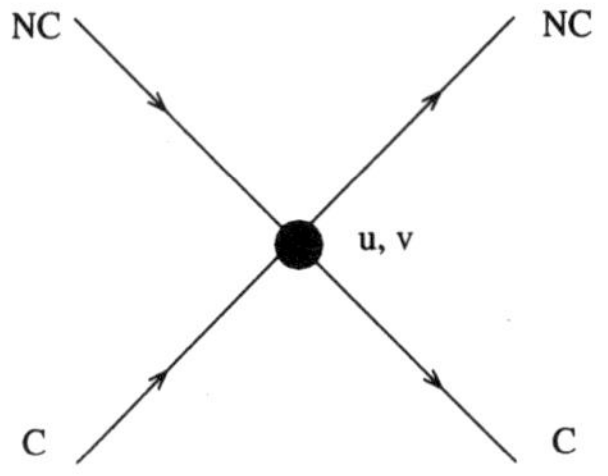

Figure 3: Elsatic two-body collisions between condensate and non-condensate particles

The first one is proportional to the occupation number of initial non-condensed states; since each occupation number is proportional to fugacity $z \equiv e^{\beta\mu}$ (where β is the inverse temperature of the thermal bath and μ its chemical potential), then the diagram is $O(z^2)$. The second one is proportional to the occupation number of the initial non-condensed state (that is, proportional to z), and energy conservation brings about another fugacity factor, so the final result is that this diagram is also $O(z^2)$ [14, 15].

- There are elastic two-body collisions that conserve the number of condensed particles, and are shown in Figure 3. These diagrams are proportional to fugacity, $O(z)$ [14, 15].

When fugacity is small (which occurs when the gap between the condensate and non-condensate single-particle levels is bigger that the thermal energy $k_{\rm B}T$), then the elastic processes dominate over the inelastic ones. The elastic collisions are also the most relevant ones in the computation of the decoherence rate, since we shall see that their contribution scales as N^2 (with N the total number of particles in the condensate), whereas the inelastic contribution scales as N. For these reasons, from now on we shall concentrate only on the elastic processes.

5 Decoherence-free pointer subspace in BEC

As we have mentioned in Section III, when $2\lambda \gg k_{\rm B}T$ the antisymmetric environmental states are nearly empty. Only the symmetric states are occupied. Since these latter states cannot distinguish between A and B, all collisions that involve symmetric non-condensed states will not destroy the quantum phase coherence between the Schrödinger cat's components. We refer to this condition as the perfect symmetrization limit. In this case, the states $|\pm\rangle \equiv (|N,0\rangle \pm |0,N\rangle)/\sqrt{2}$ span a decoherence-free pointer subspace (also known in the literature as decoherence-free subspace or, for short, DFS) of the Hilbert space, since they have degenerate eigenvalues of the interaction hamiltonian V [16, 17, 18]. Any state of that subspace can be written as $\alpha|N,0\rangle + \beta|0,N\rangle$, with α and β complex numbers. If $\mathcal{P}_{[\alpha|N,0\rangle+\beta|0,N\rangle]}$ denotes a projector onto that subspace, then

$$[V, \mathcal{P}_{[\alpha|N,0\rangle+\beta|0,N\rangle]}] = 0, \tag{9}$$

which means that any quantum superposition $\alpha|N,0\rangle + \beta|0,N\rangle$ in the subspace is an eigenstate of the interaction hamiltonian (a perfect pointer state), and as such will retain its phase coherence and last forever. The interaction Hamiltonian between the condensate and the thermal cloud is a sum of products of condensate operators and environmental operators. Only terms with symmetric environmental operators are relevant because the antisymmetric states are empty. The total Hamiltonian is symmetric with respect to $A \leftrightarrow B$ so, to preserve this symmetry, the relevant terms with symmetric environmental operators also contain symmetric condensate operators. The argument simplifies a lot for small fugacity where there is only one leading term with the $N_A + N_B$ condensate operator. The states $|\pm\rangle$ are its eigenstates with the same eigenvalue N. They are also (almost) degenerate eigenstates of the condensate Hamiltonian build out of $N_{A,B}$. The coherent transitions $A \leftrightarrow B$ break this degeneracy of $|\pm\rangle$ but the difference of their eigenfrequencies is negligible as compared to the usual condensate lifetime of $\sim$ 10s. In the next-to-leading order in fugacity there are symmetric interaction terms which change the number of condensed atoms. These terms drive the $|N,0\rangle$ and $|0,N\rangle$ states into slightly "squeezed-like" states $|S,0\rangle$ and $|0,S\rangle$ respectively [22]. There are also terms which exchange A with B. They give each state a small admixture of the opposite component. Superpositions of these are still decoherence-free pointer subspaces - there are no relevant antisymmetric operators to destroy their quantum coherence.

6 Decoherence comes into the scene

When the antisymmetric environmental states begin to be occupied, then the commutation relation (9) is only approximate and states within the subspace will decohere. Hence we must face the problem of calculating decoherence rates due to interactions with the environment.

We shall first consider the case of $\lambda = 0$, which corresponds to no Josephson-coupling being applied, and for which the two states $|\pm\rangle$ are exactly degenerate with respect to the condensate hamiltonian. In this case there is no difference between symmetric and antisymmetric environmental states. After some long but straightforward calculation which involves computing the master equation for the reduced density matrix of the condensed particles (see [12] for details), we find a lower bound for the decoherence rate

$$t_{\rm dec}^{-1} > 16\pi^3 \left(4\pi a^2 \frac{N_{\rm E}}{V} v_T\right) N^2, \tag{10}$$

where N is a number of condensed atoms, $v_T = \sqrt{2k_{\rm B}T/m}$ is a thermal velocity in the noncondensed thermal cloud, a is a scattering length, V is a volume of the trap, and $N_{\rm E}$ is a number of atoms in the thermal cloud,

$$N_{\rm E} \approx e^{\frac{\mu}{k_{\rm B}T}} \left(\frac{k_{\rm B}T}{\hbar\omega}\right)^3 . \tag{11}$$

This is a lower estimate since we have only considered terms to leading order in fugacity $O(z)$ and to leading order in condensate size $O(N^2)$. Next-to-leading order terms are $O(z^2)$ and $O(N)$, in agreement with Refs. [14, 15], so they were neglected here. Even without going into details of our derivation it is easy to understand where a formula like Eq.(10) comes from. N^2 is the main factor which makes the decoherence rate large. It comes from the master equation of the Bloch-Lindblad form $\dot{\rho} \sim -[N_{\rm A} - N_{\rm B}, [N_{\rm A} - N_{\rm B}, \rho]]$, with A and B the two internal states of the atoms. N^2 is the distance squared between macroscopically different components of the superposition $(|N,0\rangle + |0,N\rangle)/\sqrt{2}$ - the common wisdom reason why macroscopic objects are classical [1]. The factor in brackets in Eq.(10) is a scattering rate of a condensate atom on noncondensate atoms - the very process by which the thermal cloud environment learns the quantum state of the system.

Let us estimate the decoherence rate for a set of typical parameters: $T = 1\mu$K, $\omega = 50$Hz, and $a = 3\ldots5$nm. The thermal velocity is $v_{\rm T} \approx 10^{-2}$m/s. The volume of the trap can be approximated by $V = 4\pi a_{\rm return}^3/3$, where $a_{\rm return} = \sqrt{2k_{\rm B}T/m\omega^2}$ is a return point in a harmonic trap at the energy of $k_{\rm B}T$. We estimate the decoherence time as $t_{\rm dec} \approx 10^5{\rm sec}/(N_{\rm E}N^2)$. For $N_{\rm E} = 10^0 \ldots 10^4$ and $N = 10\ldots10^7$ it can

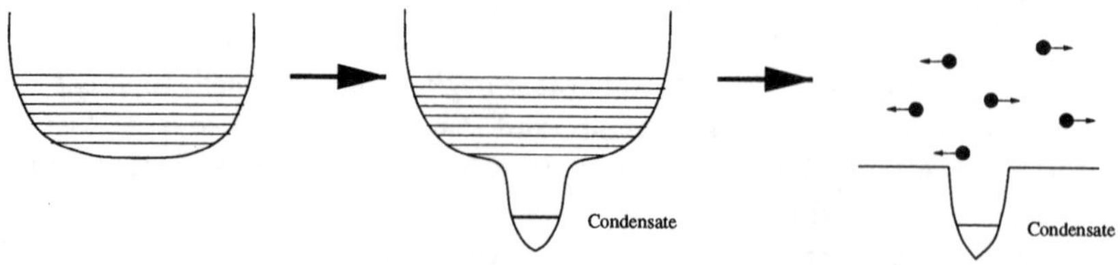

Figure 4: First step in engineering the trap: an optical dip is superimposed on a big magnetic trap, which is later opened so that most of the thermal atoms disperse away.

range from 1000s down to 10^{-13}s. For $N = 10$ our (over-)estimate for $t_{\rm dec}$ is large. However, already for $N = 1000$ and $N_{\rm E} = 10$ (which are still within the limits of validity of the two-mode approximation) $t_{\rm dec}$ shrinks down to milliseconds. Given that our $t_{\rm dec}$ is an upper estimate and that big condensates are more interesting as Schrödinger cats, it is clear that for the sake of cat's longevity, one must go beyond the standard harmonic trap setting.

7 Trap engineering

From Eqs.(10,11) it is obvious that the decoherence rate depends a lot on temperature and on chemical potential. The two factors strongly influence both $N_{\rm E}$ and $v_{\rm T}$. Both can be improved by the following scenario, which is a combination of present day experimental techniques. In the experiment of Ref.[19] a narrow optical dip was superposed at the bottom of a wide magnetic trap. The parameters of the dip were tuned so that it had just one bound state. The gap between this single condensate mode and the first excited state was 1.5μK, which at $T = 1\mu$K gives a fugacity of $z = \exp(-1.5)$. We need the gap so that we can use the single mode approximation. At low temperatures, the gap results in a small fugacity, which is convenient for calculations. We propose to prepare a condensate inside a similar combination of a wide magnetic and a narrow optical trap (or more generally: a wide well plus a narrow dip with a single bound mode) and then to open the magnetic trap and let the noncondensed atoms disperse. The aim is to get rid of the thermal cloud as much as possible. A similar technique was used in the experiment of Ref.[20] (see Figure 4).

Let us estimate the ultimate limit for the efficiency of this technique. At the typical initial temperature of 1μK the thermal velocity of atoms is 10^{-2}m/s. An atom with this velocity can cross a 1μm dip in 10^{-4}s. If we wait for, say, 1s after opening the wide trap, then all atoms with velocities above 10^{-6}m/s will disperse away from the dip. A thermal velocity of 10^{-6}m/s corresponds to the temperature of $10^{-8}\mu$K. As the factor $N_{\rm E} v_T \sim T^{7/2}$ in Eq.(10), then already 1s after opening the wide trap the decoherence rate due to non-condensed atoms is reduced by a factor of 10^{-28}!

It is not realistic to expect such a "cosmological" reduction factor. The "dip" which is left after the wide harmonic trap is gone could be, for example, a superposition of an ideal dip plus a wide shallow well (which was a negligible perturbation in presence of the wide harmonic trap). The well would have a band of width ΔE of bound states which would not disperse but preserve their occupation numbers from before the opening of the wide trap. They would stay in contact with the condensate and continue to "monitor" its quantum state. Even if such a truncated environment happens to be already relatively harmless, there are means to do better than that.

Further reduction of the decoherence rate can be achieved by "symmetrization" of the environmental states. To this end we propose to turn on the Josephson-like coupling λ in order to induce coherent transitions from states A and B. This has already been achieved experimentaly [21]; in their case $\lambda \approx$ 1kHz. When $\Delta E \ll 2\lambda$ we are in the perfect symmetrization limit, and hence no decoherence takes place. Indeed, the symmetric and antisymmetric ΔE-bands of states can be visualized as two ladders shifted with respect to each other by 2λ. In other words, the two sets of states feel the same, but shifted, trapping potentials. When $\Delta E \ll 2\lambda$, then the antisymmetric O_s's are nearly empty since they can evaporate into symmetric states and then leave the trap . The symmetric S_s's cannot distinguish between A and B so they do not destroy the quantum coherence between the Schrödinger cat's components. After symmetrization, $N_{\rm E}$ in Eq.(10) has to be replaced by the final number of atoms in the antisymmetric states only;

$$N_{\rm E}^{\rm O} \approx \begin{cases} n_\lambda \exp((\mu - \lambda)/k_{\rm B}T), & \text{for } 2\lambda < \Delta E \\ 0, & \text{for } 2\lambda > \Delta E \end{cases} \tag{12}$$

Here T is the temperature before opening the wide trap. n_λ is the number of antisymmetric bound states which remain within the ΔE-band of symmetric states. Atoms in these antisymmetric bound states cannot disperse away. For $2\lambda > \Delta E$ this number n_λ is zero and there is no decoherence from the thermal cloud.

In Figure 5 we make a schematic graph of the steps we propose to engineer the trap in order to reduce decoherence due to the thermal cloud.

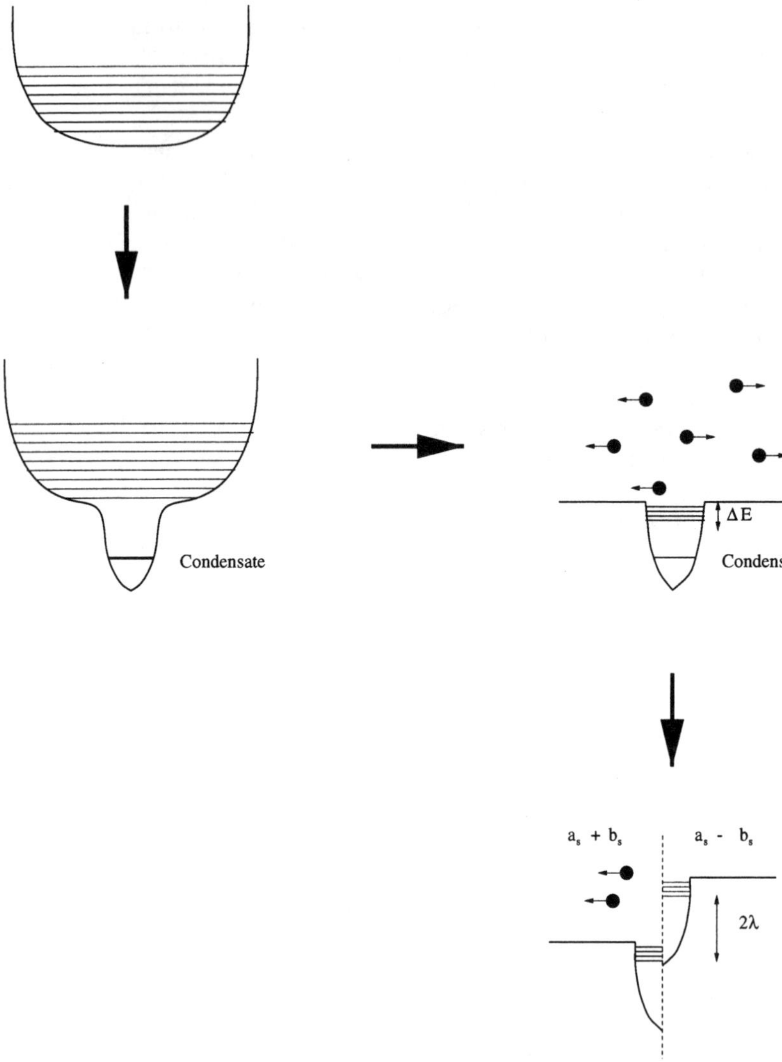

Figure 5: Second step in engineering the trap: After the magnetic trap is opened, there is still a band ΔE of thermal states in the "mouth" of the dip. Turning on the Josephson coupling λ shifts the energy levels of symmetric and antisymmetric thermal states, and the latter become empty.

8 Other sources of decoherence

Interaction with the thermal atoms is not the only source of decoherence for the quantum superposition state of the condensate. Among other possible sources we can mention:

- Ambient magnetic fields: the condensed atoms have magnetic moments. If the magnetic moments of A and B were different the magnetic field would distinguish between them and would introduce an unknown phase into the quantum superposition, thus rendering its underlying coherent nature undetectable. Fortunately the much used $|F, m_F\rangle = |2,1\rangle, |1,-1\rangle$ states of ^{87}Rb have the same magnetic moments. For them the magnetic field is a "symmetric" environment.

- Different scattering lengths: there is a typical $\approx 1\%$ difference between the $A - A$ and $B - B$ scattering lengths. The condensate Hamiltonian is not perfectly symmetric under $A \leftrightarrow B$. Even for a perfectly symmetrized environment, symmetric environmental operators couple to not fully symmetric condensate operators. This means that for the 1% difference of scattering lengths symmetrization can improve decoherence time by at most two orders of magnitude as compared to the unsymmetrized environment.

- Three-body losses: due to collisions involving three particles, the condensate is not stable but loses atoms via recombination into molecular states. Hence a condensate has a finite lifetime, of the order of $10 \ldots 20$s. The atoms which escape from the condensate carry information about its quantum state. They destroy its quantum coherence. In the experiment of Ref.[19] the measured loss rate per atom was 4/s for $N = 10^7$ or around 1 atom per 10^{-7}s. The last rate scales like N^3 so already for $N = 10^4$ just one atom is lost per second; decoherence time is 1s. One possibility for increasing this decoherence time is to increase slightly the dip radius. The loss rate scales like density squared so an increase in the dip width by a factor of 2 reduces the loss rate by a factor of $2^6 = 64$.

9 Discussion

The aim of this paper was to discuss the "longevity" of Schrödinger cats in BEC's. We have shown that while in the standard traps decoherence rates are significant enough to prevent long-lived macroscopic superpositions of internal states of the condensate, the strategy of trap engineering and symmetrization of the environment will be able to deal with that issue.

What remains to be considered is how one can generate such macroscopic quantum superposition, and how one can detect it. The issue of generation was already touched upon in Refs. [8, 9]. We have little to add to this. However, in the context of the Gordon and Savage proposal, it is fairly clear that the time needed to generate the cat state would have to be short compared to the decoherence time. If our estimates of Eq.(10) are correct, symmetrization procedure appears necessary for the success of such schemes.

Detection of Schrödinger cat states is perhaps a more challenging subject. In principle, states of the form $(|N,0\rangle + |0,N\rangle)/\sqrt{2}$ have a character of GHZ states, and one could envision performing measurements analogous to those suggested in [27] and carried out in [28], where a 4-atom entangled state was studied. However, this sort of parity-check strategy, appropriate for $N \leq 10$, is likely to fail when N is larger, or when (as would be the case for the "quasi-squeezed" states anticipated here [22]) N is not even well defined.

A strong circumstantial evidence can be nevertheless obtained from two measurements. The first one would consist of a preparation of the cat state, and of a measurement of the internal states of the atoms. It is expected that in each instance all (to within the experimental error) would turn up to be in either A or B states. However, averaged over many runs, the number of either of these two alternatives would be approximately equal. Decoherence in which the environment also "monitors" the internal state of the atoms in the A versus B basis would not influence this prediction. We need to check separately whether the cat state was indeed coherent. To do this, one could evolve the system "backwards". However, this is not really necessary. For, as Gordon and Savage point out, when, in their scheme, we let the system evolve unitarily for more or less twice the time needed for the generation of the cat state, it will approximately return to the initial configuration. Thus, we can acquire strong evidence of the coherence of the cat provided that this unitary return to the initial configuration can be experimentally confirmed.

These are admittedly rather vague ideas, which serve more as a "proof of principle" rather than as a blueprint for an experiment. Nevertheless, they may, we hope, encourage more concrete investigation of such issues with a specific experiment in mind.

10 Acknowledgements

We are indebted to E.Cornell, R.Onofrio, E.Timmermans, and especially to J.Anglin for very useful comments. This research was supported in part by NSA.

References

[1] W.H. Zurek, Physics Today **44**, 36 (1991); Phil.Trans.Roy.Soc. London **A 356**, 1793 (1998); D. Guilini *et al*, Decoherence and the Appearance of a Classical World in Quantum Theory (Springer, Berlin, 1996).

[2] M. Brune *et al*, Phys.Rev.Lett.**77**, 4887 (1996).

[3] M. Arndt *et al*, Nature **401**, 680 (1999).

[4] C. J. Myatt *et al*, Nature **403**, 269 (2000).

[5] J. R. Friedman *et al*, cond-mat/0004293.

[6] For a review see W. Ketterle *et al*, cond-mat/9904034. See also E.A. Cornell *et al*, J. Low Temp.Phys. **113**, 151 (1998).

[7] J. Stenger *et al*, Nature **396**, 345 (1998).

[8] J.I. Cirac *et al*, Phys.Rev.**A57**, 1208 (1998)

[9] D. Gordon and C.M. Savage, Phys.Rev.**A59**, 4623 (1998).

[10] J. Ruostekoski *et al.*, Phys.Rev.**A57**, 511 (1998).

[11] J. Ruostekoski, cond-mat/0005469.

[12] D. A. R. Dalvit, J. Dziarmaga, and W. H. Zurek, Phys.Rev.**A62**, 13607 (2000).

[13] M.J. Steel and M.J. Collett, Phys.Rev.**A57**, 2920 (1998).

[14] J. Anglin, Phys.Rev.Lett.**79**, 6 (1997).

[15] J. Ruostekoski and D.F. Walls, Phys.Rev.**A58**, R50 (1998).

[16] W.H. Zurek, Phys.Rev.**D24**, 1516 (1981); *ibid.* **26**, 1862 (1982).

[17] P. Zanardi and M. Rasetti, Phys.Rev.Lett.**79**, 3306 (1997).

[18] D.A. Lidar *et al*, Phys.Rev.Lett.**81**, 2594 (1998).

[19] D.M. Stamper-Kurn *et al*, Phys.Rev.Lett.**81**, 2194 (1998).

[20] D.M. Stamper-Kurn *et al*, Phys.Rev.Lett. **80**, 2027 (1998).

[21] D.S. Hall *et al*, Phys.Rev.Lett.**81**, 1539 (1998); D.S. Hall *et al*, Phys.Rev.Lett.**81**, 1543 (1998).

[22] P. Navez, Mod.J.Phys.Lett. **B12**, 705 (1998). A.I. Salomon, Phys.Rev.**B60**, 3044 (1999).

[23] E.Cornell, private communication.

[24] D.F. Walls and G. Milburn, Quantum Optics, Springer Verlag 1995.

[25] J.C. Eilbeck *et al*, Physica D **16**, 318 (1985); G.J. Milburn *et al*, Phys.Rev.A**55**, 4318 (1997).

[26] C.W. Gardiner and P. Zoller, cond-mat/9905087.

[27] N.D.Mermin, Phys.Rev.Lett. **65**, 1838 (1990).

[28] C.A.Sackett *et al*, Nature **404**, 256 (2000).

Decoherence and its Implications in Quantum Computation and Information Transfer
T. Gonis and P.E.A. Turchi (Eds.)
IOS Press, 2001

Mixed-state entanglement and quantum information theory

Ryszard HORODECKI
Institute of Theoretical Physics and Astrophysics,
University of Gdańsk, 80–952 Gdańsk, Poland

Abstract

The basic aspects of the structure of noisy entanglement are presented in the context of positive maps and quantum communication. The discontinuity in the structure due to bound entanglement is emphasised. The fundamental irreversibility of the distillation process is also considered in the context of thermodynamical analogies.

1 Introduction

Entanglement [1, 2] is one of the most intriguing phenomena leading to various quantum effects with possible applications in quantum theory. It is a property of compound quantum system by virtue of which the local subsystem exhibit correlations that cannot be explained in classical terms. Then, the pure entanglement state cannot be written as a product of states of the individual subsystems. It is a direct consequence of tensorial description of the system in terms of Hilbert space and superposition principle.

From pragmatic point of view entanglement is the key to performing tasks that cannot be realized classically [3, 4]. A common source of entanglement is phase-matched parametric down conversion producing entangled two-photon states. It has lead to an explosion of the number of experiments studying various aspects of entanglement, in particular, in the context of the quantum information and quantum communication.

The notorious problem is that decoherence effects due to environment convert the pure entangled state into statistical mixture and degrade quantum entanglement [5, 6]. For example suppose that environment represented by the one-half spin particle in the state $|\uparrow_E\rangle$ "attacks" the B-particle from AB pair in the singlet state. Then the controlled-not gate U_{BE}^{C-NOT} generates entanglement between AB system and environment E.

$$|\Psi_{AB}^{singlet}\rangle|\uparrow_E\rangle \stackrel{I_A\otimes U_{BE}^{C-NOT}}{\longrightarrow} |\Psi_{ABE}^{out}\rangle = \frac{1}{\sqrt{2}}(|\uparrow_A\downarrow_B\downarrow_E\rangle - |\downarrow_A\uparrow_B\uparrow_E\rangle) \neq |\Phi\rangle_{AB}|\tilde{\phi}_E\rangle$$

In result, after averaging over environment we obtain statistical mixture of product states.

$$\varrho_{AB} = Tr_E(|\Psi^{out}_{ABE}\rangle\langle\Psi^{out}_{ABE}|) = \frac{1}{2}(|\uparrow_A\downarrow_B\rangle\langle\uparrow_A\downarrow_B| + |\downarrow_A\uparrow_B\rangle\langle\downarrow_A\uparrow_B|).$$

Thus here initial entanglement has been completely destroyed. In general, however, some residual noisy entanglement can be kept by final (mixed) state. It opens a package of fundamental problems:

- What does it mean that a given state ϱ is entangled (or inseparable)?
- Are there any operational separability criteria?
- What is the structure of the noisy entanglement?
- Is the noisy entanglement useful for quantum communication?
- Are there basic measures of entanglement?
- Are there any fundamental laws governing processing of entanglement?

Obviously it is not possible to present here the panorama of the results related to those problems. Therefore I will restrict myself to the presentation of only some of them.

2 Scalar separability criteria

Clearly, the first question comes to providing a suitable definition of separability. It was introduced by Werner for bipartite systems [7]. More generally we have the following

Definition 1. A state belonging to parties A,B,C, ... is said to be inseparable (entangled) if it cannot be written in separable form

$$\varrho = \sum_i p_i \varrho_i^A \otimes \varrho_i^B \otimes \varrho_i^C \ldots \tag{1}$$

for any possible probabilities p_i with $\sum_i p_i = 1$ and set of density matrices $\{\varrho_i^A, \varrho_i^B, \varrho_i^C, \ldots\}$, where ϱ_i^A operates on the Hilbert space $\mathcal{H}_A$ and so on.

Unfortunately the above definition does not provide operational criterion of separability. In fact given a state it is very hard to say if the state is separable or not. In this context it is natural to ask: Are there any *operational* separability criteria? By term "separability criteria" we mean

Definition 2. A necessary condition for separability (i.e.. a condition satisfied by all separable states) is called *separability criterion.*

Historically, the first separability criteria were scalar ones. In 1989 Werner has proved that separable state ϱ satisfies all possible Bell inequalities [7]. Moreover Werner has provided spectacular criterion in terms of the "flip" operator V. Than separable state ϱ satisfies inequality.

$$\mathrm{Tr} V\varrho \geq 0, \tag{2}$$

where V defined as $V\phi\otimes\psi = \psi\otimes\phi$ acts on the $\mathcal{H} = C^d \otimes C^d$, $dim\mathcal{H} = d^2$.

Another scalar criteria of separability was introduced in terms of entropic inequalities [8]

$$S_\alpha(\varrho_A) \leq S_\alpha(\varrho), \quad S_\alpha(\varrho_B) \leq S_\alpha(\varrho). \tag{3}$$

Here $\varrho_{A(B)} = \mathrm{Tr}_{B(A)}\varrho$ and $S_\alpha = (1-\alpha)^{-1}\log \mathrm{Tr}\varrho^\alpha$ is quantum α-entropy. For $\alpha = 0, 1, 2, \infty$ we have

$$\begin{aligned} S_0 &= \log R(\varrho) \\ S_1 &= -\mathrm{Tr}\varrho\log\varrho \\ S_2 &= -\log \mathrm{Tr}\varrho^2 \\ S_\infty &= -\log||\varrho||\,, \end{aligned}$$

where $R(\varrho)$ denotes the rank of the state ϱ (number of non-vanishing eigenvalues). and $||\cdot||$ is operator norm. It has been shown [8], that for any separable state on the finite dimensional Hilbert space inequalities are satisfied for α=0,1,2,∞. In some cases entropic criteria are stronger than the Bell ones. For example it is easy to check that the following orthogonal mixture [9]

$$\varrho = \sum_i p_i|\psi_i\rangle\langle\psi_i|, \quad \sum_i p_i = 1 \tag{4}$$

where

$$\begin{aligned} |\psi_1\rangle &= ae_1\otimes e_1 + be_2\otimes e_2; \quad a, b > 0 \\ |\psi_2\rangle &= ae_1\otimes e_2 + be_2\otimes e_1; \quad \{e_i\} \text{ a standard basis in } C^2. \end{aligned}$$

satisfies Clauser-Horne-Shimony-Holt (CHSH) inequality [10] whereas it violates the 2-entropic inequality. In general however the above scalar criteria are independent. As we will see further, they are weaker than structural ones being related to the structure of the density operator itself.

3 Structural separability criteria

3.1 Partial transposition criterion

At that time the meaning of the Bell and entropic inequalities was more or less clear, the Werner's condition was quite puzzling though. While some physicists were dealing with scalar criteria they were surprised by Peres' structural criterion [11]. It says that

Theorem 1. *If a state ϱ is separable, it has positive partial transposition i.e. the eigenvalues of the partial transposition are nonnegative.*

Here ϱ^{T_B} is defined by the matrix elements.

$$\langle e_m\otimes e_\mu|\,\varrho^{T_B}\,|e_n\otimes e_\nu\rangle = \langle e_m\otimes e_\nu|\,\varrho\,|e_n\otimes e_\mu\rangle. \tag{5}$$

Although the matrix ϱ^{T_B} depends on the used bases, its eigenvalues do not. Consequently for any state the condition can be checked using an arbitrary orthonormal basis. It has been shown [11], that partial transposition criterion is essentially stronger than the scalar ones. Thus there was a question: why transposition is such a good "detector" of entanglement? In the same time somewhere on the shelves of libraries there were papers on positive maps, ignored by physicists as not useful. The key observation was that the transposition is positive map.It has lead to the revealing a fundamental connection between quantum entanglement and the theory of positive maps [12, 13, 14, 15].

3.2 Positive and completely positive maps.

Peres' partial transposition criterion strongly suggested that some positive maps can play a role of detectors of entanglement and that the necessary and sufficient condition of separability should be expressed in terms of positive maps. Let $\mathcal{B}(\mathcal{H}_n), \mathcal{B}^+(\mathcal{H}_m)$ and Λ: $\mathcal{B}(\mathcal{H}_n) \rightarrow \mathcal{B}(\mathcal{H}_m)$ are set of linear operators on n-dimensional complex Hilbert space, set of positive linear operators belonging to $\mathcal{B}(\mathcal{H}_n)$ and linear map respectively.

Definition 3. A map Λ is said to be positive if it maps positive operators into positive ones $\Lambda : \mathcal{B}^+(\mathcal{H}_n) \rightarrow \mathcal{B}^+(\mathcal{H}_m)$.

As I mentioned, positive maps were not of interest for physicists. The relevant maps were the so-called *completely* positive ones for them. The latter satisfy stronger condition. Namely, they must remain positive if tensored with identity.

Definition 4. Λ is completely positive (CP) when induced map

$$I_k \otimes \Lambda : B(\mathcal{H}_k \otimes \mathcal{H}_n) \rightarrow B(\mathcal{H}_k \otimes \mathcal{H}_m)$$

is positive for all $k = 1, 2, 3...$

However, in the present context the CP maps are *not* interesting, as they do not feel entanglement! We will need yet another notion. Namely, by *decomposable* map, $\Lambda : B(\mathcal{H}_n) \rightarrow B(\mathcal{H}_m)$ we mean the map that can be written in the form [12, 14]

$$\Lambda = \Lambda_1^{CP} + \Lambda_2^{CP} \circ T$$

where Λ_1^{CP} and Λ_2^{CP} are completely positive maps and T is transposition. It is remarkable, that there are positive maps that are not completely positive. An example is transposition. It is positive because if P is a one-dimensional projector so is P^T. However $(I \otimes T)P_+$ is not positive map. Namely the operator

$$(I \otimes T)P_+, \quad \text{where } P_+ = |\psi_+\rangle\langle\psi_+|, \quad \psi_+ = \frac{1}{\sqrt{d}}\sum_i |i\rangle \otimes |i\rangle$$

is not positive despite P_+ is a positive operator.

4 Separability of mixed states: necessary and sufficient condition

4.1 Theorems "Λ" and "A"

There was the idea that if a state ϱ is entangled, then there is a positive map Λ such that $(I \otimes \Lambda)\varrho$ is no longer a state, i.e. it is not positive. This idea has materialised in two complementary theorems.

Theorem "Λ"		Theorem "A"
A state ϱ acting on $\mathcal{H}\otimes\mathcal{H}$ is separable iff $(I\otimes\Lambda)\varrho\geq 0$ for all positive maps $\Lambda : B^{+}(\mathcal{H})\rightarrow B^{+}(\mathcal{H})$	$\Leftrightarrow$	*A state ϱ acting on $\mathcal{H}\otimes\mathcal{H}$ is separable iff $TrA\varrho\geq 0$ for all $A\in B(\mathcal{H}\otimes\mathcal{H})$ such that for any separable σ $TrA\sigma\geq 0$*

Note that the theorem "Λ" involves a set of *structural* conditions while complementary theorem "A" represents a set of *scalar* conditions. Both the theorems are equivalent if considered collectively. In particular the operator "A" was called "entanglement witness". The link is formed by an isomorphism between the set of "entanglement witnesses" and the set of positive maps which can be expressed in the form [16]

$$A = I\otimes\Lambda^{\dagger}\tilde{P}_{+}. \tag{6}$$

where $\tilde{P}_{+}=|\tilde{\psi}_{+}\rangle\langle\tilde{\psi}_{+}|$ and $\tilde{\psi}_{+}=\sum_i|i\rangle\otimes|i\rangle$. Then having A one can obtain Λ and vice versa.

4.2 Flip operator V as "entanglement witness"

Please note that the above theorems together with the isomorphism (6) solve a puzzle of Werner' scalar condition of separability i.e. $TrV\varrho\geq 0$ for any separable ϱ. Now it is clear that "flip" operator V is nothing as "entanglement witness". Indeed for some states σ we have $TrV\sigma<0$, hence from theorem "A" it follows that σ is entangled. Note however that V is not universal. In fact there are entangled states for which still $TrV\varrho\geq 0$, for example ψ_+. Notwithstanding we can apply isomorphism. In result we obtain

$$V \quad\rightarrow\quad \Lambda_V = T \quad \text{(transposition!)}$$

In this way we obtained much stronger condition

$$(I\otimes T)\varrho\geq 0,$$

being nothing as structural partial transposition condition. Thus we see, that though characterisation "Λ" and "A" are both equivalent if considered collectively, the single conditions (i) $Tr(\varrho A)\geq 0$ (ii) $(I\otimes\Lambda_A)(\varrho)\geq 0$ are not.

4.3 Operational characterizations of entanglement in low dimensions ($2\otimes 2$ and $2\otimes 3$ systems)

As I already mentioned above, not all positive maps are good "detectors" of entanglement. For example CP maps don't feel entanglement at all. In this context it is convenient to use the following lemma.

Lemma. If Λ is decomposable, then the condition $(I \otimes \Lambda)\varrho \geq 0$. is equivalent to the condition $(I \otimes T)\varrho \geq 0$.

Fortunately, Størmer [13] and Woronowicz [14] showed that *all* positive linear maps $\Lambda : B^+(\mathcal{H}_2) \to B^+(\mathcal{H}_2)$ and $\Lambda : B^+(\mathcal{H}_2) \to B^+(\mathcal{H}_3)$ are decomposable: $\Lambda = \Lambda_1^{CP} + \Lambda_2^{CP} \circ T$. Then from this fact and the above lemma it follows [11, 17]

Theorem 2. *A state of* $2\otimes 2$ *or* $2\otimes 3$ *system is separable iff its partial transposition is positive operator.*

This theorem provides very simple operational criterion of separability of mixed states for low dimensional systems. It is stronger than the scalar criteria. In particular for states both criteria 2-entropic and partial transposition ones are equivalent. However it is not hard to see that for the following states

$$\varrho = p|\psi_-\rangle\langle\psi_-| + (1-p)|\uparrow\uparrow\rangle\langle\uparrow\uparrow|$$

with $|\psi_-\rangle = 1/\sqrt{2}(e_1 \otimes e_2 - e_2 \otimes e_1)$, $|\uparrow\uparrow\rangle = e_1 \otimes e_2$, the partial transposition criterion is much stronger than the 2-entropic one. Indeed, applying the above theorem one finds that the state ϱ is separable only for $p = 0$. In this case the 2-entropic criterion gives $p \geq 1/3$

5 Entangled states with positive partial transposition.

5.1 Range criterion

Since Størmer-Woronowicz characterisation of positive maps applies only to low dimensional systems, it follows that for higher dimensions positive partial transposition (PPT) will not constitute necessary and sufficient condition of separability. It implies that there exist PPT entangled states. Thus for low dimensions PPT states are separable, and vice versa. On the other hand, the non-positive partial transposition states (NPT) are entangled and vice versa. However this doesn't hold for higher dimensions (see fig.1)

Now a basic question arises: how to check whether given PPT is entangled? Needless to say, we cannot use - the strongest so far - Peres criterion, as the state is to be PPT! Fortunately independent *range criterion* was found [18], which appeared to be stronger than PPT in some cases.

Theorem 3. *If a state acting on Hilbert space* $\mathcal{H}_A \otimes \mathcal{H}_B$ *is separable, then there exists a family of product vectors* $\{\phi_i \otimes \psi_i\}$ *such that they span range (or support) of* ϱ *and the vectors* $\{\phi_i \otimes \psi_i^*\}$ *span the support of* ϱ^{T_B}.

Here the star denotes the complex conjugation (the conjugation is taken in the same basis as the transposition). First explicit exampla of PPT entangled states was constructed in 1997 [18]. The following matrix represents a state $2 \otimes 4$ system and is written in the standard product basis $\{|ij\rangle\}$

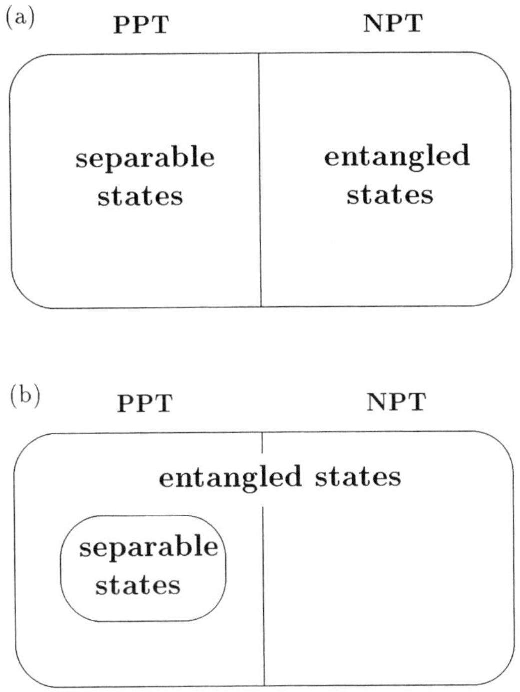

Figure 1: Structure of entanglement of mixed states for $2 \otimes 2$ and $2 \otimes 3$ system (a) and for higher dimensions (b)

$$\varrho_b = \frac{1}{7b+1}\begin{bmatrix} b & 0 & 0 & 0 & 0 & b & 0 & 0 \\ 0 & b & 0 & 0 & 0 & 0 & b & 0 \\ 0 & 0 & b & 0 & 0 & 0 & 0 & b \\ 0 & 0 & 0 & b & 0 & 0 & 0 & 0 \\ 0 & 0 & 0 & 0 & \frac{1+b}{2} & 0 & 0 & \frac{\sqrt{1-b^2}}{2} \\ b & 0 & 0 & 0 & 0 & b & 0 & 0 \\ 0 & b & 0 & 0 & 0 & 0 & b & 0 \\ 0 & 0 & b & 0 & \frac{\sqrt{1-b^2}}{2} & 0 & 0 & \frac{1+b}{2} \end{bmatrix}, \tag{7}$$

where $0 < b < 1$. One can show that $\varrho_b^{T_B} \geq 0$. But one can check that *none* of the product vectors belonging to the range of ϱ_b, if partially conjugated (as stated in theorem), belongs to the range of $\varrho_b^{T_B}$. One can see now that the state ϱ_b is PPT, and it violates the range criterion extremely.

Note that apart from the partial transposition and the range criteria there is another structural reduction criterion based on the positive maps $\Lambda(\varrho) = I\, Tr\varrho - \varrho$ [19, 20]. For dimensions higher than $2 \otimes 2$ and $2 \otimes 3$ this criterion appeares to be weaker than the above criteria. The importancy of the reduction criterion follow from the fact that any state violating the criterion can be distilled [19] (see in this context Ref.[21]).

5.2 Systematic construction of the PPT entangled states.

First systematic construction of the PPT entangled states was based on the concept of so-called unextendible product basis (UPB) [22]. It is defined as a set of product vectors in $\mathcal{H}_{AB}$ such that it has less elements than the dimension of the space $\mathcal{H}_{AB}$ and there does not exist any product vector orthogonal to all of them. A well known example of UPB basis for $3 \otimes 3$ system is given by

$$\begin{gathered} |v_0\rangle = \frac{1}{\sqrt{2}}|0\rangle|0-1\rangle, \quad |v_2\rangle = \frac{1}{\sqrt{2}}|2\rangle|1-2\rangle, \\ |v_1\rangle = \frac{1}{\sqrt{2}}|0-1\rangle|2\rangle, \quad |v_3\rangle = \frac{1}{\sqrt{2}}|1-2\rangle|0\rangle \\ |v_4\rangle = \frac{1}{3}|0+1+2\rangle|0+1+2\rangle\ . \end{gathered} \tag{8}$$

It can be seen, that any subset of three vectors spans the three-dimensional space of that party preventing any new vector from being orthogonal to all of them.

Having the UPB $\{w_i = \phi_i \otimes \psi_i\}_{i=1}^k$ constitute UPB for $d \otimes d$ system. One can construct PPT entangled states as follows First we write projector P onto the space $\mathcal{H}$ spanned by this UPB

$$P = \sum_i |w_i\rangle\langle w_i|.$$

Then we consider the state distributed on the complementary subspace $\mathcal{H}^\perp$ ($\dim \mathcal{H}^\perp = d^2 - k$):

$$\varrho = \frac{1}{d^2-k}P^\perp = \frac{1}{d^2-k}(I-P)$$

Now since w_i constitute UPB, the support of ϱ contains no product vectors. Then by the range criterion the state is entangled. It is easy to check that it is PPT. Although UPB's provide simple method to construct a wide class of PPT entangled states, not all such states can be constructed by this method. For example the above $2 \otimes 4$ PPT entangled states (5) cannot be constructed in this way (see in this context Refs. [23, 24]). Remarkably the presented UPB method developed first by Bennett *et al* [22]. has lead to first systematic construction of non-decomposable positives maps! [25]

6 Distillation of noisy entanglement

So far we have considered mathematical characterization of noisy entanglement. Let us now pass to the more physical description. Namely, we will check if the noisy entanglement is useful for quantum communication.

Imagine that Alice and Bob share n pairs of entangled particles, each of them in the same mixed state ϱ. Of course they cannot use those pairs for teleportation, as the state is mixed and fidelity of transmission would be poor. They however can reverse the process of degradation of entanglement, namely they can obtain the maximally entangled pairs by means of classical communication and local quantum operations (LQCC). This process called *distillation*, was first considered in pioneering paper by Bennett *et al.* [26]

Obviously it is not possible to get *ideal* singlet pairs. Instead one can require that the obtained pairs become more and more similar to maximally entangled ones if we consider large amount of initial pairs. More precisely if Alice and Bob share n pairs each in the state ϱ_{AB} than they can obtain m_n pairs in the joined state ϱ' (see fig.(2)).

Figure 2: Distillation scheme.

Then the similarity condition is expressed by the asymptotic condition for fidelity between the real state ϱ' of output pairs and desired product pure state

$$(\psi_+)^{\otimes m_n} : F_n(\varrho') \stackrel{n\to\infty}{\to} 1,$$

where

$$F_n(\varrho') = \underbrace{\langle\psi_+|\dots\langle\psi_+|}_{m_n} \varrho' \underbrace{|\psi_+\rangle\dots|\psi_+\rangle}_{m_n},$$

It appears that it is possible to obtain nonzero asymptotic yield of distilled maximally entangled pairs:

$$D_{\mathcal{P}} = \lim_{n\to\infty} \frac{m_n}{n}$$

Then distillable entanglement E_D is defined as the maximal possible yield that Alice and Bob can obtain

$$E_D(\varrho) = \max_{\mathcal{P}} D_{\mathcal{P}}.$$

Definition 5. The state ϱ_{AB} is distillable if there exist a protocol P that satisfies the above conditions.

7 Bound entanglement

In this context a fundamental question arises: *Can all mixed states be distilled ?* Clearly the separable states can be distilled as they do not contain entanglement at all. But what about noisy entanglement? The two following theorems provide the answer to this question [27].

Theorem 4. *Quantum state with positive partial transposition are non distillable.*

This means that PPT states are entangled but their entanglement is *bound* - it cannot be recovered by means of LQCC operations In this case the external noise was too weak to destroy entanglement completely. But it was strong enough to bind it.

It is important result, as it implies the existence of a new type of entanglement, that is masked in more subtle way than known so far, and the role of which is far from clear. It is natural to ask: how to characterise the distillable (or free) entanglement? The answer is given by the following

Theorem 5. *A state ϱ on the space $\mathcal{H}$ is distillable iff there exists n and two-qubit subspace of the space $\mathcal{H}^{\otimes n}$ such that $\varrho^{\otimes n}$ restricted to this subspace is entangled.*

The above theorem implies: *distillable (or free) entanglement is two-qubit entanglement. It is compatible with the earlier result* that any entangled (inseparable) $2 \otimes 2$ system represent entanglement which, however small, can be distilled to singlet form.

It must be emphasized here that the set of distillable states has not yet been characterised *operationally*. Thus it is not clear so far, whether there exists bound entangled NPT states. Finally the relations between the entanglement and distillability are presented on the diagrams (see fig.(3))

So, for low dimensions, all PPT states are separable and non-distillable, while all NPT states are free entangled. For higher dimensions the situation is less clear. Apart from (non-distillable) separable states they are bound entangled states. The slanted pattern indicates the hypothetical set of bound entangled NPT states.

7.1 Quantifying bound entanglement

To quantify the bound entanglement, let us recall definition of entanglement formation [28]. It is the minimal number of maximally entangled pairs needed to produce the state ϱ. Thus we have quite opposite situation to that of distillation: One starts with singlet

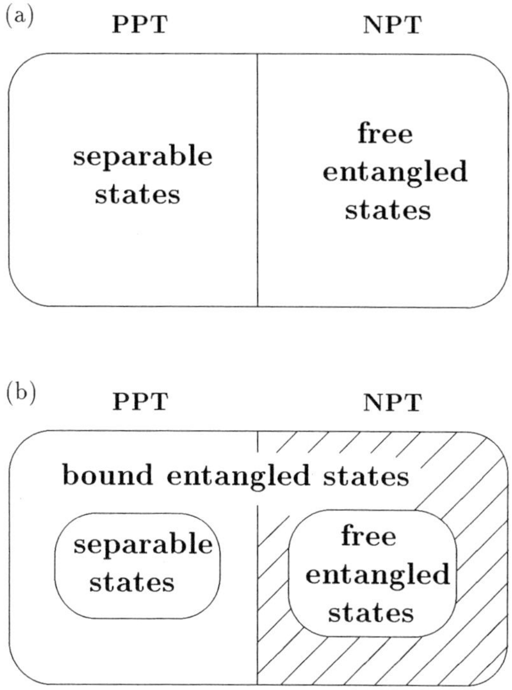

Figure 3: Entanglement and distillability of mixed states for $2 \otimes 2$ and $2 \otimes 3$ system (a) and for higher dimensions (b). The slanted pattern denotes the hypothetical set of bound entangled NPT states.

pairs, and wants to produce the largest number of pairs in the state ϱ (see fig.(4)). Again Alice and Bob cannot produce the pairs exactly in the states ϱ. But for larger and larger number of input pairs, they can obtain more and more similar state to ϱ. This is expressed by the condition for fidelity

$$F(\varrho^{\otimes n}, \varrho') \stackrel{n\to\infty}{\to} 1.$$

Now entanglement of formation E_F is defined as minimal yield of needed singlets.

$$E_F(\varrho) = \min\left\{\frac{\text{number of singlets}}{\text{number of output pairs}}\right\} = \min\left\{\lim_{n\to\infty}\frac{m_n}{n}\right\}$$

Having defined E_F it is natural to define the bound entanglement as a difference between E_F and free entanglement E_D [27]

$$E_B(\varrho) = E_F(\varrho) - E_D(\varrho).$$

This is reasonable definition which includes extremal cases. Namely, for pure entangled states entanglement of formation equals to free entanglement, $E_B = 0$. For bound entangled states entanglement of formation equals to bound entanglement, while $E_D =$

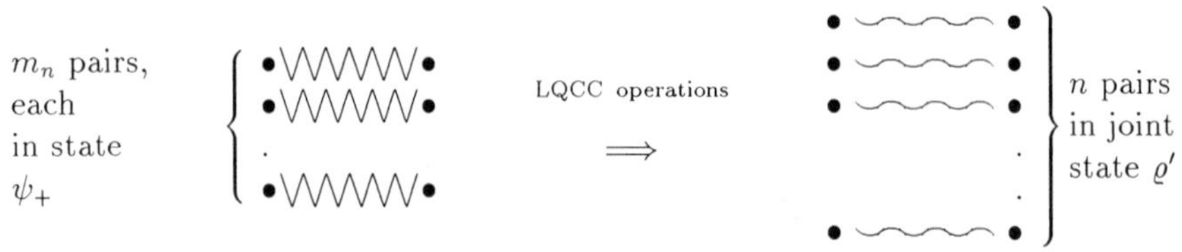

Figure 4: Formation scheme.

0. As we will see in a moment, in general, the mixed state can contain both kinds of entanglement, i.e. E_D, $E_B \neq 0$

In conclusion: There is discontinuity in the structure of noisy entanglement. Beginning with two spin 1 systems quantum mechanics implies existence of two qualitatively different types of entanglement: *free entanglement* E_D, that can be distilled and *bound entanglement*, that cannot be converted by LQCC operations to the singlet form.

8 Irreversibility of asymptotic entanglement manipulations.

Note that if bound entanglement is non-zero then to produce the state we need more singlets than we can obtain *back* from it. Thus in the process of formation some entanglement was irreversibly bound. Ironically, it has not yet been rigorously shown if the bound entangled states have non-zero bound entanglement!

Now it is reasonable to ask: can distillation process be irreversible for some states? Equivalently: can we have strict inequality between entanglement of formation and free entanglement, so that the bound entanglement is non-zero?

Quite recently, this sharp inequality was proven [29] for some subfamily of the Werner states of the form

$$\varrho = (1-p)P_{SYM} + pP_{ASYM}$$

where P_{SYM} and P_{ASYM} are projectors onto symmetric and antisymmetric space respectively. The very idea of the proof is clear: find functions $E_1(\varrho)$ and $E_2(\varrho)$ such that the following inequality is satisfied for some ϱ

$$E_D \leq E_1 < E_2 \leq E_F.$$

Fortunately, one can find such functions that for some Werner states satisfy the inequality. In particular $E_1(\varrho) = \log(2N+1)$ is built on the basis of the so-called *negativity* N [30, 31]. The latter is defined as the absolute value of the sum of the negative eigenvalues of partially transposed state.One can show that this measure is upper bound for distillable entanglement $E_D(\varrho)$.

In particular for the above Werner states the E_1 is given by

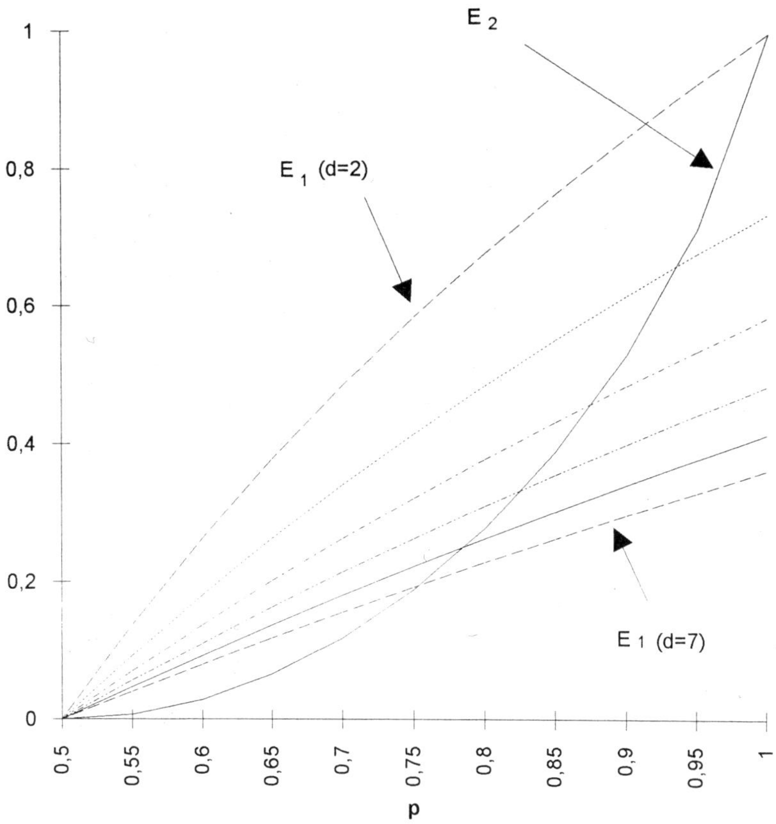

Figure 5: Irreversibility of distillation.

$$E_1(\varrho_W(p,d)) = \begin{cases} \log\left[\frac{2}{d}(2p-1)+1\right] & \text{for } p > \frac{1}{2} \\ 0 & \text{for } p \leq \frac{1}{2} \end{cases} \tag{9}$$

The function E_2 is a modification of relative entropy of entanglement in the context of bound entanglement $E_2(\varrho) = \inf_{\mathcal{S}} S(\varrho|\sigma)$ where the infimum is taken over all PPT states σ. It appears that entanglement of formation is upper bound for E_1. Moreover, for Werner states, E_2 is a lower bound for entanglement of formation.

$$E_2(\varrho_W(p,d)) = \begin{cases} 1 - H(p) & \text{for } p > \frac{1}{2} \\ 0 & \text{for } p \leq \frac{1}{2} \end{cases} \tag{10}$$

The relations between E_1 and E_2 for the family of Werner states of $d \otimes d$ system are shown on the fig.(5). Note that E_1 depends both on p and on the dimension d, while E_2 does not depend upon d. Now we see clearly, that for some subfamily of the Werner states (with larger d and p) there is a definite gap between distillable entanglement and entanglement of formation implying that bound entanglement is sharply greater than 0. Then we come to important conclusion : *Asymptotic entanglement manipulations exhibit genuine irreversibility.* It can be seen as a new kind of irreversibility due to quantum information processing. It must be noted here, that such irreversibility was discovered earlier in the case of pure states of the multiparticle systems [32, 33].

9 Thermodynamical analogies

The irreversibility of the distillation process comes to mind the analogy to irreversible thermodynamics processes. Some physicists believe that there should exist the lows governing entanglement processing in quantum communication analogous to those in the thermodynamics.

In particular entanglement-energy analogy was developed and the information counterpart of the first low of thermodynamics was proposed: *entanglement of compound system does not change under unitary processes on one of the subsystems.* Here we present some thermodynamical analogies as follows

(entanglement of formation) E_F	$\longleftrightarrow$	U (internal energy)
(free entanglement) E_D	$\longleftrightarrow$	F (free energy)
(bound entanglement) E_B	$\longleftrightarrow$	TS (bound energy)
sending qubits	$\longleftrightarrow$	work
$E_F - E_D > 0$	$\longleftrightarrow$	$U - F > 0$
$E_F = E_D + E_B$	$\longleftrightarrow$	$U = F + TS$ (Gibbs-Helmholtz equation)

Note, that the inequality $E_F - E_D > 0$ can be viewed as an analogy to irreversible thermodynamical analogue to irreversible thermodynamical processes where only the free energy (which is not equal to the total energy) can be converted to useful work. This strongly support the view according to which the equation connecting the quantities E_F, E_B and E_D can be regarded as quantum information counterpart of the thermodynamical Gibbs-Helmholtz equation where T and S are the temperature and the entropy of the system.

Recently basing on the postulates that (i) entanglement is a form of quantum information corresponding to internal energy (ii) sending qubits corresponds to work, it has been shown [34] that in the closed bipartite quantum communication systems the information is conserved. In this context it would be interesting to consider ontological aspects of quantum information, however [34] it goes beyond the scope of this paper.

10 Some aspects of bound entanglement

The above entanglement-energy analogy albeit formal, suggested in particular that entanglement can be pumped from one to other system. In particular it allowed to find a new quantum effect - activation of the bound entanglement that correspond to chemical activation process [35].

As we have seen, bound entanglement if applied directly remains useless in quantum communication. Then there was a question: can be bound entanglement somehow activated to produce any effect useful for practical purposes. The answer is "yes". Suppose that Alice and Bob have a huge number of bound entangled pairs of $d \otimes d$ system. Clearly, as bound entanglement cannot be distilled, Alice and Bob cannot obtain even a single pair in the state with fidelity greater than classical one, that is fidelity F greater than (1/d). Here F is overlap with the maximally entangled state.

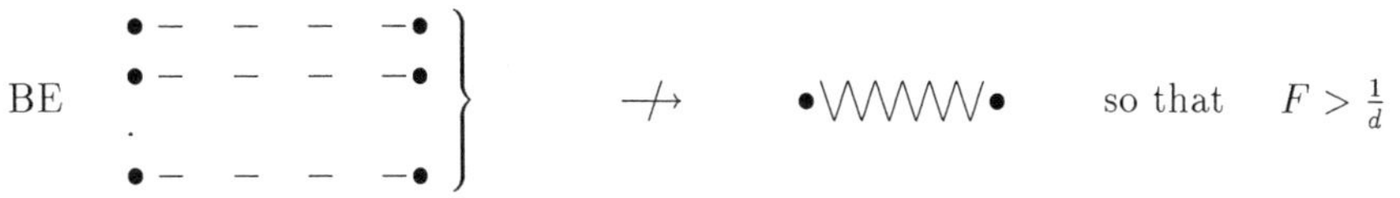

Suppose, now that Alice and Bob have only one free entangled pair, such that it is impossible to obtain fidelity greater than given fidelity F_0. So, both from large number of bound entangled pairs, and one free entangled pair, if alone, it is impossible to get arbitrarily high fidelity.

Now, it appears, if we put the two things together, then we *can* obtain the pair of arbitrarily high fidelity.

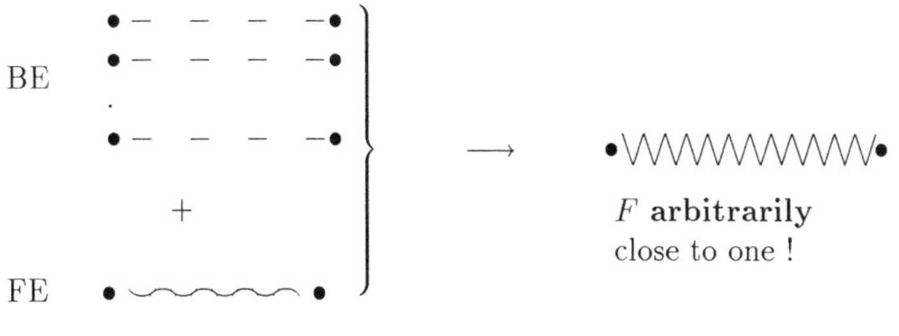

Thus, we can say that bound entanglement has been activated by the single free entangled pair.

It has been suggested [35] that distillable entanglement E_D may be superadditive. Quite recently Shor *et al* [36] proved that in the case of more than two parties the bound entanglement can be truly activated in the sense of being superadditivity of distillable entanglement. Namely it appeared that differ bound enanglement state, tensored together can make a distillable state. Thus new nonclassical effect was called *superactivation*. On the other hand Murao and Vedral [37] showed that Smolin's four-particle unlocked bound entangled state can be used for remotely concentrating of information.

In above context it would be interesting to know whether the similar effects can be obtained for bipartite systems.

References

[1] A. Einstein, B. Podolsky and N. Rosen, Phys. Rev. **47**, 777 (1935).

[2] E. Schrödinger, Proc. Cambridge Philos. Soc. **31**, 555 (1935).

[3] A. K. Ekert, Phys. Rev. Lett. **67** (1991) 661; C. H. Bennett, S. J. Wiesner: Phys. Rev. Lett. **69** (1992) 2881; C. H. Bennett, G. Brassard, C. Crepeau, R. Jozsa, A. Peres and W. K. Wootters, Phys. Rev. Lett. **70**, 1895 (1993); P. W. Shor: in *Proceedings of the 35th Annual Symposium on the Foundations of Computer Science* (IEEE Computer Society, Los Alamos, CA, (1994) p.124; R. Cleve and H. Burhmann, Phys. Rev. A **56**, 1201 (1997); L. K. Grover, Phys. Rev. Lett. **79**, 4709-4712 (1997).

[4] C. H. Bennett *et all.*, J. Cryptol. **5**, 3 (1992); P. D. Townsend, J. G. Rarity, and P. R. Tapster, Electron. Lett. **29**, 1291 (1993); D. S. Naik *et all.*, quant-ph/9912105; T. Jaunewein *et all.*, quant-ph/9912117; N. A. Gershenfeld and I. L. Chuang, Science **275**, 350 (1997); D. G. Cory, A. F. Fahmy and T. Havel, Proc, Natl. Acad. Sci. USA. **94**, 1634 (1998); I. L. Chuang, L. M. K. Vandersypen, X. Zhou, D. W. Leung and S. Lloyd, Nature **393**, 143 (1998); J. A. Jones, M. Mosca, and M. H. Hansen *ibid* **393**, 344 (1998); D. Bouwmeester, J.W. Pan, K. Mattle, M. Elbl, H. Weinfurter and A. Zeilinger, Nature (London) **390**, 575 (1997); D. Boschi, S. Brance, F. De Martini, L. Hardy and S. Popescu, Phys. Rev. Lett. **80**, 1121 (1998); A. Furusawa *et all.*, Science **282**, 706 (1998).

[5] W. H. Żurek, Physics Today **44**, 36 (1991).

[6] R. Alicki, K. Lendi, *Quantum dynamical semigroups and applications*, Springer-Verlag 1987.

[7] R. F. Werner, Phys. Rev. A **40**, 4277 (1989).

[8] R. Horodecki and M. Horodecki, Phys. Rev. A **54**, 1838 (1996).

[9] R. Horodecki, Phys. Lett. A **210**, 223 (1996).

[10] J. F. Clauser, M. A. Horne, A. Shimony and R. A. Holt, Phys. Rev. Lett. **23**, 880 (1969).

[11] A. Peres, Phys. Rev. Lett. **77**, 1413 (1996).

[12] M. D. Choi, Linear Algebra Appl. **12**, 95 (1975).

[13] E. Størmer, Acta. Math. **110**, 233 (1963); E Størmer, Proc. Amer. Math. Soc. **86**, 402 (1982).

[14] S.L. Woronowicz, Rep. Math. Phys. **10**, 165 (1976).

[15] H. Osaka, Linear Algebra Appl. **135**, 73 (1991).

[16] A. Jamiołkowski, Rep. Math. Phys. **3**, 275 (1972).

[17] M. Horodecki, P. Horodecki and R. Horodecki, Phys. Lett. A **223**, 1 (1996).

[18] P. Horodecki, Phys. Lett. A **232**, 333 (1997).

[19] M. Horodecki and P. Horodecki, Phys. Rev. A **59**, 4206 (1999).

[20] N. Cerf, C. Adami and R. M. Gingrich, Phys. Rev. **60**, 898 (1999).

[21] G. Giedke, L. Duan, J. I. Cirac and P. Zoller, quant-ph/0007061.

[22] C. H. Bennett, D. DiVincenzo, T. Mor, P. Shor, J. Smolin and B. Terhal, Phys. Rev. Lett. **82**, 5385 (1999).

[23] D. Bruß and A. Peres, Phys. Lett. A **61**, 030301R (2000).

[24] J. Smolin, quant-ph/0001001.

[25] B. Terhal, Quantum Algorithms and Quantum Entanglement, PhD Thesis, Universiteit van Amsterdam, Amsterdam (1999).

[26] C. H. Bennett, G. Brassard, S. Popescu, B. Schumacher, J. Smolin and W. K. Wootters, Phys. Rev. Lett. **76**, 722 (1996).

[27] M. Horodecki, P. Horodecki and R. Horodecki, Phys. Rev. Lett. **80**, 5239 (1998).

[28] C. H. Bennett, D. P. DiVincenzo, J. Smolin, and W. K. Wootters, Phys. Rev. A **54**, 3824 (1996).

[29] M. Horodecki, P. Horodecki and R. Horodecki, Phys. Rev. Lett. **84**, 4260 (2000).

[30] Życzkowski, P. Horodecki, A. Sanpera and M. Lewenstein, Phys. Rev. A **58**, 883 (1998).

[31] G. Vidal, private communication.

[32] C. H. Bennett, S. Popescu, D. Rohrlich, J. A. Smolin, and A. V. Thapliyal quant-ph/990807.

[33] N. Linden, S. Popescu, B. Schumacher and M. Westmoreland quant-ph/9912039.

[34] R. Horodecki, M. Horodecki and P. Horodecki, quant-ph/0002021.

[35] P. Horodecki, M. Horodecki and R. Horodecki, Phys. Rev. Lett. **82**, 1046 (1999).

[36] P. W. Shor, J. A. Smolin and A. V. Thapliyal, quant-ph/0005117.

[37] M. Murao and V. Vedral, quant-ph/0008078.

Decoherence and its Implications in Quantum Computation and Information Transfer
T. Gonis and P.E.A. Turchi (Eds.)
IOS Press, 2001

Prospects for Macroscopic Quantum Coherence

Anupam Garg

Department of Physics and Astronomy,

Northwestern University,

Evanston, Illinois 60208, U.S.A.

Abstract

Whether macroscopically distinct states can exist in a superposition is a long standing problem in quantum theory. Two recent experiments, one on a magnetic macromolecule, and another on an rf SQUID, provide dramatic instances of such superpositions. These experiments do not, however, enable us to test a macrorealistic Bell inequality, whose violation would rule out classically-minded theories in which a macroscopic system *is* in a definite state at all times.

1. Introduction

Right from the birth of the subject, the quantum measurement problem has been recognized as the most difficult to unravel in quantum mechanics. It leads to a violent clash with our common sense, as evidenced by the language and thought experiments that have been constructed in the attempts to come to grips with it. Thus we have the paradoxes of Schrödinger's cat and Wigner's friend, and Einstein's question 'Is the moon there when no body looks?', which are disturbing because we do not really *know* the outcome of an actual experimental test, and we are unwilling to *believe* that the cat really could be in a superposition of states, or that the moon really might not be there when no one is looking. Since even contemplating experiments with cats or the moon is ridiculous, we are at an impasse.

This situation has driven a large amount of work over the last two decades, both experimental and theoretical, to study and find other systems, not as fantastic as a cat, but still nevertheless such that they might reasonably be termed "macroscopic", in which superpositions of macroscopically distinct states might be observed. Still, it does not appear to this author that the fundamental conundrum has gone away despite these studies, or the fascinating new directions for research opened up by the idea of quantum computation. To the question asked by Leggett [1–4], 'Do superpositions of macroscopically distinct states exist?' the experimental answer seems to be a fairly loud 'yes', though one can still legitimately ask whether the systems which have been prepared in superposition states, a single ion in a trap, an atom plus a photon in a microcavity, a single molecule of Fe_8, or even an rf SQUID with 10^{10} electrons, really deserves the appelation of Schrödinger's cat. (None of these systems seems to evoke the same sense of revulsion as the example of the cat [5].) The experiments are still very far from the realm of everyday experience, and we still do not know if there is a boundary that separates the microscopic and macroscopic worlds, or what physical qualities—spatial extent, mass, number of particles, complexity—would characterize this boundary, or whether such a boundary should be sharp or fuzzy. The new fields of quantum computing and quantum information processing have provided us with an extremely exciting and new way to think about wavefunctions in terms of their information content, but they have not so far softened the affront to our gut feeling of what it means for a thing to exist.

Thus it seems that the best we can do is to continue studying more and varied examples of quantum superpositions. Leggett's question is really only answerable by experiment, and the role of the theoretician is to suggest systems in which the question can be posed sharply and in a way that offers the best hope for clear-cut answers, and to provide a framework in which to analyze the experimental results. However, the only framework we have at present is quantum theory, so one is in the paradoxical position of relying on the very theory that one suspects might be at fault.

It is with this motivation that in this article I discuss some recent work to look for large scale quantal superpositions. It would be impossible for me to improve on the lucidity of Leggett's reviews [1-4] for the conceptualization of macroscopic quantum phenomena, and I will not attempt to do so. Nor will I survey the progression of theoretical and experimental work over two decades. In addition to the original papers, there now exist many reviews [6,7], and a monograph [8] that more than do justice to the field. Rather, I will confine myself to much more focussed treatments of two specific systems that show fairly dramatic instances of macroscopically distinct superpositions.

The two systems that I will focus on are a magnetic molecule, $[Fe_8O_2(OH)_{12}\text{-}(tacn)_6]^{8+}$, generally abbreviated to Fe_8, in which the states in question involve the simultaneous switching of 40 or so electronic spins, and an rf-SQUID, in which the states involved differ in the sense of circulation of approximately 10^9 Cooper pairs. The Fe_8 system is discussed in Sec. 2, and the pertinent experiment is one by Wernsdorfer and Sessoli [9]. The SQUID system is discussed in Sec. 3, with special reference to a very recent experiment of Lukens and coworkers [10]. Both experiments are in agreement with quantum theory, but of course this does not automatically rule out alternative theories. In Sec. 4, therefore, I will reprise an argument due to Leggett [11] that examines the consequences of a certain point of view termed "macrorealism." This argument is based on deriving Bell-like inequalities which would be violated if quantum mechanics applies to macroscopic objects. Neither of the two systems discussed herein is presently capable of putting these inequalities to the test, but that may well change in the future!

FIGURES

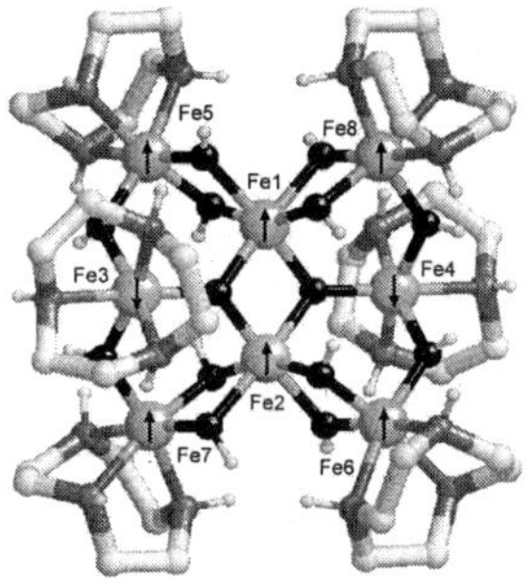

FIG. 1. The Fe_8 molecule. Figure courtesy of Dr. W. Wernsdorfer.

2. The Fe_8 System

The molecule Fe_8 (which is an abbreviation for the formula $[Fe_8O_2(OH)_{12}(tacn)_6]^{8+}$) is magnetic, and forms good single crystals. It has no strict symmetry, but an approximate D_2 symmetry can be recognized (see Fig. 1). In its ground manifold it is found to have a total spin of 10, which is believed to arise from intramolecular antiferromagnetic interactions that result in six of the eight Fe^{3+} ions ($S = 5/2$) being parallel to each other and antiparallel to the remaining two [12–15]. (See arrows in Fig. 1.) However, the net spin is not free to rotate at will; spin-orbit and spin-spin interactions lead to a magnetic anisotropy. A variety of experimental techniques (electron spin resonance, ac susceptibility, magnetic relaxation, Mössbauer spectroscopy, neutron scattering) indicates that a single molecule can be described by the Hamiltonian

$$\mathcal{H} = k_1 J_x^2 + k_2 J_y^2 - g\mu_B \mathbf{J} \cdot \mathbf{H}, \tag{1}$$

with $J = 10$, $k_1 \approx 0.33$ K, and $k_2 \approx 0.22$ K. We have added a term describing the interaction with an externally applied magnetic field $\mathbf{H}$. The g factor is very close to 2, and μ_B is the Bohr magneton. The spin is dimensionless, i.e., it is measured in units of $\hbar$.

The Hamiltonian (1) as written presumes that the low temperature dynamics of the molecule can be described by a single macrovariable, the total spin **J**. How good an assumption is this? The most obvious weakness is that the intramolecular exchange might be so weak that other spin multiplets are very close in energy to the $J = 10$ multiplet. EPR experiments do not show any evidence of another multiplet, and so while a definite number is not avaliable, one can reasonably guess that any other multiplet (with $J = 9$, say) will lie at least tens of Kelvin above the $J = 10$ multiplet. At least as far as tunneling dynamics are concerned, the other multiplets can then be ignored.

The second assumption is that the macrovariable **J** is indeed decoupled from other degrees of freedom. How good is this assumption, and what has been left out? First, interactions between different molecules. However, the molecules are quite large and well separated from each other in the solid, Thus it is a good approximation to ignore the dipolar interactions between the molecules, or somewhat less approximately, to treat these in the form of a weak inhomogeneous field. Second, we have ignored the interactions with phonons. These are indeed very important for magnetic relaxation [14,15], but the phenomena we are interested in occur at very low temperature (indeed the theory presented here is only for zero temperature), where phonons can safely be neglected. Third, the hyperine interactions have been ignored. The only isotope of iron with a magnetic nucleus, ^{57}Fe, has a natural abundance of 2–2.5%. Thus about 75% of the molecules have no nuclear spin in the magnetic species at all, and for those that do, the hyperfine field seen by the electronic spin is about 10–100 Oe. The net effect of the nuclear spins can be lumped into a weak inhomogeneous field along with the dipolar fields. The width of this field distribution is measured to be about 200 Oe [16], which is well below the equivalent anisotropy field of $\sim$ 2.5 T. (Alternatively, we could say that the energy scales associated with dipolar interactions and magnetic anisotropy are 0.2 K and 25 K, respectively.)

It is therefore a good first approximation to study the tunneling of the molecular spin using Eq. (1), and to include interactions with the other degrees of freedom later if necessary. The spin described by this Hamiltonian has, in the absence of an applied field, easy or minimum energy directions $\pm\hat{\mathbf{z}}$, hard or maximum energy directions $\pm\hat{\mathbf{x}}$, and medium directions (saddle points of the energy surface) $\pm\hat{\mathbf{y}}$. If the field $\mathbf{H}$ is exactly zero, the $\pm\hat{\mathbf{z}}$ directions are classicaly exactly degenerate, and it is possible to resonate between them by quantum tunneling, in exact analogy with the 24 GHz masing resonance of the NH_3 molecule. This state of affairs continues to hold if a magnetic field is applied anywhere in the xy plane. As long as $H_z = 0$, and $\mathbf{H}$ is small enough, there are two degenerate minima in the energy separated by an energy barrier, setting the stage for resonance or coherence.

One way to understand the tunneling is as follows. We rewrite Eq. (1) after subtracting out a term $k_2\mathbf{J}\cdot\mathbf{J}$ (which is a constant, since $\mathbf{J}\cdot\mathbf{J} = J(J+1) = 110$) and explicitly putting $H_z = 0$:

$$\mathcal{H} = -k_2 J_z^2 + (k_1 - k_2) J_x^2 - g\mu_B (J_x H_x + J_y H_y). \tag{2}$$

If we regard the last three terms as perturbations, then the J_x^2 term gives rise to $\Delta m = 2$ transitions, and thus mixes -10 and $+10$ via $-8, -6, \ldots, +8$, and the H_x and H_y terms allow $\Delta m = 1$ transitions, and also mix -10 and +10 (see Fig. 2a). This picture allows

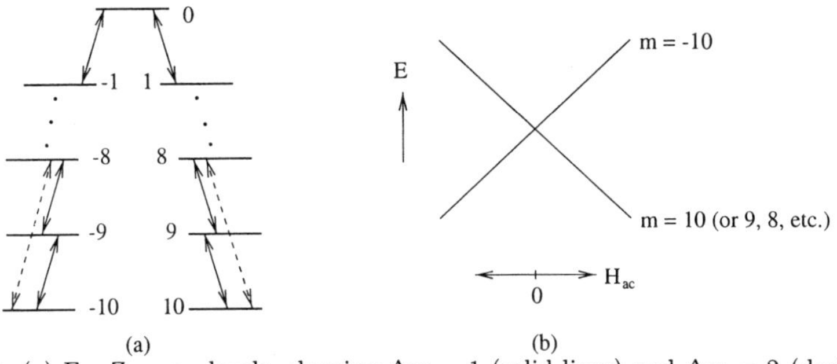

FIG. 2. (a) Fe_8 Zeeman levels, showing $\Delta m = 1$ (solid lines) and $\Delta m = 2$ (dashed lines) transitions. (b) The Landau-Zener-Stückelberg process.

us to think of spin tunneling in direct analogy to a particle tunneling through an energy barrier.

Next, let us ask how the tunnel splitting, Δ, is to be measured. Direct numerical diagonalization of the 21×21 Hamiltonian matrix with the experimental numbers for k_1 and k_2 shows that $\Delta \sim 10^{-8}$ K. We have already said that dipolar fields give rise to a non degeneracy or bias, ϵ, of about 0.1 K between the $m = \pm 10$ states. Since ϵ is enormous compared to Δ, left to itself, a spin in one of the levels will have almost no chance of tunneling to the other level. [This point can be easily seen by considering a two-level Hamiltonian

$$\begin{pmatrix} \epsilon & \Delta \\ \Delta & -\epsilon \end{pmatrix}. \tag{3}$$

It is now directly verified that if we start at $t = 0$ in the $+\epsilon$ level, say, the probability of finding the system in the $-\epsilon$ level at a later time is of order $\Delta^2/(\epsilon^2 + \Delta^2)$.]

Wernsdorfer and Sessoli get around this problem by applying an ac magnetic field along the z axis. Then as H_z changes with time, the energies of the ± 10 levels move in opposite directions, and at some point in the cycle they will cross, giving rise to a Landau-Zener-Stückelberg (LZS) process [17] (see Fig. 2b). The magnitude $H_{\rm ac}$ of the ac field is chosen to be comfortably larger than the dipolar fields, but not so large as to make the -10 state ever cross the $+9$ state. In the so-called non-adiabatic limit, where the crossing is passed rapidly, the probability of a transition from -10 to $+10$ is very small in each passage. Further, different passages may be assumed to be uncorrelated because the stray dipolar fields, which are after all fluctuating, randomize the phase of each molecule. The net result is that there is a rate or transition probability per unit time, $\Gamma_{\rm LZS}$, for $-10 \leftrightarrow 10$ transitions, given by

$$\Gamma_{\rm LZS} \sim \frac{\Delta^2}{\omega H_{\rm ac}}, \tag{4}$$

where ω is the frequency of the ac field. If the sample is first saturated to the -10 state, say, by slow field cooling, and the field is then removed, and the ac field applied,

LZS transitions will be induced, giving rise to a relaxation of the net magnetization of the sample. One can determine $\Gamma_{\rm LZS}$ by measuring the time dependence of the magnetization relaxation. From this measurement and the known values of ω and $H_{\rm ac}$, one can then obtain Δ. One of the nice features of this method is that nearly all molecules in the sample undergo a crossing at some point in the cycle, as long as $H_{\rm ac}$ is large enough to encompass the range of biases.

It is apparent that the above protocol can be carried out in the presence of a transverse dc field (in the xy plane) $H_\perp$, enabling one to measure Δ as a function of $H_\perp$. This ability is yet another very nice feature of Fe_8, since it provides one with a way to study systematic dependences on an in situ controllable variable other than the temperature. Temperature, in fact, turns out to be a rather inconvenient variable. The most obvious criterion for evidence of quantum mechanics is that as temperature is lowered, one should see a flattening or bottoming out of the transition rate since thermally activated processes are frozen out. However, lowering the temperature invariably lengthens equilibration and heat transport times, and one is always left with nagging questions about possible heat leaks, and whether the sample and the thermometer temperatures are the same, or if the former is hung up. The experience of the last two decades has been that it is very difficult to settle such questions. Changing the magnetic field, by contrast, involves no such complications, at least a priori.

What, then, do we expect for Fe_8 as $H_\perp$ is changed? Naively, increasing $H_\perp$ decreases both the energy barrier and the angle through which the spin needs to tunnel, suggesting that Δ will increase monotonically with $H_\perp$. What Wernsdorfer and Sessoli actually see is rather different (see Fig. 3a). When $\mathbf{H}\|\hat{\mathbf{y}}$ ($\phi = 90°$), the behavior is of the monotonic kind, but when $\mathbf{H}\|\hat{\mathbf{x}}$ ($\phi = 0$), one finds that Δ oscillates with H_x!

The oscillations in Δ with H_x were in fact predicted for the model (1) [18], without knowing of its relevance to Fe_8, or even the existence of this system. The most visual explanation is in terms of Feynman paths for the spin [19]. Since the magnitude of the

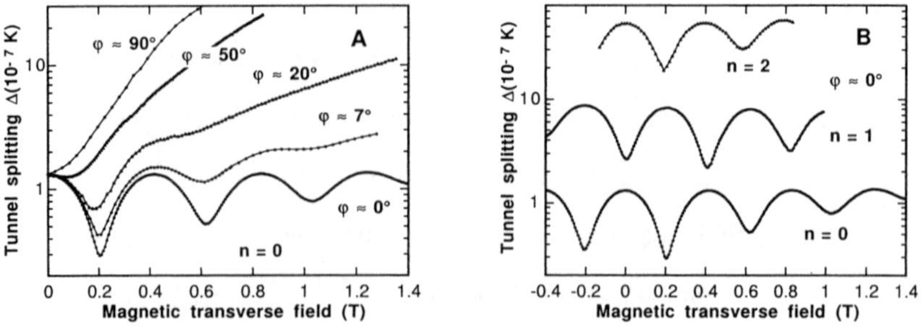

FIG. 3. Measured splittings for Fe_8 for (a) $-10 \leftrightarrow 10$ transitions for various orientations of $\mathbf{H}$ in the xy plane, and (b) for $\mathbf{H}\|\hat{\mathbf{x}}$ between $m = -10$ and $m = 10 - n$ states. Figure courtesy of Dr. W. Wernsdorfer.

spin is fixed, and only its direction $\hat{\mathbf{n}}$ is variable, it can be thought of as living on the unit sphere, and the paths can be described by giving its orientation. Howver, just as in the tunnleing of a massive particle, the path integral is dominated by the least action paths, and just as in the particle case the momentum must be imaginary, now $\hat{\mathbf{n}}$ must be allowed to become complex. (It still remains a unit vector, though, i.e., $\hat{\mathbf{n}} \cdot \hat{\mathbf{n}} = 1$.) When one finds these least action paths, or instantons for Fe_8, with $\mathbf{H}\|\hat{\mathbf{x}}$, one finds that there are two instantons which wind about $\hat{\mathbf{x}}$ in opposite directions (see Fig. 4). That this is so can be made plausible as follows. When $\mathbf{H} = 0$, the energy on the equator is

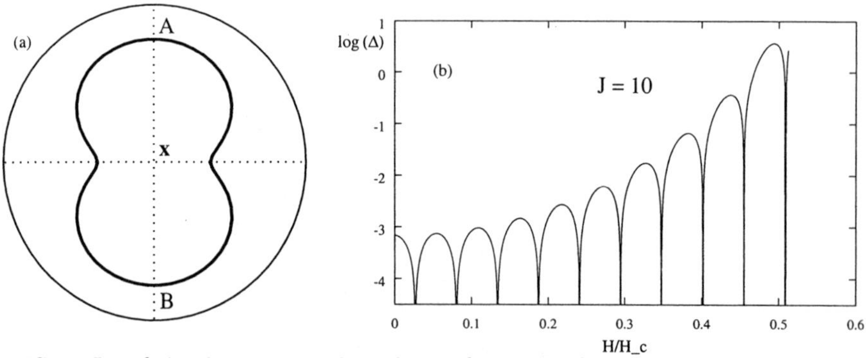

FIG. 4. Interfering instanton trajectories, and numerically computed $-10 \leftrightarrow 10$ tunnel splitting when $\mathbf{H}\|\hat{\mathbf{x}}$.

least when $\hat{\mathbf{n}} = \pm\hat{\mathbf{y}}$, and the instanton projection onto the real unit sphere passes through these points. Now if H_x is turned on, the instantons evolve smoothly, and there are two of them.

The reason for the oscillations is that there is a Berry-like phase in the path integral for spin [20,21] which for a closed path is proportional to the solid angle enclosed by it. If we describe the tunneling by a Euclidean action, $\Delta \sim e^{-S_E}$, the real part S_R of this Euclidean action is equal for the two paths, but the imaginary parts differ by $J\Omega$, where Ω is the area of the loop formed by the two instantons. Thus,

$$\Delta \propto \exp(-S_R) \cos\Phi, \tag{5}$$

where $\Phi = J\Omega/2$. As H_x is increased, the minima of the energy, $\hat{\mathbf{n}}_1$ and $\hat{\mathbf{n}}_2$, both move toward the x axis, and the area Ω decreases. In the simplest model for Fe_8, Eq. (1), Φ sweeps through an odd multiple of $\pi/2$ a total of $2J$ times counting zero and negative values of H_x. Each time this happens, Δ vanishes. One can view the plot of Δ versus H_x, or even better Δ^2 versus H_x, which is what Wernsdorfer and Sessoli measure directly, as a plot of intensity versus position on the screen of a standard two-slit interfernce experiment. That one does not observe perfect destructibve intereference is due to the inhomogenieties and fluctuating fields mentioned above, and is largely a technicality. There is much to be gained from studying the precise value and shape of the minima in the experimental curve, but that would take us too far away from the aims of this article.

What about the behaviour when $\mathbf{H}$ is in the xy plane but has a y component as well? If the angle ϕ with the x axis is small, there are still two instantons, but there is now no symmetry to ensure equality of the real parts. Thus just as in two-slit interference with unequal slit widths, the contrast in intensity is weaker, now the oscillations in Δ^2 vs. $H_\perp$ are less pronounced. The trend seen with increasing ϕ is as expected in this picture.

Much more interesting are the curves marked $n = 2$ and $n = 3$ in Fig. 3b. These are obtained by adding a z component to the dc magnetic field that is such as to bring into degeneracy the -10 and $+9$ ($n = 2$) or the -10 and $+8$ ($n = 3$) levels, and inducing LZS transitions between the degenerate levels using the same protocol as in the $-10 \leftrightarrow 10$ case. Again, one sees that Δ oscillates with H_x, but the oscillations are shifted in phase by half a period for each successive excited level in the positive J_z hemisphere.

The new oscillations can also be understood in terms of the Berry-like phase, but this is not easy to see immediately. In fact, since a vanishing of Δ means that the levels in question are exactly degenerate, these oscillations appear at first sight to violate the von-Neumann Wigner theorem which says that energy levels repel unless they transform differently under some symmetry of the Hamiltonian. Let us briefly recall the argument. Suppose two nearly degenerate levels have energies E_1 and E_2. Under an incremental perturbation V, the secular matrix is

$$\begin{pmatrix} E_1 + V_{11} & V_{12} \\ V_{21} & E_2 + V_{22} \end{pmatrix}, \tag{6}$$

with $V_{21} = V_{12}^*$. The eigenvalues of this matrix are equal only if

$$E_1 + V_{11} = E_2 + V_{22}, \qquad V_{12} = V_{12}^* = 0. \tag{7}$$

This represents three conditions (two if V is real), which cannot generally be satisfied by variation of only one parameter. The exception is if V is diagonal from the outset because of some symmetry. If there is no symmetry, then one must vary two parameters (assuming V to be real) to get a level crossing.

When $\mathbf{H}\|\hat{\mathbf{x}}$, Eq. (1) is invariant under a 180° rotation about $\hat{\mathbf{x}}$, but this property is lost when H_z and H_x are both nonzero [22]. The oscillations when $\mathbf{H}\|\hat{\mathbf{x}}$ can be understood as symmetry allowed level crossings, but the others can not, and were not predicted. They are nontrivial instances of diabolical points, or what used to be called conical instersections [23]. The term diabolical originates in the fact that the energy surface from Eq. (6) is

$$E = \text{constant} \pm \sqrt{X^2 + Y^2}, \tag{8}$$

where $X = E_1 + V_{11} - E_2 - V_{22}$, and $Y = V_{12}$. This is a double cone or *diabolo* in the XY plane, or equivalently, in the plane of parameters in the Hamiltonian.

What is surprising about the Fe_8 system is that the diabolical points lie on such nice straight lines in the H_x-H_z plane. Theoretical expressions for the locations of these points have been found by the author [24] and independently by Villain and Fort [25] using a discrete analog of the WKB or phase integral method. The positions found by them using the model Eq. (1) display exactly the half-period shift per excited level seen experimentally, but the precise positions and numbers of diabolical points differ from those measured, because the model Eq. (1) omits certain higher order anisotorpies that while small are quantitatively important for the tunneling. The influence of these terms is understood, and does not require any modification of the basic concepts.

We should also note that the model (1) appears to be special in that if energy levels with ordinal numbers n and $n+1$ are degenerate at a given field point, then at the same field point, $n+2$ and $n+3$ are degenerate, $n+4$ and $n+5$ are degenerate, and so on, until all $2S+1$ levels are exhausted. The pattern of diabolical points forms concentric rhombi in the H_x-H_z plane, and all in all, is highly suggestive of the presence of some higher dynamical symmetry in the Hamiltonian. The location of the diabolical points has recently been established rigorously by E. Keçecioğlu and the author [26], but we have been unable to find a higher symmetry. These features are destroyed by the fourth order terms mentioned above, but since these terms are small, it would still be useful to find the hidden symmetry, assuming it exists.

In the rest of this section, we provide a somewhat more technical discussion of spin tunneling in terms of spin coherent states, in view of its relative unfamiliarity as compared to tunneling of massive particles. Readers uninterested in this should skip to Sec. 3.

We describe the spin in terms of spin-coherent states $|\hat{\mathbf{n}}\rangle$, which is the state with maximal spin projection along the unit vector $\hat{\mathbf{n}}$:

$$\mathbf{J} \cdot \hat{\mathbf{n}} |\hat{\mathbf{n}}\rangle = J |\hat{\mathbf{n}}\rangle. \tag{9}$$

The spin-coherent states are minimum uncertainty states, and are the closest that one can get to a classical state in which the orientation of $\mathbf{J}$ is exactly specified. We then consider a transition amplitude (in imaginary time) such as

$$U_{21} = \langle \hat{\mathbf{n}}_2 | e^{-\mathcal{H}T} | \hat{\mathbf{n}}_1 \rangle, \tag{10}$$

where $\hat{\mathbf{n}}_{1,2}$ are the minima of the classical energy

$$E_{\mathrm{cl}}(\hat{\mathbf{n}}) = \langle \hat{\mathbf{n}} | \mathcal{H} | \hat{\mathbf{n}} \rangle. \tag{11}$$

We now expand the amplitude in Eq. (10) in terms of the eigenstates of $\mathcal{H}$. As $T \to \infty$, only the lowest two states will contribute, and since these differ in energy by Δ, we expect that

$$U_{21} \sim \sinh \Delta T, \quad T \to \infty. \tag{12}$$

If we can calculate U_{21}, comparison with Eq. (12) will yield Δ.

To calculate U_{21}, we write it as a Feynman path integral

$$U_{21} = \int_{\hat{\mathbf{n}}_1}^{\hat{\mathbf{n}}_2} [d\hat{\mathbf{n}}] e^{-S[\hat{\mathbf{n}}(\tau)]}, \tag{13}$$

where the paths $\hat{\mathbf{n}}(\tau)$ all run from $\hat{\mathbf{n}}_1$ at $-T/2$ to $\hat{\mathbf{n}}_2$ at $T/2$, and $S[\hat{\mathbf{n}}(\tau)]$ is the imaginary time or Euclidean action for this path. The key to the oscillations of Δ lies in the nature of this action. This can be written as

$$S[\hat{\mathbf{n}}(\tau)] = iJ\mathcal{A}[\hat{\mathbf{n}}(\tau)] + \int_{-T/2}^{T/2} E_{\mathrm{cl}}[\hat{\mathbf{n}}(\tau)]\, d\tau, \tag{14}$$

$$\mathcal{A}[\hat{\mathbf{n}}(\tau)] = \oint_{C=\hat{\mathbf{n}}(\tau)-\hat{\mathbf{n}}_R} d\Omega. \tag{15}$$

The term $\mathcal{A}[\hat{\mathbf{n}}(\tau)]$ is the kinetic term in the action, and has the mathematical structure of a Berry phase. (The same term is responsible for the Haldane gap in antiferromagnetic spin chains [20,21]). One of the key properties of such a term is that it can not be made manifestly gauge invariant, and this fact has given rise to misstatements in

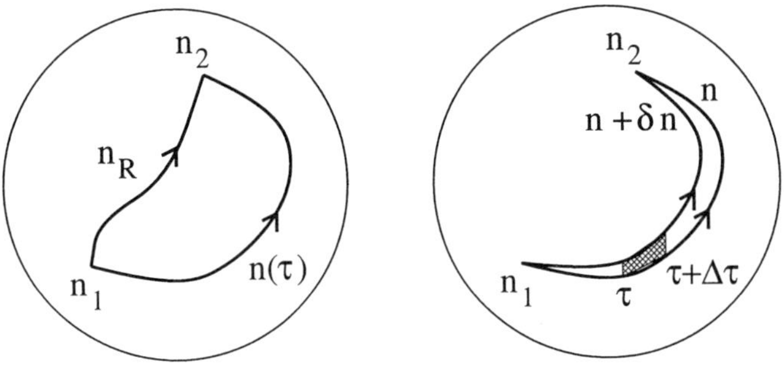

FIG. 5. The kinetic term in the spin action, and its variation.

the literature in which a representation arising from a particular gauge choice and the singularities of the spherical polar coordinate system is said to be responsible for the oscillations in Δ. We have written it in a way that is designed to avoid these pitfalls. In words, Eq. (15) says that $\mathcal{A}$ is the solid angle enclosed by the closed curve $\mathcal{C}$ formed by $\hat{\mathbf{n}}(\tau)$ and a reference path $\hat{\mathbf{n}}_R$ (taken backwards) running from $\hat{\mathbf{n}}_1$ to $\hat{\mathbf{n}}_2$ (Fig. 5). The reference path $\hat{\mathbf{n}}_R$ is arbitrary, but must be the same for all $\hat{\mathbf{n}}$ in the path integral. Its choice is equivalent to fixing the gauge. However, because we have defined it as a geometrical object, an area, it does not depend on how we choose coordinates on the sphere, and it is clearly nonsingular.

One way to see that $\mathcal{A}[\hat{\mathbf{n}}(\tau)]$ is indeed a kinetic term is to derive the least action path, i.e., the classical equations of motion. Suppose we vary the path $\hat{\mathbf{n}}(\tau)$ by $\delta\hat{\mathbf{n}}(\tau)$. The variation $\delta\mathcal{A}$ is given by the area of the thin sliver on the sphere enclosed by the curves $\hat{\mathbf{n}}(\tau)$ and $\hat{\mathbf{n}}(\tau) + \delta\hat{\mathbf{n}}(\tau)$ in Fig. 5. The part of this area due to the segments between τ and $\tau + \Delta\tau$ is easily seen to be given by

$$\begin{aligned}\Delta(\delta\mathcal{A}) &= \Big[\delta\hat{\mathbf{n}}(\tau) \times [\hat{\mathbf{n}}(\tau+\Delta\tau) - \hat{\mathbf{n}}(\tau)]\Big] \cdot \hat{\mathbf{n}}(\tau) \\ &= \left(\delta\hat{\mathbf{n}}(\tau) \times \frac{d\hat{\mathbf{n}}}{d\tau}\right) \cdot \hat{\mathbf{n}}(\tau)\, \Delta\tau. \end{aligned} \tag{16}$$

Thus the total change is given by

$$\delta\mathcal{A} = \int \delta\hat{\mathbf{n}}(\tau) \cdot \left(\frac{d\hat{\mathbf{n}}}{d\tau} \times \hat{\mathbf{n}}(\tau)\right) d\tau. \tag{17}$$

The variation of the second term in Eq. (15) is elementary: $\int \delta\hat{\mathbf{n}}(\tau)\cdot(\partial E_{\rm cl}/\partial\hat{\mathbf{n}})\, d\tau$. Thus the condition for the action to be an extremum is

$$iJ\frac{d\hat{\mathbf{n}}}{d\tau} \times \hat{\mathbf{n}}(\tau) + \frac{\partial E_{\rm cl}}{\partial \hat{\mathbf{n}}} = 0, \tag{18}$$

or, taking the cross product with $\hat{\mathbf{n}}$ and noting that $\hat{\mathbf{n}} \cdot (d\hat{\mathbf{n}}/d\tau) = 0$,

$$iJ\frac{d\hat{\mathbf{n}}}{d\tau} = -\left(\hat{\mathbf{n}} \times \frac{\partial E_{\rm cl}}{\partial \hat{\mathbf{n}}}\right). \tag{19}$$

This is the imaginary time equation for Larmor precession in the effective magnetic field $\partial E_{\rm cl}(\hat{\mathbf{n}})/\partial\hat{\mathbf{n}}$. If written in terms of spherical polar coordinates, it will be seen to have a Hamiltonian structure with ϕ and $J\cos\theta$ as conjugate variables, and $E_{\rm cl}$ as the Hamiltonian. One consequence is that $E_{\rm cl}$ is conserved along the classical path, and this is directly seen from Eq. (19):

$$\begin{aligned}\frac{dE_{\rm cl}}{d\tau} &= \frac{\partial E_{\rm cl}}{\partial \hat{\mathbf{n}}}\cdot\frac{d\hat{\mathbf{n}}}{d\tau} \\ &= \frac{i}{J}\frac{\partial E_{\rm cl}}{\partial \hat{\mathbf{n}}}\cdot\left(\hat{\mathbf{n}} \times \frac{\partial E_{\rm cl}}{\partial \hat{\mathbf{n}}}\right) \\ &= 0.\end{aligned} \tag{20}$$

Returning to the evaluation of U_{21}, we perform the integral in Eq. (13) by steepest descents, which is justfied when $T \to \infty$. The dominant paths, or instantons, are just those which minimize the action. Since these paths must conserve $E_{\rm cl}$, and since $\hat{\mathbf{n}}_{1,2}$ are minima of $E_{\rm cl}$, it follows that the paths must be complex, and the area $\mathcal{A}$ must be defined on the complexified unit sphere. We can always take $E_{\rm cl}$ to be zero along the instanton by adjusting the zero of energy, so that the action for an instanton is just $iJ\mathcal{A}$. The actual evaluation of $\mathcal{A}$ for the instanton path is usually more simply done by transforming Eq. (15) to a line integral via Gauss's theorem. Deliberately using notation from electrodynamics, let us write $d\Omega = \mathbf{B}\cdot\hat{\mathbf{n}}\, ds$, where $\mathbf{B} = \hat{\mathbf{n}}$, and ds is an area element. The line integral is $\int \mathbf{A}\cdot d\hat{\mathbf{n}}$, with $\mathbf{B} = \nabla \times \mathbf{A}$. Since $\mathbf{B} = \hat{\mathbf{n}}$, it is a magnetic monopole field, for which it is known that any vector potential $\mathbf{A}$ must necessarily have a singularity somewhere. If the singularity is concentrated into a Dirac string at the south pole, one can write

$$\mathcal{A}[\hat{\mathbf{n}}(\tau)] = \oint_{\mathcal{C}} [1 - \cos\theta(\phi)]\, d\phi, \tag{21}$$

as long as $\mathcal{C}$ does not pass through the south pole, and as long as one agrees to increment ϕ by $\pm 2\pi$ every time one crosses the date line. For the instanton path, $\theta(\phi)$ can be found by energy conservation, so one never has to solve the equation of motion (19).

For Fe_8 with $\mathbf{H} \| \hat{\mathbf{x}}$, there are, as already stated, *two* instanton paths, which we label A and B. Denoting the respective actions by S_A and S_B, symmetry yields

$$\mathrm{Re} S_A = \mathrm{Re} S_B \equiv S_R. \tag{22}$$

The imaginary parts, on the other hand, are necessary unequal, since by the interpretation of $\mathcal{A}$ as an area, we have

$$S_B - S_A = iJ \times \Omega[A - B], \tag{23}$$

where $\Omega[A-B]$ is the area of the closed loop formed by the two instantons. To evaluate Δ one must add the amplitudes for both instantons, so that the total amplitude is as in Eq. (5).

The evaluation of the integrals for S_R and Φ is straightforward [27,18]. We provide the results for completeness

$$S_R = J \left[\ln \left(\frac{\sqrt{1-u_0^2} + \sqrt{\lambda}}{\sqrt{1-u_0^2} - \sqrt{\lambda}} \right) - \frac{u_0}{\sqrt{1-\lambda}} \ln \left(\frac{\sqrt{(1-u_0^2)(1-\lambda)} + u_0\sqrt{\lambda}}{\sqrt{(1-u_0^2)(1-\lambda)} - u_0\sqrt{\lambda}} \right) \right], \tag{24}$$

$$\Phi = J\pi(1 - u_0/\sqrt{1-\lambda}), \tag{25}$$

with $u_0 = g\mu_B H / 2k_1 J$.

3. The rf-SQUID

An rf-SQUID is a superconducting loop interrupted by a Josephson junction. When an external magnetic flux Φ_x is applied to the loop, the flux actually trapped in the loop Φ need not equal Φ_x, since there could be a supercurrent I flowing in the loop. The Josephson relation $I = I_c \sin(\phi)$ connects this current to ϕ, the phase difference of the order parameter across the junction. (Here, I_c is the critical current of the junction.) The phase difference ϕ is related to the flux Φ by the usual fluxoid quantization argument:

$$\phi = 2\pi \frac{\Phi}{\Phi_0} \mod 2\pi, \tag{26}$$

where $\Phi_0 = hc/2e$ is the superconducting flux quantum. Thus the SQUID acts esentially as a nonlinear inductance. In addition to the energy due to the geometrical inductance of the loop, there is another contribution due to the Josephson effect, which is periodic in Φ with period Φ_0. The potential energy can be written as

$$U(\Phi) = \frac{1}{2L}(\Phi - \Phi_x)^2 - \frac{I_c \Phi_0}{2\pi} \cos(2\pi\Phi/\Phi_0). \tag{27}$$

There is also an inertia associated with the flux because the junction has a capacitance C, so the voltage across the junction can not change instantaneously. Thus, in the absence of any dissipative effects, the equation of motion can be written as

$$C\ddot{\Phi} = -\frac{dU}{d\Phi}. \tag{28}$$

How does this description fit in with quantum mechanics? The flux through the loop can certainly be thought of as an operator in terms of microscopic variables: the electron coordinates, and the electromangetic fields. Then by demanding that the equations of motion have the correct classical from in the correspondence limit, we can argue that the momentum canonically conjugate to Φ is

$$p_\Phi = \frac{\partial L}{\partial \dot{\Phi}} = C\dot{\Phi} = -Q, \tag{29}$$

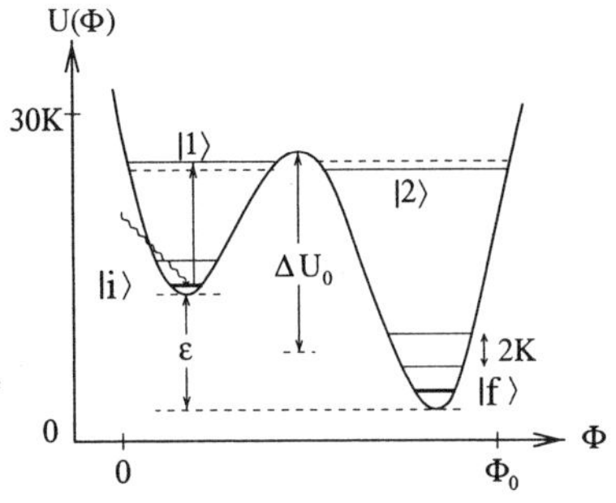

FIG. 6. The SQUID potential.

where Q is now the charge across the junction. The quantum mechanical description must therefore be obtained by writing $p_\Phi = -i\hbar\partial/\partial\Phi$, and the associated Schrödinger equation

$$-\frac{\hbar^2}{2C}\frac{\partial^2\Psi(\Phi)}{\partial\Phi^2} + U(\Phi)\Psi(\Phi) = i\hbar\frac{\partial\Psi(\Phi)}{\partial t}. \tag{30}$$

We refer readers to Leggett's article [3] and to the paper by Ambegaokar, Eckern, and Schön [28] for a more comprehensive discussion of this quantization procedure.

Equation (30) is exactly like the Schrödinger equation for a massive particle of mass C moving in the potential $U(\Phi)$. This description of the system leaves out other degrees of freedom such as the quasiparticles, but these are essentially not excited at all at sufficiently low temperatures (comapred to the gap energy scale), so it is at least qualitatively correct. In fact, the experimental results, obtained at 40 mK, seem to indicate that even quantitatively it may not be too far off.

The shape of this potential is shown in Fig. 6. If $\beta_L \equiv 2\pi L I_c/\Phi_0 > 1$, then it has more than one minimum. The minima are separated on the scale of Φ_0, and the barrier energy ΔU_0 is on the scale of $\Phi_0^2/4\pi^2 L$. If $\Phi_x = \Phi_0/2$, the potential is exactly symmetric about $\Phi_0/2$. By varying Φ_x, the potential can be made asymmetric, and one can introduce a bias ϵ between the minima. The experiment by Lukens and coworkers [10] makes use of a SQUID in which the single junction is replaced by a smaller loop containing two junctions. The flux through this small loop, $\Phi_{x,\mathrm{dc}}$ can be varied independently of Φ_x. This control is a crucial element in the experiment, since it essentially allows one to modulate I_c and thus vary the barrier height U_0 largely independently of ϵ. The experimental values of various parameters are $L \simeq 240\,\text{pH}$, $C \simeq 0.1\,\text{pF}$, and $\beta_L \simeq 2.3$.These yield $U_0 \simeq 30\,\text{K}$, and a small oscillation frequency about the two potential minima of about 40 GHz. It is quite remarkable that one should have an energy scale of a few tens of Kelvin (or a few meV) controlling the dynamical behavior of a macrovariable that collective describes 10^{10} electrons.

The experimental procedure consists of starting the system in the left well, which can be done by suitably cycling Φ_x and $\Phi_{x,\text{dc}}$. Since the thermal energy $k_B T \ll U_0$, we can be sure that the system is in the ground state in this well. A 96 GHz microwave pulse one millisecond long and of fixed power is then applied. One then measures the flux in the SQUID to see if the system has made a transition to the right well. This protocol is repeated several times, and the transition probability is measured as a function of both ΔU_0 and ϵ.

Let us now discuss how this experiment would be interpreted quantum mechanically. Let the lowest states in the left and right wells be denoted by $|i\rangle$ and $|f\rangle$, and let there be an energy eigenstate $|1\rangle$ near the top of the barrier. The transition path $|i\rangle \to |1\rangle \to |f\rangle$ will have a high probability when two conditions are met:

(1) $E_1 - E_i = \hbar\omega_{\text{rf}}$.

(2) The state $|1\rangle$ has a large amplitude in *both* wells.

Condition (1) is necessary to make sure that there is a state which can resonantly absorb the microwaves. Condition (2) is necessary because unless this state has a large amplitude in the left well, the matrix element for the $1 \leftrightarrow i$ transition is small. Likeweise, unless this state has a large amplitude in the right well, the $1 \leftrightarrow f$ transition element is small.

Condition (2) is equivalent to saying that in the vicinity of the energy E_1, the flux should be able to tunnel from the left to the right well. If we can find such a state that is below the top of the barrier, then this automatically implies that there is another state $|2\rangle$ separated from the first by an amount that is on the order of Δ, the tunnelling amplitude. [This can be seen by modelling the two states in terms of a two level system Hamiltonian as in Eq. (3).]

Suppose that one could vary the frequency $\omega_{\rm rf}$ of the microwave radiation. Then by studying the transition probability as a function of $\omega_{\rm rf}$ one could map out the energy levels near the top of the barrier. However, the method would be effective only for states with appreciable tunnelling between wells. It is experimentally more effective to keep $\omega_{\rm rf}$ fixed and vary Φ_x and $\Phi_{x,{\rm dc}}$, effectively varying ΔU_0 and ϵ.

The point of the experiment is that the energy levels of $|1\rangle$ and $|2\rangle$ undergo an anticrossing as ΔU_0 and ϵ are varied. (Unlike Fe_8, there is now no diabolical point, as the Schrödinger equation for Φ is one-dimensional.) Suppose that we keep ΔU_0 fixed, and vary ϵ. Loosely speaking, we can think of this as sweeping E_1 and E_2 through the resonance one after the other. In general, we should expect unequal transition probabilities, which we will show up as two unequal peaks in the plot of transition probability versus ϵ, or the equivalent experimental variable Φ_x. If we now repeat the procedure for several different values of ΔU_0, we will map out the anticrossing. The peaks in the plot of transition probability versus Φ_x will move closer together, and then diverge again. At the same time, the asymmetry in the transition probability will reverse its sense, and the weight will shift from the left peak to the right.

This is exactly what is seen in the experiments (Fig. 7). The inset in the figure shows the positions of a few of the energy levels as deduced from the transition probability data. Two anticrossings are evident. The changes in the current associated with the two states are of the order of $2\,\mu$A, and in the flux are on the order of $\Phi_0/4$. Although the states are not the ground states in the respective wells, they are the most macroscopically distinct to show a superposition to date.

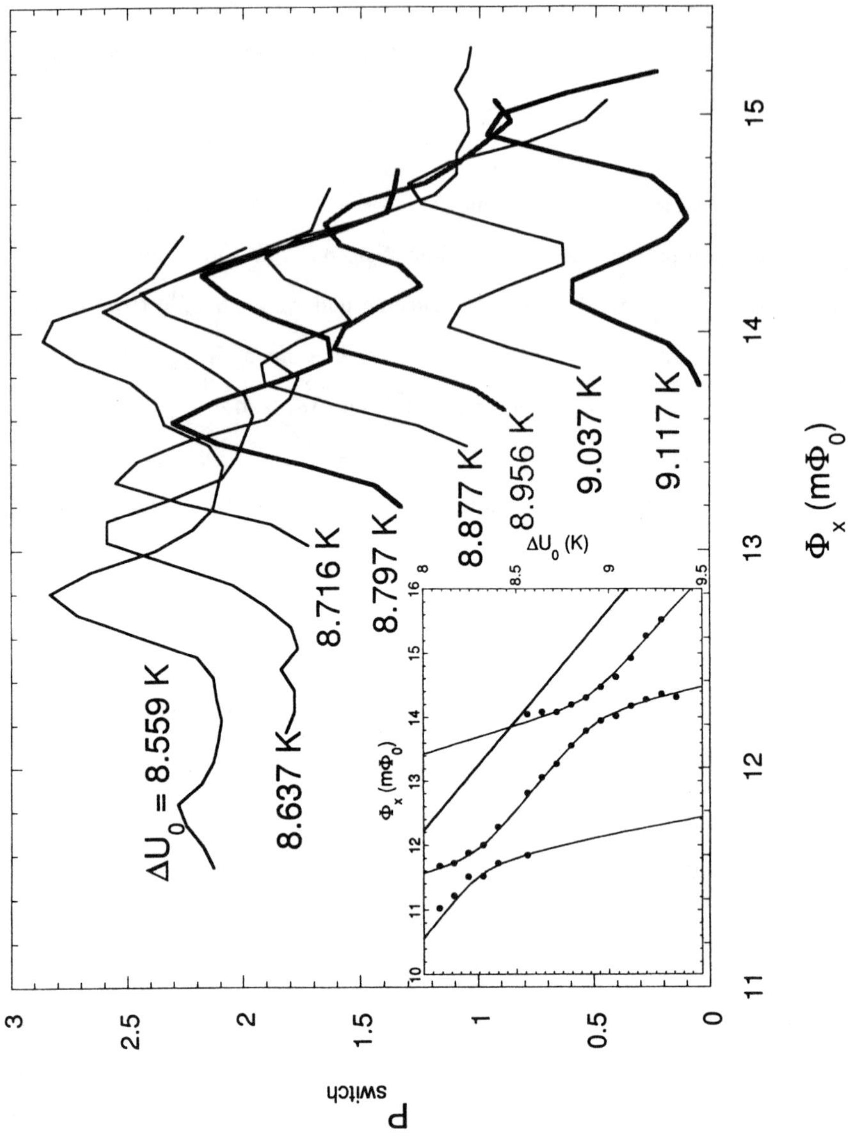

FIG. 7. Transition probability measured in the Stony Brook experiment. The inset shows the observed peak positions in the ΔU_0-Φ_x plane. Figure courtesy of Dr. Lukens.

5. Macrorealism

The experiments on Fe_8 and the rf SQUID, as we have seen, are well described by quantum theory. The latter experiment, in particular, gives fairly strong evidence of a superposition of macroscopic distinct states. Is it clear, however, that our everyday sense of reality is absolutely untenable?

The situation in this regard is analogous to that which prevailed prior to Bell's inequality in 1964. Some older textbooks on quantum mechanics argued that the two-slit interference experiment made it very hard to entertain hidden variable theories, since it was very hard to see how one could get destructive interference with them. Yet, Bohm's pilot wave theory did just that, and it was only with Bell's inequality and its experimental tests that one recognized locality as a key ingredient, and was able to exclude local hidden variable theories with any degree of logical rigour.

One can in fact make an argument very similar to that of Bell which makes definite predictions starting from a view point termed "macrorealism" [11]. The experiments to date are not in a potition to make significant tests of these predictions, but that might change in the future. It is therefore useful to present this argument.

The macrorealistic point of view is based on two fundamental assumptions: (A1) A macroscopic system *is* in a definite macroscopic state at all times. (A2) It is possible to determine the state of the system with arbitrarily small perturbation on the system's subsequent evolution. This viewpoint is patently in conflict with quantum mechanics. Assumption A1 is basic to this viewpoint, and is very much classical in spirit. On its basis we argue that any macroscopic physical property Q of the system must have definite values at all times. The dynamical behavior of Q may be very complex and unknown, and even stochastic. When the system is such that it has available to it two or more states that are *macroscopically* distinct, then this assumption seems especially unimpeachable from a classical world view. Assumption A2 explicitly contradicts what we learn from quantum mechanincs and thought experiments like the Heisenberg gamma ray microscope, but those concepts pertain to the microscopic world, and it ia questionable whether they should be imported into the macroscopic world if one is willing to accept classical notions of existence for macro objects. Indeed, it would be extremely strange to adopt a position where one accepts A1 but rejects A2, as there would then be no way to give an operational meaning to A1. In the words of a leading contemporary historical figure, it would cast into doubt "what the meaning of *is* is".

Let us suppose then that we have a macroscopic system for which Q is a dichotomic varaible. The values that it takes can always be set equal to ± 1. In the case of the SQUID, e.g., we could define Q to be -1 if the flux is found withing some range $\Delta\Phi$ of the left minimum (region L), and +1 if it is found within $\Delta\Phi$ of the right minimum (region R). The width $\Delta\Phi$ is chosen so that the regions L and R do not overlap, and yet the system is almost never found in the excluded central region C. The selection of $\Delta\Phi$ could be based on a subsidiary experiment and would constitute "prior knowledge" of the system. In other words, we would begin the main experiment knowing that one's system is "almost never" found in the region C, or that the probability for finding it was less than some number $\epsilon \ll 1$. In the subsequent argument, we will set ϵ to zero for simplicity. Setting it to a nonzero value would modify the averages to be calculated by small amounts, and would not alter the final conclusion that quantum mechanical predictions are incompatible with macrorealism.

Likewise for the Fe_8 system, we could define $Q = \pm 1$ to be the states in which $|m| \geq 7$, say, and C to be the region with $|m| \leq 6$.

Let us now prepare a large ensemble of such systems at some time $t = 0$, with all the members of the ensemble prepared in the same way. We divide this ensemble into three subensembles. On the first, we measure Q at times t_1 and $t_2 > t_1$. On the second, we measure Q at t_1 and $t_3 > t_2$, and on the third, we measure Q at t_2 and t_3 (see Fig. 8). We now measure the averages $\langle Q(t_1)Q(t_2)\rangle$, $\langle Q(t_2)Q(t_3)\rangle$ etc. Strictly speaking, the

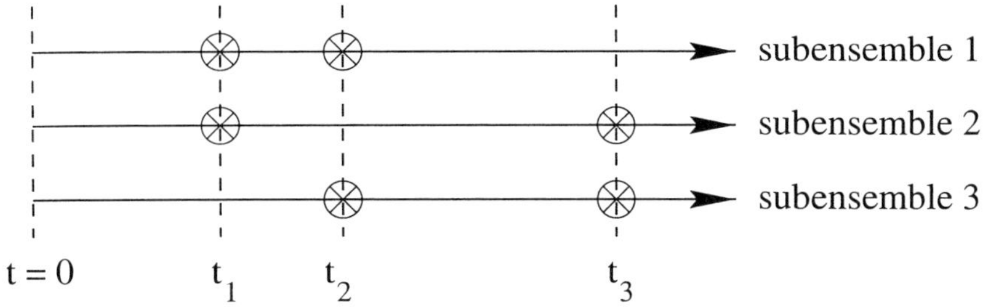

FIG. 8. Subensembles and measurement times for deriving macrorealistic Bell inequality.

averages with different time arguments have been measured on different subensembles, but assumption A2 allows us to argue that at least in principle the measurement at t_1 could have been made in a way that did not affect the subsequent time evolution, so that subsequent measurements of Q at times t_2 and t_3 on subensemble 1 would have given the same results as those obtained from subensemble 3. Thus the subensembles actually yield averages with respect to the same ensemble.

The rest of the argument is closely parallel to that of Bell. We have the obvious inequalities which hold for each member of the ensemble.

$$[Q(t_1) + Q(t_2) + Q(t_3)]^2 \geq 1$$
$$Q(t_1)Q(t_2) + Q(t_1)Q(t_3) + Q(t_2)Q(t_3) \geq -1.$$

If we average the second inequality, we obtain

$$\langle Q(t_1)Q(t_2)\rangle + \langle Q(t_1)Q(t_3)\rangle + \langle Q(t_2)Q(t_3)\rangle \geq -1. \tag{31}$$

This is the simplest macrorealistic Bell inequality.

Quantum mechanics, on the other hand, makes predictions that disagree with this inequlity. For both the Fe_8 and rf-SQUID, one could, in the ideal situation of being able to study a perfectly isolated system, arrange matters so that

$$\langle Q(t_i)Q(t_j)\rangle = \cos\Delta(t_j - t_i), \tag{32}$$

where Δ is the relevant resonance frequency. Substituting in the inequality, and choosing $t_3 - t_2 = t_2 - t_1 = 2\pi/3\Delta$, the left hand side would equal $-3/2$, which is less than -1!

The present experiments on Fe_8 do not allow a test of this inequality because the measurments are not done on a single molecule, and it is rather difficult to conceive of how this might be done at present. The experiments on the rf-SQUID run into the problem that it is difficult to turn the mesurement SQUID off and on on the $1/\Delta$ time scale. Perhaps an old proposal of Tesche [29] could be revived, but for the moment we must be content with marveling at how the flux in a SQUID with 10^{10} electrons can display energy level quantization in accord with quantum mechanics in a regime where the the wavefunction is delocalized over two widely separated regions of flux.

Acknowledgments

The research described in Sec. 2 has been supported by the U. S. National Science Foundation, and that in Sec. 4 by the John D. and Catherine T. MacArthur Foundation. I am indebted to many people for discussions and collaborations that have informed me and shaped my views over the years, but I would especially like to thank my experimental colleagues, Siyuan Han, Jim Lukens, and Wolfgang Wernsdorfer.

REFERENCES

[1] A. J. Leggett, Prog. Theor. Phys. Suppl. No. **69**, 80 (1984).

[2] A. J. Leggett, in *Essays in Theoretical Physics in Honour of Dirk ter Haar*, edited by W. J. Parry (Pergamon, Oxford, 1984).

[3] A. J. Leggett, in *Percolation, Localization, and Superconductivity*, NATO ASI Series B: Physics, Vol. 109, edited by A. M. Goldman and S. A. Wolf (Plenum, New York, 1984).

[4] A. J. Leggett, in *Chance and Matter* (Proceedings of the Les Houches Summer School, Course XLVI, June-Aug. 1986), edited by J. Souletie, J. Vannimenus, and R. Stora (North-Holland, Amsterdam, 1987).

[5] Thus, I very much doubt that if the example of the cat had not been thought of, and these were the only systems that we knew about, R. G. Griffiths, J. Stat. Phys. **36**, 219 (1984), would have used the term "grotesque" to describe their superposition states.

[6] J. Clarke, A. N. Cleland, M. H. Devoret, D. Esteve, and J. M. Martinis, Science **239**, 992 (1988).

[7] Leon Gunther and Bernard Barbara, Eds., *Quantum Tunneling of Magnetization - QTM '94*, NATO ASI Series E: Applied Sciences, Vol. 301, Proceedings of the NATO Advanced Research Workshop on Quantum Tunneling of Magnetization - QTM '94, Grenoble and Chichilianne, France, June 27 - July 2, 1994 (Kluwer Academic, Dordecht, 1995).

[8] Ulrich Weiss, *Quantum Dissipative Systems* (World-Scientific, Singapore, 1993).

[9] W. Wernsdorfer and R. Sessoli, Science **284**, 133 (1999).

[10] J. Friedman, V. Patel, W. Chen, S. K. Tolpygo, and J. E. Lukens, Nature **406**, 43 (2000).

[11] A. J. Leggett and Anupam Garg, Phys. Rev. Lett. **54**, 857 (1985).

[12] K. Wieghardt, K. Pohl, I. Jibril, and G. Huttner, Angew. Chem. Int. Ed. Engl. **23**, 77 (1984).

[13] C. Delfs et al., Inorg. Chem. **32**, 3099 (1993).

[14] A. -L. Barra et al, Europhys. Lett. **35**, 133 (1996).

[15] C. Sangregorio et al., Phys. Rev. Lett. **78**, 4645 (1997).

[16] T. Ohm, C. Sangregorio, and C. Paulsen, Eur. Phys. J. B **6**, 195 (1998).

[17] L. D. Landau and E. M. Lifshitz, *Quantum Mechanics*, 3rd edition (Pergamon, Oxford, 1977), Sec. 90.

[18] Anupam Garg, Europhys. Lett. **22**, 205 (1993).

[19] D. Loss, D. P. DiVincenzo, and G. Grinstein, Phys. Rev. Lett. **69**, 3232 (1992); J. von Delft and C. L. Henley, *ibid.* **69**, 3236 (1992).

[20] E. Fradkin, *Field Theories of Condensed Matter Systems* (Addison-Wesley, Redwood City, 1991), Chap. 5.

[21] F. D. M. Haldane, Phys. Lett. A **93**, 464 (1983); Phys. Rev. Lett. **50**, 1153 (1983); J. Appl. Phys. **57**, 3359 (1985).

[22] Anupam Garg, Phys. Rev. B **5**, 15 161 (1995).

[23] M. V. Berry and M. Wilkinson, Proc. Roy. Soc. Lond. A **392**, 15 (1984).

[24] Anupam Garg, Phys. Rev. Lett. **83**, 4385 (1999); math-ph/0003005; cond-mat/0003114; cond-mat/0003156.

[25] J. Villain and Anna Fort, to appear in Eur. Phys. J. B.

[26] Ersin Keçecioğlu and Anupam Garg, cond-mat/0003319.

[27] Anupam Garg, Phys. Rev. B **60**, 6705 (1999).

[28] V. Ambegaokar, U. Eckern, and G. Schön, Phys. Rev. Lett **48**, 1745 (1982).

[29] C. Tesche, in *Proceeeedings of the New York Conference on Quantum Measurement Theory*, edited by D. Greenberger (New York Academy of Sciences, New York, N.Y., 1986).

Decoherence and its Implications in Quantum Computation and Information Transfer
T. Gonis and P.E.A. Turchi (Eds.)
IOS Press, 2001

Quantum Computing and Error Correction

Andrew M. Steane
Centre for Quantum Computation,
Department of Physics, University of Oxford,
Clarendon Laboratory, Parks Road, Oxford OX1 3PU, England

Abstract

The main ideas of quantum error correction are introduced. These are encoding, extraction of syndromes, error operators, and code construction. It is shown that general noise and relaxation of a set of 2-state quantum systems can always be understood as a combination of Pauli operators acting on the system. Each quantum error correcting code allows a subset of these errors to be corrected. In many situations the noise is such that the remaining uncorrectable errors are unlikely to arise, and hence quantum error correction has a high probability of success. In order to achieve the best noise tolerance in the presence of noise and imprecision throughout the computer, a hierarchical construction of a quantum computer is proposed.

1 Introduction

Quantum error correction (QEC) is one of the basic components of quantum information theory. It arises from the union of quantum mechanics and classical information theory, and leads to some powerful and surprising possibilities in the quantum context. For example, we may use a quantum error correcting code to allow us to store a general, unknown state of a two-state system in a set of five two-state systems (eg two-level atoms), in such a way that if any unknown change (arbitrarily large and including the case of relaxation) should occur in the state of any one atom, then it is still possible to recover the originally stored state precisely: the seemingly large perturbation had no effect at all on the stored information.

QEC is by now sufficiently well established that it has chapters devoted to it in text books, for example [1, 2, 3]. Nevertheless, an introduction to the ideas is still useful for people new to the subject, so much of the discussion here will be devoted to this. I will then discuss recent research in section 7.

QEC may be considered to consist of two separate parts. The first part is the theory of quantum error correcting codes, syndromes and quantum error operators. This is essentially a collection of physical insights and mathematical ideas which follow directly from the principles of quantum mechanics and are uncontroversial. The second part is the physics of noise, which must include the theory of how an open quantum system interacts with its environment. The success of QEC methods in practice will depend on our having the correct understanding of noise in open quantum systems. There is still some controversy over which methods can be correctly applied when we are considering large-scale systems such as a 'classical' environment, but the success of QEC is found not to depend on special pleading in this area. Nevertheless, it is necessary to be careful to examine what assumptions are required, so the presentation here will focus on this, particularly in sections 4 and 6.

2 Background

The first quantum error correcting codes were discovered independently by Shor [4] and Steane [5]. Shor proved that 9 qubits could be used to protect a single qubit against general errors, while Steane described a general code construction whose simplest example does the same job using 7 qubits (see section 5.1). A general theory of quantum error correction dates from subsequent papers of Calderbank and Shor [6] and Steane [7] in which general code constructions, existence proofs, and correction methods were given. Knill and Laflamme [8] and Bennett *et. al.* [9] provided a more general theoretical framework, describing requirements for quantum error correcting codes, and measures of the fidelity of corrected states.

The important concept of the stabilizer (section 5.1) is due to Gottesman [10] and independently Calderbank *et. al.* [11]; this yielded many useful insights into the subject, and permitted many new codes to be discovered [10, 11, 12]. Stabilizer methods will probably make a valuable contribution to other areas in quantum information physics. The idea of recursively encoding and encoding again was explored by several authors [13, 14, 15], this uses more quantum resources in a heirarchical way, to permit communication over arbitrarily long times or distances. Van Enk *et. al.* [16, 17] have discussed quantum communication over noisy channels using a realistic model of trapped atoms and high-quality optical cavities, and recursive techniques for systems in which two-way classical communication is possible were described by Briegel *et al.* [18].

Quantum error correction is discussed in reviews of quantum information theory, see for example [19, 20].

3 Three bit code

We will begin by analysing in detail the workings of the most simple quantum error correcting code. Exactly what is meant by a quantum error correcting code will become apparent.

Suppose a source A wishes transmit quantum information via a noisy communication channel to a receiver B. Obviously the channel must be noisy in practice since no channel is perfectly noise-free. However, in order to do better than merely sending quantum bits down the channel, we must know something about the noise. For this introductory section, the following properties will be assumed: the noise acts on each qubit independently, and for a given qubit has an effect chosen at random between leaving the qubit's state unchanged (probability $1-p$) and applying a Pauli σ_x operator (probability $p < 1/2$). This is a very artificial type of noise, but once we can correct it, we will find that our correction can also offer useful results for much more realistic types of noise.

The simplest quantum error correction method is summarised in fig. 1. We adopt the convention of calling the source Alice and the receiver Bob. The state of any qubit which Alice wishes to transmit can be written without loss of generality $a\,|0\rangle + b\,|1\rangle$.

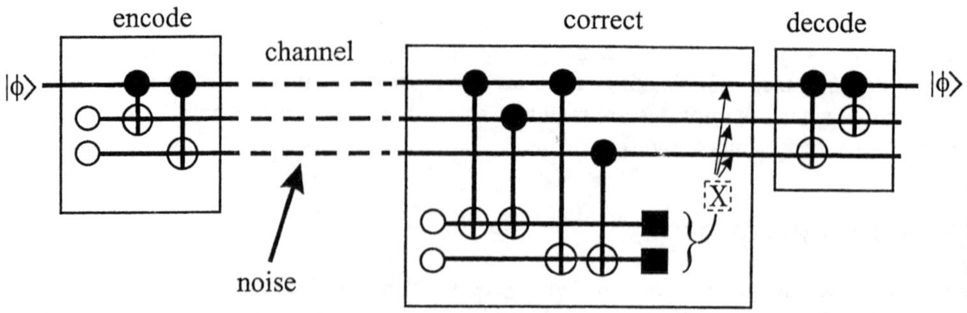

Figure 1: Example of quantum error correction which illustrates the main features. A single qubit is encoded into three, and these are sent down the noisy channel. The receiver introduces further qubits (the ancilla) and uses them to extract a syndrome by applying controlled-not gates and then measuring the ancilla bits. The received state can then be corrected. Finally, the single qubit of information may be re-extracted (decoded) from the three.

Alice prepares two further qubits in the state $|0\rangle$, so the initial state of all three is $a|000\rangle + b|100\rangle$. Alice now operates a controlled-NOT gate from the first qubit to the second, producing $a|000\rangle + b|110\rangle$, followed by a controlled-NOT gate from the first qubit to the third, producing $a|000\rangle + b|111\rangle$. Finally, Alice sends all three qubits down the channel.

The states $|000\rangle$ and $|111\rangle$ are called quantum codewords. Only codewords, or superpositions of them, are sent by Alice.

Bob receives the three qubits, but they have been acted on by the noise in the channel. Their state is one of the following:

$$\begin{array}{ll} \text{state} & \text{probability} \\ a|000\rangle + b|111\rangle & (1-p)^3 \\ a|100\rangle + b|011\rangle & p(1-p)^2 \\ a|010\rangle + b|101\rangle & p(1-p)^2 \\ a|001\rangle + b|110\rangle & p(1-p)^2 \\ a|110\rangle + b|001\rangle & p^2(1-p) \\ a|101\rangle + b|010\rangle & p^2(1-p) \\ a|011\rangle + b|100\rangle & p^2(1-p) \\ a|000\rangle + b|111\rangle & p^3 \end{array} \tag{1}$$

Bob now introduces two more qubits of his own, prepared in the state $|00\rangle$. This extra pair of qubits, referred to as an *ancilla*, is not strictly necessary, but makes error correction easier to understand and becomes necessary when fault-tolerant methods are needed. Bob uses the ancilla to gather information about the noise. He first carries out controlled-NOTs from the first and second received qubits to the first ancilla qubit, then from the first and third received qubits to the second ancilla bit. The total state of all five qubits is now

$$\begin{array}{cc} \text{state} & \text{probability} \\ (a\,|000\rangle + b\,|111\rangle)\,|00\rangle & (1-p)^3 \\ (a\,|100\rangle + b\,|011\rangle)\,|11\rangle & p(1-p)^2 \\ (a\,|010\rangle + b\,|101\rangle)\,|10\rangle & p(1-p)^2 \\ (a\,|001\rangle + b\,|110\rangle)\,|01\rangle & p(1-p)^2 \\ (a\,|110\rangle + b\,|001\rangle)\,|01\rangle & p^2(1-p) \\ (a\,|101\rangle + b\,|010\rangle)\,|10\rangle & p^2(1-p) \\ (a\,|011\rangle + b\,|100\rangle)\,|11\rangle & p^2(1-p) \\ (a\,|000\rangle + b\,|111\rangle)\,|00\rangle & p^3 \end{array} \tag{2}$$

Bob measures the two ancilla bits in the basis $\{|0\rangle, |1\rangle\}$. This gives him two classical bits of information. This information is called the *error syndrome*, since it helps to diagnose the errors in the received qubits. Bob's next action is as follows:

measured syndrome	action
00	do nothing
01	apply σ_x to third qubit
10	apply σ_x to second qubit
11	apply σ_x to first qubit

Suppose for example that Bob's measurements give 10 (i.e. the ancilla state is projected onto $|10\rangle$). Examining eq. (2), we see that the state of the received qubits must be either $a\,|010\rangle + b\,|101\rangle$ (probability $p(1-p)^2$) or $a\,|101\rangle + b\,|010\rangle$ (probability $p^2(1-p)$). Since the former is more likely, Bob corrects the state by applying a Pauli σ_x operator to the second qubit. He thus obtains either $a\,|000\rangle + b\,|111\rangle$ (most likely) or $a\,|111\rangle + b\,|000\rangle$. Finally, to extract the qubit which Alice sent, Bob applies controlled-NOT from the first qubit to the second and third, obtaining either $(a\,|0\rangle + b\,|1\rangle)\,|00\rangle$ or $(a\,|1\rangle + b\,|0\rangle)\,|00\rangle$. Therefore Bob has either the exact qubit sent by Alice, or Alice's qubit operated on by σ_x. Bob does not know which he has, but the important point is that the method has a probability of success greater than $1-p$. The correction is designed to succeed whenever either no or just one qubit is corrupted by the channel, which are the most likely possibilities. The failure probability is the probability that at least two qubits are corrupted by the channel, which is $3p^2(1-p) + p^3 = 3p^2 - 2p^3$, i.e. less than p (as long as $p < 1/2$).

To summarise, Alice communicates a single general qubit by expressing its state as a joint state of three qubits, which are then sent to Bob. Bob first applies error correction, then extracts a single qubit state. The probability that he fails to obtain Alice's original state is $O(p^2)$, whereas it would have been $O(p)$ if no error correction method had been used. We will see later that with more qubits the same basic ideas lead to much more powerful noise suppression, but it is worth noting that we already have quite an impressive result: by using just three times as many qubits, we reduce the error probability by a factor $\sim 1/3p$, i.e. a factor ~ 30 for $p = 0.01$, ~ 300 for $p = 0.001$, and so on.

3.1 Phase errors

The channel we just considered was rather artificial. However, it is closely related to a more realistic type of channel: one which generates random rotations of the qubits about the z axis. Such a rotation is given by the operator

$$P(\epsilon\phi) = \begin{pmatrix} e^{i\epsilon\phi} & 0 \\ 0 & e^{-i\epsilon\phi} \end{pmatrix} \equiv \cos(\epsilon\phi)I + i\sin(\epsilon\phi)\sigma_z$$

where I is the identity, ϵ is a fixed quantity indicating the typical size of the rotations, and $0 < \phi < 2\pi$ is a random angle. We may understand such an error as a combination of no error (I) and a phase flip error (σ_z). We can deal with this situation by using exactly the same quantum error correcting code as before, only now at either end of the channel we apply to each qubit the Hadamard rotation

$$H \equiv \frac{1}{\sqrt{2}} \begin{pmatrix} 1 & 1 \\ 1 & -1 \end{pmatrix}$$

The combined effect of phase error and these Hadamard rotations is

$$HPH = \cos(\epsilon\phi)I + i\sin(\epsilon\phi)\sigma_x \tag{3}$$

which is easy to derive using $HH = I$ and $H\sigma_z H = \sigma_x$.

We now have a situation just like that considered in the previous section, only instead of a σ_x 'bit flip' being applied randomly with probability p to each qubit, every qubit certainly experiences an error which is a combination of the identity and σ_x. When we put this type of error through the analysis of the correction network shown in figure 1, modelling the measurements on the ancilla qubits as standard Von-Neumann projective measurements, the outcome is exactly as before, with the quantity p equal to the average weight of the terms in the quantum state which have single bit-flip errors (before correction). This average is over the random variable ϕ, thus $p = \left\langle \sin^2 \epsilon\phi \right\rangle \simeq (2\pi\epsilon)^2/3$ for $\epsilon \ll 1$, where we have considered the worst case, in which the states being transmitted are taken to orthogonal states by the action of σ_z. A complete calculation of the 3-qubit system, with the random rotations and the syndrome measurement described, confirms that the fidelity of the final corrected state is $f = 1 - 3p^2$ for small p, in agreement with the results of the previous section.

3.2 Projective errors

It is a familiar feature of quantum mechanics that a set of particles where each is in an equal superposition of two states $(|0\rangle + \exp(i\phi)\,|1\rangle)/\sqrt{2}$, with the relative phase ϕ of the two terms random, is indistinguishable from a statistical mixture, i.e. a set of particles where each is either in the state $|0\rangle$ or in $|1\rangle$, randomly. This follows immediately from the fact that these two cases have the same density matrix. With this in mind, it should not be surprising that another type of error which the three-bit code can correct is an error consisting of a projection of the qubit onto the $|0\rangle, |1\rangle$ basis. To be specific, imagine the channel acts as follows: for each qubit passing through, either no change occurs (probability $1 - 2p$) or a projection onto the $|0\rangle, |1\rangle$ basis occurs (probability $2p$). The projection $|0\rangle\langle 0| = (I + \sigma_z)/2$ and $|1\rangle\langle 1| = (I - \sigma_z)/2$. Such errors are identical to phase errors (equation(3)), except for the absence of the factor i before the σ_z term. However, this factor does not affect the argument, and once again the analysis in equation (2) applies (once we have used the Hadamard 'trick' to convert phase errors to bit-flip errors) in the limit of small p. Note that we define p in terms of the effect of the noise on the state: it is not the probability that a projection occurs, but the probability that a σ_z error is produced in the state of any single qubit when quantum codewords are sent down the channel.

Another way of modeling projective errors is to consider that the projected qubit is first coupled to some other system, and then we ignore the state of the other system. Let such another system have, among its possible states, two states $|\alpha\rangle_e$ and $|\beta\rangle_e$ which are close to one another, $\langle\alpha|\beta\rangle = 1 - \epsilon$. The error consists in the following coupling between qubit and extra system:

$$(a|0\rangle + b|1\rangle)|\alpha\rangle_e \to a|0\rangle|\alpha\rangle_e + b|1\rangle|\beta\rangle_e \tag{4}$$

this is an entanglement between the qubit and the extra system. A useful insight is to express the entangled state in the following way:

$$a|0\rangle|\alpha\rangle + b|1\rangle|\beta\rangle \equiv \frac{1}{\sqrt{2}}\left[(a|0\rangle + b|1\rangle)|+\rangle_e + (a|0\rangle - b|1\rangle)|-\rangle_e\right] \tag{5}$$

where $|\pm\rangle_e = (|\alpha\rangle_e \pm |\beta\rangle_e)/\sqrt{2}$. Hence the error on the qubit is seen to be a combination of identity and σ_z, and is correctable as before. The probability that the σ_z error is produced is calculated by finding the weights of the different possibilities in equation (2) after the tracing over the extra system. We thus obtain $p = (1 - \mathrm{Re}(\langle\alpha|\beta\rangle))^2/2 = \epsilon^2/2$.

4 Most general possible error, digitization of noise

The most general possible error that a qubit can undergo is a general combination of the ones we have listed, that is, general rotations in the two-dimensional Hilbert space, combined with a possible projection onto any axis. In what follows we will model a projection by using the method just outlined, in which the qubit is entangled with another system, and the error consists in the fact that the entangling process can not be undone, because the other system is not under our control. In QEC it has become standard practice to refer to coupling to 'the environment' in this way. It might seem that this depends on modelling the environment as if it is a quantum system with unitary evolution, but in fact the technique is only a mathematical convenience for dealing with projections. If one is uncomfortable with writing down quantum states of the environment, then one should return to the language of projection, and the conclusions concerning the efficacy of QEC will remain unchanged.

With these considerations in mind, we now come to one of the basic insights of QEC: a completely arbitrary change in a general state $|\phi\rangle$ of a qubit can be written

$$|\phi\rangle|\psi_0\rangle_e \to \sum_{i=\{I,x,y,z\}} (\sigma_i|\phi\rangle)|\psi_i\rangle_e \tag{6}$$

where we have written $I = \sigma_I$ and the second system with states $|\psi_i\rangle_e$ (typically called 'states of the environment') is introduced simply to allow us to write using the language of kets a general change in the density matrix of the qubit. This density matrix of the qubit is the reduced density matrix of the pair of systems, after tracing over the second system. We can thus reproduce arbitrary changes in the density matrix of the qubit as long as we allow the states $|\psi_i\rangle_e$ to be unconstrained, that is, they are not necessarily either normalised or orthogonal.

Generalising this insight, we find that any transformation of the state of a set of qubits can be expressed

$$|\phi\rangle|\psi_0\rangle_e \to \sum_i (E_i|\phi\rangle)|\psi_i\rangle_e \tag{7}$$

where each 'error operator' E_i is a tensor product of Pauli operators acting on the qubits, $|\phi\rangle$ is the initial state of the qubits, and $|\psi_i\rangle_e$ are states of a second system, not necessarily orthogonal or normalised.

We thus express general noise and/or decoherence in terms of Pauli operators σ_x, σ_y, σ_z acting on the qubits. These will be written $X \equiv \sigma_x$, $Z \equiv \sigma_z$, $Y \equiv -i\sigma_y = XZ$. The remarkable fact embodied by (7) is that arbitrary changes, which form a continuous set, can be 'digitized' by expressing them as a weighted sum of discrete errors. This arises from the fact that the Pauli matrices with the identity form a complete set.

To write tensor products of Pauli matrices acting on n qubits, we introduce the notation $X_u Z_v$ where u and v are n-bit binary vectors. The non-zero coordinates of u and v indicate where X and Z operators appear in the product. For example,

$$X \otimes I \otimes Z \otimes Y \otimes X \equiv X_{10011} Z_{00110}. \tag{8}$$

The *weight* of an error operator is the number of terms in the tensor product which are not the identity, this is 4 in the example just given.

Error correction is a process which takes a state such as $E_i |\phi\rangle$ to $|\phi\rangle$. Correction of X errors takes $X_u Z_v |\phi\rangle$ to $Z_v |\phi\rangle$; correction of Z errors takes $X_u Z_v |\phi\rangle$ to $X_v |\phi\rangle$. Putting all this together, we discover the highly significant fact that to correct *the most general possible* noise (eq. (7)), it is sufficient to correct just X and Z errors. See eq. (9) and following for a proof. It may look like we are only correcting unitary 'bit flip' and 'phase flip' rotations, but this is a false impression: actually we will correct everything, including non-unitary relaxation processes!

5 Correction of general errors

A general quantum error correcting code will be an orthornormal set of n-qubit states (quantum codewords) which allow correction of all members of a set $S = \{E_i\}$ of *correctable errors.* We are usually interested in the case where the correctable errors include all errors (including X, Y or Z and combinations thereof for different qubits) of weight up to some maximum w. Such a code is called a w-error correcting code. It can be shown that such a code allows perfect correction of any change in the n-qubit quantum state which affects at most w qubits [7, 6]. The correction method is comparable to that shown in figure 1 for the simple 3-bit code, where first a syndrome is extracted, and then the state is corrected accordingly.

The main steps in the proof of this claim are as follows. Let $|\phi\rangle_L$ be a state consisting of a general superposition of codewords of some QEC code. A general noise process (7) will then produce the state

$$\sum_i (E_i |\phi\rangle_L) |\psi_i\rangle_e \tag{9}$$

The syndrome extraction can be done most simply by attaching an $n-k$ qubit ancilla a to the system (this is not the only method, but is convenient), and storing in it the parity checks which are satisfied by the quantum codewords, using controlled-not and Hadamard operations. Another way of saying the same thing is that we store into the ancilla the eigenvalues of a set of simultaneously commuting operators (the *stabilizer*) acting on the noisy state. The quantum codewords themselves are all simultaneous eigenstates with eigenvalue 1 of all the operators in the stabilizer, so such a process does not have any effect on a general noise-free state. In the case of a noisy state, the effect is to couple system and environment with the ancilla as follows:

$$|0\rangle_a \sum_i \left(E_i |\phi\rangle_L\right) |\psi_i\rangle_e \rightarrow \sum_i |s_i\rangle_a \left(E_i |\phi\rangle_L\right) |\psi_i\rangle_e . \tag{10}$$

The s_i are $(n-k)$-bit binary strings, the syndromes. A projective measurement of the ancilla will collapse the sum to a single syndrome taken at random: $|s_i\rangle_a \left(E_i |\phi\rangle_L\right) |\psi_i\rangle_e$, and will yield s_i as the measurement result. To keep the argument clear, we have assumed that every E_i will have a different syndrome. This assumption will be dropped in a moment. Since there is only one E_i with the syndrome we have measured, we can deduce the operator E_i^{-1} which should now be applied to correct the error. The resulting state is $|s_i\rangle_a |\phi\rangle_L |\psi_i\rangle_e$, and now we can ignore the state of environment and ancilla without affecting the system, since they are in a product state. The system is therefore "magically" disentangled from its environment, and perfectly restored to $|\phi\rangle_L$!

This remarkable process can be understood as first forcing the general noisy state to 'choose' among a discrete set of errors, via a projective measurement, and then reversing the particular discrete error 'chosen' using the fact that the measurement result tells us which one it was. Alternatively, the correction can be accomplished by a unitary evolution consisting of controlled gates with ancilla as control, system as target, effectively transferring the noise (including entanglement with the environment) from system to ancilla.

In general the true situation is that more than one error operator has the same syndrome, just as in equation (2). If two error operators E_1, E_2 have the same syndrome s for a given QEC code, then upon obtaining s in the syndrome extraction we apply to the qubits the inverse of whichever of E_1 and E_2 was most likely to have been produced by the noise process. The other error operator is then an uncorrectable error. This will be considered further in section 6.

5.1 Example codes

The above abstract treatment is clarified by some specific examples. In section 3 we considered a simple example where *either* an X error *or* a Z error (on any single qubit) was correctable. The most important generalisation is to the case of a code which allows *both* an X error *and* a Z error to be corrected. We will then further generalise to the case where errors affecting more than one qubit are correctable.

We have already noted that $HZH = X$ and this means that if the set of states (quantum error correcting code) $\{|u\rangle\}$ are to be correctable for X and Z errors, it is sufficient and necessary that $\{|u\rangle\}$ and $\{\mathbf{H}|u\rangle\}$ both be correctable for X errors, where $\mathbf{H} = H \otimes H \otimes H \cdots \otimes H$ signifies the Hadamard rotation applied to all n qubits. This was how the first general class of QEC codes was discovered [7, 6]. A general formalism in which such ideas can be handled succinctly has been developed by Gottesman [10] and Calderbank *et al.* [21]. The simplest (though not the shortest) single-error correcting code is the 7-bit code having two codewords given by

$$\begin{aligned} |0\rangle_L &= |0000000\rangle + |1010101\rangle + |0110011\rangle + |1100110\rangle \\ &\quad + |0001111\rangle + |1011010\rangle + |0111100\rangle + |1101001\rangle , \\ |1\rangle_L &= X_{1111111} |0\rangle_L . \end{aligned} \tag{11}$$

This code is constructed by combining two classical error correcting codes: the states occurring in $|0\rangle_L$ are all the members of a classical error correcting code C which has minimum distance 4, and the states occuring in $\mathbf{H}\,|0\rangle_L$ are all the members of another classical error correcting code $C^{\perp}$ (the dual of C) which has minimum distance 3 and which contains C. Furthermore, any superposition of $|0\rangle_L$ and $|1\rangle_L$ only involves members of $C^{\perp}$, and this is also true for any superposition of $\mathbf{H}\,|0\rangle_L$ and $\mathbf{H}\,|1\rangle_L$. It is not my purpose to present this code construction fully, this may be found in the references given in the introductory sections above. The aim is to give enough information here to enable an understanding of the main themes to be gained.

I referred just now to the concept of 'minimum distance'. For a set of binary words, the minimum distance is the minimum number of bits which must be flipped to convert one binary word in the set into another in the set. A classical error correcting code of distance d can correct any set of bit flips as long as they affect no more than $(d-1)/2$ bits. In the quantum case it can be shown that codes can be constructed which allow correction of any error operator E of weight less than or equal to $(d-1)/2$, where d is now the minimum distance of the quantum code.

The code given by equation (11) has distance 3 and is therefore a single-error-correcting quantum code. Since there are two orthogonal states in the code, only a single qubit of information can be stored in the 7 qubits employed. These parameters are summarised by the notation $[[n,k,d]] = [[7,1,3]]$. The code construction methods are now quite advanced. Infinite sets of codes of given distance are known, as well as a large set of quite powerful codes [21, 12]. Some parameters of known codes include, for example, $[[5,1,3]]$, $[[8,3,3]]$, $[[21,6,5]]$, $[[23,1,7]]$, $[[127,29,15]]$, $[[127,50,15]]$. The 5-bit $[[5,1,3]]$ code has the minimum number of bits needed to allow correction of a single general error.

6 Success probability

As I have remarked in the introduction, the properties of QEC codes, such as the set of correctable errors and the syndrome extraction, follow immediately from the basic principles of quantum mechanics. If a channel truly produces noise of the form (7) where all the error operators are in the correctable set for some code, then that code will certainly allow error-free communication down the channel. However, any realistic channel will produce not only correctable errors, but also uncorrectable ones. The essence of evaulating the success of QEC is to calculate the probability that the error correction will work, and this is identical to calculating the probability P_u that the channel will produce an uncorrectable error. The fidelity of the final state of the system of qubits, after passing through the channel and having correction applied, will be $1 - P_u$. For the rather artificial channel considered in section 3, $P_u = 3p^2 - 2p^3$, and for the more realistic case of random rotations through angles $\epsilon\phi$, considered in section 3.1, we obtained $P_u \simeq 3p^2 = (2\pi\epsilon)^4/3$ in the limit of small ϵ.

More general types of noise can be modelled in a number of ways. As remarked before, by 'noise' we mean simply any unknown or unwanted change in the density matrix of our system. One approach to modelling it is to consider the Hamiltonian describing a coupling between the set of qubits and their environment. For example, if the qubits are two-level atoms, we might consider the effects of spontaneous emission of photons, adopting a master equation approach.

The statement (7) about digitization of noise is equivalent to the statement that any interaction between a system of qubits and its environment can be written in the form

$$H_I = \sum_i E_i \otimes H_i^e \tag{12}$$

where the operators H_i^e act on the environment. Under the action of this coupling, the density matrix of the system (after tracing over the environment) evolves from ρ_0 to $\sum_i a_i E_i \rho_0 E_i^\dagger$. QEC returns all terms of this sum having correctable E_i to ρ_0. Therefore, the fidelity of the corrected state, compared to the noise-free state ρ_0, is determined by the sum of all coefficients a_i associated with uncorrectable errors.

For a mathematically thorough analysis of this problem, see [13, 8, 22]. The essential ideas are as follows. Noise is typically a continuous process affecting all qubits all the time. However, when we discuss QEC, we can always adopt the model that the syndrome is extracted by a projective measurement. Any statement such as 'the probability that error E_i occurs' is just a short-hand for 'the probability that the syndrome extraction projects the state onto one which differs from the noise-free state by error operator E_i'. We would like to calculate such probabilities.

To do so, it is useful to divide up (12) into a sum of terms having error operators of different weight:

$$H_I = \sum_{\mathrm{wt}(E)=1} E \otimes H_E^e + \sum_{\mathrm{wt}(E)=2} E \otimes H_E^e + \sum_{\mathrm{wt}(E)=3} E \otimes H_E^e + \dots \tag{13}$$

There are $3n$ terms in the first sum, $3^2 n!/(2!(n-2)!)$ terms in the second, and so on. The strength of the system-environment coupling is expressed by coupling constants which appear in the H_E^e operators. In the case that only the weight 1 terms are present, we say the environment acts independently on the qubits: it does not directly produce correlated errors across two or more qubits. This language has lead to some confusion, since clearly the environment is acting on all the qubits at once, and sometimes the term 'spatially correlated' is used to refer to any situation where one degree of freedom (such as a mode of the environment) can interact with all the qubits. Nevertheless, if the interaction only contains error operators of weight 1, then in the terminology of QEC we have a situation of independent or uncorrelated errors. The important point is that in this case, although errors of all weights will still appear in the density matrix of the noisy system, the size of the terms corresponding to errors of weight w will scale as $O(\epsilon^{2w})$, where ϵ is a parameter giving the system-environment coupling strength.

As an example, consider dephasing produced by coupling between a set of qubits and a bath of oscillators:

$$H_I(t) = \hbar \sum_{q=1}^{n} \sum_k \sigma_{z,q} \left(g_{kq}^* e^{-i\omega_{kq} t} a_{kq} + g_{kq} e^{i\omega_{kq} t} a_{kq}^\dagger \right)$$

where g_{kq} are the coupling coefficients, a_{kq} are the lowering operators for the modes. Each mode here couples to all the qubits, but the error operators σ_z in the Hamiltonian are all of weight 1, and the resulting system density matrix will be of the form $\sum_i a_i E_i \rho_0 E_i^\dagger$ where for small times t,

$$a_i \sim O\left(\left| \sum_{q=1}^{n} \sum_k g_{kq} \right|^{2\,\mathrm{wt}(E_i)} \right).$$

Since QEC restores all terms in the density matrix whose errors are of weight $\le t = (d-1)/2$, the overall failure probability P_u is the probability $P(t+1)$ for the noise to generate an error of weight $t+1$ or more. In the case of uncorrelated noise, this probability is approximately

$$P(t+1) \simeq \left(3^{t+1}\binom{n}{t+1}\epsilon^{t+1}\right)^2 \tag{14}$$

when all the single-qubit error amplitudes can add coherently (i.e. the qubits share a common environment), or

$$P(t+1) \simeq 3^{t+1}\binom{n}{t+1}\epsilon^{2(t+1)} \tag{15}$$

when the errors add incoherently (i.e. either seperate environments, or a common environment with couplings of randomly changing phase). The significance of (14) and (15) is that they imply QEC works extremely well when t is large and $\epsilon^2 < t/3n$. Since good codes exist, t can in fact tend to infinity while t/n and k/n remain fixed. Therefore as long as the noise per qubit is below a threshold around $t/3n$, almost perfect recovery of the state is possible. The ratio t/n constrains the rate k/n of the codes which can be constructed.

As a numerical example, consider the [[127, 29, 15]] code. Let us suppose we employ this code to store a general 29-qubit state in a collection of 127 two-level atoms, and we envisage that the excited state of each atom is subject to spontaneous decay with a lifetime of 1 second. If the sequence of operations required for syndrome extraction can be completed in 1 millisecond, then the error probability per atom is of order $p = 10^{-3}$. The probability that an uncorrectable error occurs is approximately $(3np)^8/8! \simeq 10^{-8}$. Therefore we could repeatedly correct the state for 10^5 s= 1 day before the fidelity of the stored state falls to a half, despite the fact that during such a process every one of the atoms will have decayed on average about 50,000 times!

Such uncorrelated noise is a reasonable approximation in many physical situations, but we need to be careful about the degree of approximation, since we are concerned with very small terms of order ϵ^d. If we relax the approximation of completely uncorrelated noise, equations (14) and (15) remain approximately unchanged, if the coupling constants in (13) for errors of weight t are themselves of order $\epsilon^t/t!$ [14].

A very different case in which QEC is also highly successful is when a set of correlated errors, also called burst errors, dominate the system-environment coupling, but we can find a QEC whose stabilzer includes all these correlated errors. This is sometimes called 'error avoiding' rather than 'error correction' since by using such a code, we don't even need to correct the logical state: it is already decoupled from the environment. The general lesson is that the more we know about the environment, and the more structure there exists in the system-environment coupling, the better able we are to find good codes.

The obvious approach to take in practice is a combined one, in which we first discover the correlated contributions to the noise in our system, and design a first layer of encoding accordingly, and then overlay a second layer optimized for minimum-distance coding. Such ideas have been studied in classical coding theory, where sometimes many layers of encoding are combined, including tricks such as not placing adjacent physical bits in the same logical block when the code is not designed for burst errors.

The process of encoding one bit in several, and then encoding each of those bits, and so on, is referred to as code *concatenation*. When used recursively, we obtain a powerful code whose behaviour is relatively simple to analyse. This is the structure which underlies the "threshold result", which states that arbitrarily long quantum computations can be made reliable by introducing more and more layers of concatenation, conditioned only that the level of noise per time step and per elementary operation on the physical hardware is below a finite threshold. In other words, we don't require a more and more precise and noiseless computer if we want to evolve longer and longer computations. However, we do need a bigger one.

7 Fault tolerant methods

There is ongoing research on constructing quantum error correcting codes, and understanding the bounds on their size. In particular, there are interesting cases of codes or correction methods which are not covered by the stabilizer formalism.

In another direction, we are interested in codes whose construction is thoroughly understood, but we wish to consider what happens when we take a further step towards a realistic model of a quantum computer. This further step is to allow not only the channel, but also all the quantum gate operations involved in the corrective process to be themselves noisy. It was far from obvious whether QEC could still succeed in such conditions, since if we 'correct' a set of qubits on the basis of a faulty syndrome, we will in fact introduce more errors.

Methods which can have a good probability of success even when all the operations involved are imperfect are called 'fault tolerant'. Fault tolerance is possible through the use of simple ideas such as repetition, and also some more subtle ideas which allow us to construct networks of operations in which the routes by which errors can propagate around a set of qubits are restricted. The main themes were proposed by Shor [23] and helpfully summarised by Preskill [24, 1]. A significant collection of further ideas have been put forward by Gottesman [25, 26], which allow fault tolerant methods to be found for a wide class of QEC codes, and the methods were further improved in [27, 28].

It was shown in [28] that a quantum computer whose size was a factor 22 larger than the number of logical qubits required for its computations could be stabilized sufficiently to allow large computations once the error probability per elementary gate operation was of order 10^{-5}, and the error probability for every other qubit not participating in the gate was of order 10^{-7} during the time of one gate. The analysis was based on a $[[127, 29, 15]]$ code. To calculate the overall failure probability it is necessary to track errors through networks of many hundreds of elementary gates, and it is only possible to do this analytically in an approximate way. I have now written a (classical) computer code to automate such error tracking calculations, and this has allowed the analytical approximate methods to be checked for the case of smaller codes such as $[[23, 1, 7]]$. This gives some confidence that the estimates in [28] are correct for the codes and networks considered there.

However, any such estimates involve assumptions not only about what codes and fault tolerant techniques are employed, but also about what degree of communication and of parallelism the physical hardware of the quantum computer allows. By communication we mean the ease with which one qubit can be coupled in a quantum gate with any other, not necessarily a neighbouring one. By parallelism we mean the degree to which operations can be carried out simultaneously on different sets of qubits. The case considered in [28] assumed that controlled multiple-not operations, in which a single control qubit is coupled to multiple target qubits, would be possible within each block of 127 qubits in a single time step, that operations in different blocks could be done in parallel, and that qubits in one block could be coupled to qubits in any one of several

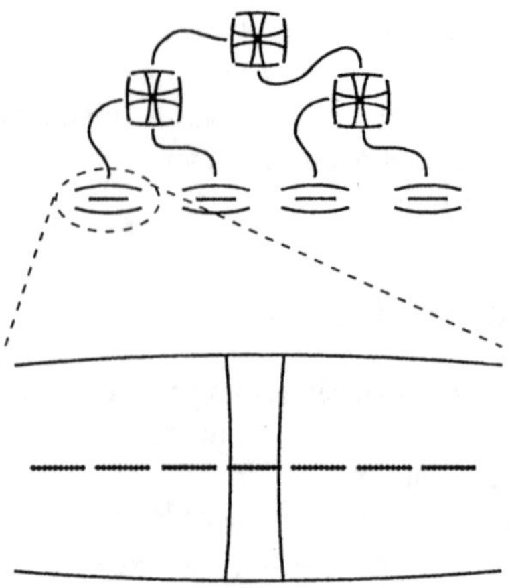

Figure 2: Hierarchical design for a quantum computer. The qubits are in groups of about 10. The physical mechanism of the computer allows gates between any pair in a group, and between the end qubits of each group and their nearest neighbours. Around 10 such groups form a block (the illustration shows 7 groups per block). Each block is coupled to the others by another physical mechanism, such as an optical method, and this allows quantum communication across long distances in the computer. To allow any one block to communicate with one of several others, switching networks may be necessary, as illustrated.

other blocks without introducing extra operations to cover communication problems. Unfortunately, it is not clear whether these assumptions are realistic.

High parallelism and ready communication tend to be mutually exclusive. At one extreme is the case of only nearest-neighbour interactions, i.e. limited communication, but any number of (logically mutually consistent) operations can be carried out in parallel. It was shown in [29] that in this case fault-tolerant methods can succeed, but no estimate of the noise threshold has yet been made. It is certainly not as high as the figures just quoted, which assumed much better communication. At the other extreme is the case of perfect communication, that is any qubit can communicate directly with any other, through a single shared degree of freedom such as a photon mode or the motion of ions in a single trap, but with no parallelism for gates involving more than one qubit. It is unlikely that this case can be stabilized sufficiently to allow large quantum computations.

Obviously the most robust physical system will be one which combines good parallelism with good communication, but the extreme case of full parallelism and direct communication between every pair of qubits is almost certainly not optimal because any process which allows such a high degree of inter-qubit coupling will have correspondingly high unwanted coupling to noisy degrees of freedom. I propose that the best construction for a quantum computer is a *hierarchical* one, as illustrated in figure 2. The most important role of the parallelism and communication inside the computer is to get the syndromes out as quickly as possible, so that qubits are not left accumulating noise for too long before they are corrected. The syndrome extraction operation, when carried out in a fault-tolerant manner, is itself hierarchical, because many two-bit logic gates are needed between qubits in each ancilla block, while coupling between one block and another is needed less frequently, and distant blocks need to communicate the least. The idea of the hierarchical quantum computer is that the physical hardware should reflect this logical state of affairs if an optimal robustness against noise is to be achieved.

Acknowledgements

I thank the organisers of the workshop. This work was supported by EPSRC and the European Community network QUBITS.

References

[1] H.-K. Lo, S. Popescu, and T. Spiller, editors. *Introduction to quantum computation and information.* World Scientific, Singapore, 1998.

[2] D. Bouwmeester, A. Ekert, and A. Zeilinger, editors. *The physics of quantum information.* Springer-Verlag, Berlin, 2000.

[3] Michael A. Nielsen and Isaac L. Chuang. *Quantum Computation and Quantum Information.* Cambridge University Press, 2000.

[4] P. W. Shor. Scheme for reducing decoherence in quantum computer memory. *Phys. Rev. A*, 52:R2493–6, 1995.

[5] A. M. Steane. Error correcting codes in quantum theory. *Phys. Rev. Lett.*, 77:793–797, 1996.

[6] A. R. Calderbank and P. W. Shor. Good quantum error–correcting codes exist. *Phys. Rev. A*, 54:1098–1105, 1996.

[7] A. M. Steane. Multiple particle interference and quantum error correction. *Proc. Roy. Soc. Lond. A*, 452:2551–2577, 1996.

[8] E. Knill and R. Laflamme. A theory of quantum error-correcting codes. *Phys. Rev. A*, 55:900–911, 1997. quant-ph/9604034,Phys.Rev.Lett. 84 (2000) 2525-2528.

[9] C. H. Bennett, D. P. DiVincenzo, J. A. Smolin, and W. K. Wootters. Mixed state entanglement and quantum error correction. *Phys. Rev. A*, 54:3825, 1996.

[10] D. Gottesman. Class of quantum error-correcting codes saturating the quantum hamming bound. *Phys. Rev. A*, 54:1862–1868, 1996.

[11] A. R. Calderbank, E. M. Rains, N. J. A. Sloane, and P. W. Shor. Quantum error correction and orthogonal geometry. *Phys. Rev. Lett.*, 78:405–409, 1997.

[12] A. M. Steane. Enlargement of Calderbank-Shor-Steane quantum codes. *IEEE Transactions on Information Theory*, 45:2492–2495, 1999.

[13] Emanuel Knill and Raymond Laflamme. Concatenated quantum codes. *quant-ph/9608012*, 1996.

[14] Dorit Aharonov and Michael Ben-Or. Fault-tolerant quantum computation with constant error rate. *SIAM Journal of Computation*, page submitted for publication. eprint quant-ph/9906129.

[15] A. Y. Kitaev. Quantum error correction with imperfect gates. In *Quantum Communication, Computing and Measurement (Proc. 3rd Int. Conf. of Quantum Communication and Measurement)*, pages 181–188, New York, 1997. Plenum Press.

[16] S. J. van Enk, J. I. Cirac, and P. Zoller. Ideal quantum communication over noisy channels: a quantum optical implementation. *Phys. Rev. Lett.*, 78:4293–4296, 1997.

[17] S. J. van Enk, J. I. Cirac, and P. Zoller. Purifying two-bit quantum gates and joint measurements in cavity QED. *Phys. Rev. Lett.*, 79:5178–5181, 1997.

[18] H.-J. Briegel, W. Dür, J. I. Cirac, and P. Zoller. Quantum repeaters: The role of imperfect local operations in quantum communication. *Phys. Rev. Lett.*, 81:5932–5935, 1998.

[19] Andrew Steane. Quantum computing. *Rep. Prog. Phys.*, 61:117–173, 1998.

[20] C. H. Bennett and P. W. Shor. Quantum information theory. *IEEE Transactions on Information Theory*, 44:2724–2742, 1998.

[21] A. R. Calderbank, E. M. Rains, P. W. Shor, and N. J. A. Sloane. Quantum error correction via codes over GF(4). *IEEE Transactions on Information Theory*, 44:1369–1387, 1998.

[22] E. Knill, R. Laflamme, and L. Viola. Theory of quantum error correction for general noise. *Phys. Rev. Lett.*, 84:2525–2528, 2000.

[23] P. W. Shor. Fault-tolerant quantum computation. In *Proc. 35th Annual Symposium on Fundamentals of Computer Science*, pages 56–65, Los Alamitos, 1996. IEEE Press. quant-ph/9605011.

[24] J. Preskill. Reliable quantum computers. *Proc. R. Soc. Lond. A*, 454:385–410, 1998.

[25] D. Gottesman. A theory of fault-tolerant quantum computation. *Physical Review A*, 57:127–137, 1998. quant-ph/9807006.

[26] D. Gottesman and I. L. Chuang. Quantum teleportation is a universal computational primitive. *Nature*, 402:390, 1999. quant-ph/9908010.

[27] A. M. Steane. Active stabilisation, quantum computation, and quantum state sythesis. *Phys. Rev. Lett.*, 78:2252–2255, 1997. quant-ph/9608026.

[28] A. M. Steane. Efficient fault-tolerant quantum computing. *Nature*, 399:124–126, 1999. quant-ph/9809054.

[29] D. Gottesman. Fault-tolerant quantum computation with local gates. 1999. quant-ph/9903099.

Decoherence and its Implications in Quantum Computation and Information Transfer
T. Gonis and P.E.A. Turchi (Eds.)
IOS Press, 2001

From entanglement witnesses to positive maps: towards optimal characterisation of separability

Paweł HORODECKI
Faculty of Technical Physics and Applied Mathematics, University of Gdańsk, 80–952 Gdańsk, Poland

Abstract

Basic aspect of both entanglement witnesses as well as positive maps methods of characterisation of bipartite separability is briefly reviewed. The operators - maps isomorphism and applied to point out that any hermitian operator with partial trace equal to identity corresponds to identity preserving linear map. Using this property and entanglement witnesses theory it is proved that *identity preserving* positive maps characterise bipartite separability completely. The possibility of simple generalisation of the result the case of multipartite case is poined out.

1 Introduction

Quantum entanglement is one of mysterious resources quantum mechanics provided us for information processing. In particular it has been shown to serve as source of save cryptographic key [1], a medium in quantum teleportation process [2]. It allows for super-dense coding [3] and takes part in one of very promising phenomena in quantum information theory i.e. quantum computing [4]. While original theoretical proposals of those phenomena involved pure states in practical implementations one deals with mixed states. Given mixed state it is important to ask whether or not it involves entanglement. In particular there is a need of methods of detection of entanglement on both mathematical and physical grounds. First, we need mathematical tools to judge whether given density matrix written on the paper represents quantum entanglement. Second, we should look for physical methods to detect entanglement of the state of given physical system.

In the case of pure states of, say, bipartite system, the problem is relatively simple. One look at the reduced density matrix and the global state represents nonzero entanglement iff the matrix is mixed. Obviously, it is not the case for mixed states - the maximally mixed state represents no entanglement (as it can be represented as equally weighted random mixture of pure product states) although it has mixed reduced states.

In this paper we shall focus first on bipartite case i.e. the case of the system composed of two quantum systems. Finally we point out that the main result can be easily generalised to the multipartite case.

First, in section 2, we recall the notion of quantum separability and ways of its characterisations. In section 3 after proving some simple properties of hermitian maps we prove the main result of the paper. Section 4 contains discussion of the results.

2 Characterisation of separability – review of needed properties

2.1 Definition

Quantum separability is defined as a lack of quantum correlations in the system. The formal definition [5] in the case of bipartite case is the following [1]

Definition 1 *The state ϱ acting on the Hilbert space $\mathcal{H} = \mathcal{H}_1 \otimes \mathcal{H}_2$ is called separable if it can be approximated in the trace norm by the states of the form*

$$\varrho = \sum_{i=1}^{k} p_i \varrho_i \otimes \tilde{\varrho}_i \tag{1}$$

where ϱ_i and $\tilde{\varrho}_i$ are states on $\mathcal{H}_1$ and $\mathcal{H}_2$ respectively.

Usually we deal with a *finite* dimensional Hilbert space $\dim\mathcal{H} = N < \infty$. For this case it is known [6, 7] that any separable state can be written as a convex combination of *finite* product pure states $k \leq N^2$, i.e. in those cases (unlike in the case of continuous variables) the "approximation" part of the definition is not needed.

What is the physical sense of the above definition? There is an intuitive interpretation in Ref. [8] which is a generalisation of the general principle [9, 10] saying that sometimes it is not possible to ascribe the state vectors to subsystems despite the fact that the wave function of the whole system is well defined. Namely the explanation of the status of any quantum state not satisfying the above definition goes as follows [8]: "As one knows, a mixed state can in general come from reduction of some pure state or from a source producing randomly pure states. If a mixed state is separable, then it produces the statistics equivalent to the one generated by an ensemble of product states. In the latter case, the nonfactorability is due to the lack of knowledge of the observer only. However if a mixed state is inseparable, then there is certainly no way to ascribe to the subsystems, even in principle, their state vectors."

2.2 Entanglement witnesses and positive maps as "detectors" of entanglement

Now we are interested in having both mathematical and physical operational tools of detecting quantum entanglement i.e. revealing possible violation of condition (1). In particular it would be good to have set of "detectors" allowing to reveal *any* violation of (1). Form the mathematical point of view in such case we say about operational *necessary and sufficient conditions* for separability. Those are some number of mathematical conditions which are satisfied simultaneously by a given sate if and only if it is separable. Otherwise there should be at least one of the conditions which is violated.

From physical point of view we should look for possible interpretations and physical of given set of (possibly reformulated) conditions.

Historically the first set of separability conditions were Bell inequalities. But it was not realised until the proof in Ref. [5] was provided. The first *scalar* condition of separability not corresponding to Bell inequality but being restriction to mean value of some observable was provided also in Ref. [5]. The other conditions based on entropy inequalities were provided [11].

[1] The present definition is due to Werner [5], who used the notion classically correlated states.

The first *structural* condition i. e. the one referring to the global properties of operator instead of using scalar inequality has been introduced by Peres [12].

This condition is sometimes called *positive partial transpose (PPT) separability condition.* It states that

Theorem I .- [12] If ϱ is separable then the matrix ϱ^{T_B} defined by the equation

$$\varrho^{T_B}_{m\mu,n\nu} \equiv \langle m, \mu | \varrho^{T_B} | n, \nu \rangle = \varrho_{m\nu,n\mu} \tag{2}$$

is still a legitimate state.

Note that the matrix ϱ^{T_B} is associated with an arbitrary fixed product orthonormal basis $\{|i,j\rangle\}$. Still, the spectrum of ϱ does not depend on the choice of the basis.

The above criterion happened to be enormously strong. There was a question: is it also sufficient ? The answer to this question was given later with aid of the general characterisation of separability has been provided in two following ways [13]:

- First it was formulated in terms of mean values of some hermitian operators:

  Characterisation 1.- The state ϱ is separable iff $\mathrm{Tr}(\varrho X) \geq 0$ *for all hermitian operators X such that*

$$\mathrm{Tr}(\sigma_{sep} X) \geq 0 \tag{3}$$

  for all separable σ_{sep} holds.

- Second it was given by means of positive maps (PM):

  Characterisation 2.- state ϱ defined on $\mathcal{H}_1 \otimes \mathcal{H}_2$ is separable iff operator

$$[I \otimes \Lambda](\varrho) \geq 0 \tag{4}$$

  for all positive Λ such that $\Lambda : \mathcal{B}(\mathcal{H}_B) \to \mathcal{B}(\mathcal{H}_A)$.

Recall that the map is called *positive* if it maps matrices with nonnegative spectrum onto the matrices of the same kind.

There is a link between those two characterisations given by Jamiołkowski [14] isomorphism:

$$X^{\Lambda} = [I \otimes \Lambda^{\dagger}](P_+) \tag{5}$$

Where $P_+ = |\Psi\rangle\langle\Psi|$, $|\Psi\rangle = \sum_{i=1}^{dim\mathcal{H}_A} |k\rangle \otimes |k\rangle$, $|\Psi\rangle \in \mathcal{H}_A \otimes \mathcal{H}_A$. Here $\Lambda^{\dagger} : \mathcal{H}_A \to \mathcal{H}_B$ is the map which is conjugated to Λ (see sect. 3.1).

Further it has been pointed out [15] that all X-s which are positive (i.e. have nonnegative spectrum) can be abandoned as the satisfy the formula (3) trivially. Basing on Characterisation 1 the definition of entanglement witness has been introduced:

Definition .- [15] Hermitian operator W is entanglement witness (EW) iff it satisfies (3) for all σ separable, but has negative mean value on some entangled state.

Of course it means (see [17]) that W *has at least one negative eigenvalue.* The final form of the characterisation of bipartite separability is [15]:

Characterisation 1'.- The state ϱ is separable iff it satisfies $Tr(W\varrho) \leq 0$ for all entanglement witnesses W.

It is interesting that there is a close connection between EW-s and Bell inequalities [16]. The counterpart of the reduction of set of hermitian "detectors" (3) to entanglement witnesses was previously known on the ground of maps [13]. Namely characterising separability one can restrict oneself the positive maps which are not completely positive or, in other words, do not represent any physical action (see [18] for discussion).

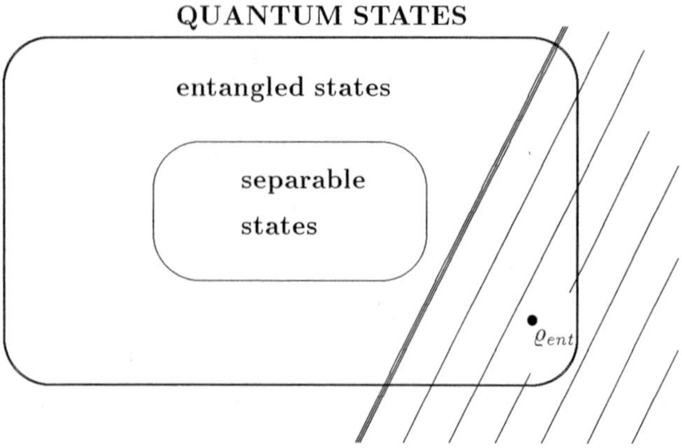

Figure 1: The geometrical interpretation of entanglement witness W. It is represented by a hyperplane (bold line). The states lying on the left hand side of the hyperplane - in particular *all* separable states - satisfy $Tr(W\varrho) > 0$. The hyperplane separates some subset of entangled states (like ϱ_{ent}) from separable states by providing the *strictly negative* mean value $Tr(W\varrho)$ on each element of the subset.

2.3 Recent progress and open questions

The characterisation 1' (resp. 1) recalled in the previous subsection is very important as it is given by means of hermitian operators i. e. *observables. The mean values of given entanglement witnesses can be measured in the laboratory.* Given a quantum system in laboratory one can try to measure some entanglement witness and possibly learn about entanglement. So the important question is

Question 1 .- Given quantum system, what is the minimal, operational set of entanglement witnesses characterising separability i.e. detecting all entanglement ?

But there is another question. Consider the case when we have in the lab the system in a state which is known via let us, say, quantum tomography method. Given the state of our system written on the paper and the question is: is it entangled or not ?

We should recall here that Characterisations 1 and 2 are completely equivalent but *the single* conditions (i) $Tr(\varrho X^{\Lambda}) \geq 0$ (ii) $[I \otimes \Lambda](\varrho) \geq 0$ are not. On mathematical ground sometimes the second one is much stronger than the first (see [19] for example). For this purposes the question were:

Question 2a.- What is the minimal, operational set of PM-s (resp. EW-s) which characterises all separable states on given Hilbert space ?

Question 2b.- Is there any common, structural property the elements of the minimal set would posses ?

First one should ask about what is the power of the very transparent PPT criterion. Note that we have $\varrho^{T_B} = (I \otimes T)(\varrho)$ where T stands for the map transposing the matrix elements of second system. This map is known to be positive (see [14, 20]). It has turned out [13] that PPT condition is necessary and sufficient for low spin systems i. e. we have:

Theorem II .- [13] The $2\otimes 2$ *(*$2\otimes 3$*)* [2] *state* ϱ *is separable iff it satisfies PPT criterion.*

[2] The system $N \otimes M$ is defined as a one defined on the Hilbert space $\mathcal{H} = \mathcal{H}_A \otimes \mathcal{H}_B$, $\dim \mathcal{H}_A = N$, $\dim \mathcal{H}_B = M$.

From the above we see that in the case of low dimensional systems the answer to the Question 2 is enormously simple: there is only *one* PM (transposition T) needed to characterise separability !

However, it is not the case in general. It was known first [13] from the fact that in the cases of higher dimensions there are nondecomposable maps. The counterexamples i. e. PPT entangled states [6, 24, 25, 26, 27] [3] were provided by techniques basing on *range criterion* independent on PPT one [6]. The criterion states that *if the state ϱ is separable exist a set of product vectors $\{|e_k, f_k\rangle\}$ that (i) they span the range $R(\varrho)$ (ii) the set of partial complex conjugated product states $\{|e_k, f_k^*\rangle\}$ spans the range of the partial transpose of ρ with respect to the second system, i.e., ϱ^{T_B}*. It has been pointed out (see [6, 15]) that to detect entanglement of such states one needs so called *nondecomposable maps* examples of which are in general hard to construct (see [20, 28]). The remarkable progress in construction of such maps has been done in Ref. [15] with the help of so called unextenable product basis (UPB) method [24, 25].

In context of the fact that for detecting entanglement of PPT type one needs nondecomposable maps the corresponding *nondecomposable witnesses* have been introduced [17] *via* the isomorphism (5). Moreover from the set of all PPT entangled states the special class of so called *edge states* violating the range criterion extremally was singled out. Basing on this class the remarkable reduction of tools characterising bipartite separability has been performed. First [17, 29] *only some special entanglement witnesses* are needed, namely those which (i) detect entanglement of "edge" states (ii) are "optimal" in well defined sense which can be verified numerically. Second [29] the corresponding set of nondecomposable PM-s generates (in sense of convex combination) the set of all nondecomposable PM-s which is a step towards characterisation of the full set of PM-s.

In this paper we shall address the question from slightly different point of view. Starting from general set of PM-s which, according to Characterisation 2, describes separable states completely, we shall look for subset which (i) would have some structural property which can be easily verified (ii) would still characterise separability completely.

We shall show that above properties are satisfied by PM-s which are *identity preserving* i. e. satisfy $\Lambda(I_B) = I_A$. Here I_A, I_B stand for identities on $\mathcal{H}_A, \mathcal{H}_B$. Subsequently we shall omit the indices A, B at identity matrices.

3 Reducing set of positive maps characterising separability

3.1 Some properties of linear hermitian maps

Here we shall briefly discuss and extend some properties of linear hermitian maps in the context of previous results [14, 18] concerning, in particular, quantum channels.

Let us recall that linear map is called *hermitian* if it preserves the hermiticity of operators. It has been shown [23] that linear Λ is hermitian iff X^Λ in (5) is hermitian operator. Λ is called *identity preserving* (IP) iff $\Lambda(I) = I$. It is called *trace preserving* iff $Tr(\Lambda(B)) = Tr(B)$ for any B. Further, given the map $\Lambda : B(\mathcal{H}_1) \to B(\mathcal{H}_2)$ we can consider it as *an operator on Hilbert space* [4] $B(\mathcal{H}_1)$ endowed with the scalar product

$$\langle A, B\rangle \equiv Tr(A^\dagger B) \tag{6}$$

[3]The matrices which can be viewed as a prototypes of such states was even known before on mathematical ground [28].

[4]It is so called Hilbert-Schmidt space.

Then the conjugated map $\Lambda^\dagger : B(\mathcal{H}_2) \to B(\mathcal{H}_1)$ is given simply by hermitian conjugate of the operator in terms of the above scalar product.

Consider the following lemma:

Lemma 1 *The linear, hermitian map Λ is identity preserving iff the conjugated map $\Lambda^\dagger$ is trace preserving*

Proof .- If $\Lambda(I) = I$ then using the properties of scalar product (6) and hermiticity of Λ we have for any hermitian B $Tr(B) = Tr(BI) = Tr(B\Lambda(I)) = Tr(\Lambda^\dagger(B)I) = Tr(\Lambda^\dagger(B))$. Thus $\Lambda^\dagger$ preserves trace of any hermitian B. Using hermitian basis in Hilbert-Schmidt space and applying linearity of trace operation, we get that $\Lambda^\dagger$ is trace preserving. On the other hand, let us suppose that map $\Lambda^\dagger$ is trace preserving. It is not difficult to see that the map conjugated to the hermitian one is hermitian itself. Using this and, again, Hilbert-Schmidt formalism we get for any hermitian B $Tr(B\Lambda(I)) = Tr(\Lambda^\dagger(B)I) = Tr(BI)$ for any hermitian B, hence $\Lambda(I) = I$. □

Now we shall prove the second lemma:

Lemma 2 *Let Λ be linear and hermitian map. Then Λ is identity preserving iff $X_A^\Lambda \equiv Tr_B(X^\Lambda) = I$ where X^Λ is given by formula (5).*

Remark .- The above lemma is extension and reformulation of the recent analysis concerning relation between quantum states and quantum channels. In fact, in the special case, when we put in place of $\Lambda^\dagger$ a quantum channel we obtain (after matching the notation) the states - channels isomorphism from Ref. [22].

Proof .- We shall first show that for any hermitian operator B

$$Tr(X_A^\Lambda B) = Tr(B^T\Lambda(I)) \tag{7}$$

In fact we have $Tr(X_A^\Lambda B) \equiv Tr(X^\Lambda B\otimes I) = Tr([I\otimes\Lambda^\dagger](P_+)B\otimes I) = Tr(P_+ B\otimes\Lambda(I)] = Tr(VB^T \otimes \Lambda(I)) = Tr(B^T\Lambda(I))$. Here we have used the fact that (i) $[I\otimes T](P_+) = V$ where V is a "flip" operator [5] $V\phi\otimes\psi = \psi\otimes\phi$, (ii) $Tr(VX\otimes Y) = Tr(XY)$ for any $X, Y \in \mathcal{H}_A$.

Now let us prove lemma 2. Suppose that $\Lambda(I) = I$. Then for any hermitian B we have $Tr(X_A^\Lambda B) = Tr(B^T\Lambda(I)) = Tr(B^T) = Tr(BI)$. As before it implies that $X_A^\Lambda = I$.

Conversely, if the latter holds then according to (7) we have $Tr(B) = Tr(B^T\Lambda(I)] = Tr(B(T\circ\Lambda)(I))$. Hence $(T\circ\Lambda)(I) = I$ and finally $\Lambda(I) = I$. □

3.2 Identity preserving positive maps characterise separability

In this section we shall prove the main result contained in the following theorem:

Theorem 1 *The state ϱ on $\mathcal{H}_A\otimes\mathcal{H}_B$ is separable if and only if the matrix $[I\otimes\Lambda](\varrho)$ has nonnegative spectrum for all positive, identity preserving maps $\Lambda : B(\mathcal{H}_B) \to B(\mathcal{H}_A)$.*

Before we shall prove the result we have to note the observation

Observation 1 *For any entanglement witness W its partial reductions W_A, W_B are positive i.e. have nonnegative spectra.*

Proof .- The proof of the above fact is immediate by considering separable states proportional to $|\phi\rangle\langle\phi| \otimes I$, $(I \otimes |\phi\rangle\langle\phi|)$.

Proof of the Theorem .- The proof of the "only if" part is trivial. Consider the second, "if" part which is equivalent to the statement that if ϱ is entangled than there exists some positive, identity preserving map Λ such that $[I \otimes \Lambda](\varrho)$ is not positive. To prove this let us take an arbitrary entangled ϱ. Then there exists entanglement witness (see. sect. II) such that

$$Tr(W\varrho) < 0. \tag{8}$$

The isomorphism (5) gives the positive map Λ_W (the one for which $W = [I \otimes \Lambda_W](P_+))$ which, obviously, "detects"entanglement of ϱ. Now, let P_{W_A} be a projection onto the range of W_A. We also use the operators $\sqrt{W_A}$, $\sqrt{W_A^{-1}}$ which obviously exist (according to Observation W_A is positive) and are pseudoinverses of each other: $\sqrt{W_A^{-1}}\sqrt{W_A} = \sqrt{W_A}\sqrt{W_A^{-1}} = P_A$.

We shall first prove that the following positive [5] map

$$\Lambda'(\cdot) \equiv \sqrt{W_A^{-1}}\Lambda(\cdot)\sqrt{W_A^{-1}} \tag{9}$$

detects entanglement of ϱ. Then, with the help of Λ' we shall construct identity preserving map which also detects entanglement of the state ϱ.

Consider the operator $Y \equiv [I \otimes \Lambda'](\varrho)$. It is hermitian as Λ' is positive, hence hermitian. To prove that Λ' detects entanglement of ϱ it is enough to show that Y must have negative eigenvalue.

To this aim let us introduce the vector $|\Psi\rangle \equiv \sqrt{W_A} \otimes I|\Psi_+\rangle$. Then we have

$$\begin{aligned} \langle\Psi|Y|\Psi\rangle &= Tr(P_{W_A} \otimes IP_+P_{W_A} \otimes I[I \otimes \Lambda](\varrho)) = \\ & Tr(P_{W_A} \otimes I[I \otimes \Lambda^\dagger](P_+)P_{W_A} \otimes I\varrho) = Tr(W\varrho) < 0 \end{aligned} \tag{10}$$

as $P_{W_A} \otimes IWP_{W_A} \otimes I = W$, because P_{W_A} is a projection onto support of W_A.

Thus we have proved that Λ' detects entanglement of ϱ. Now we shall consider the operator

$$\tilde{W} = [I \otimes \Lambda'^\dagger](P_+) + P_{W_A}^\perp \otimes I \tag{11}$$

where $P_{W_A}^\perp$ is projector perpendicular to P_{W_A}.

From positivity of Λ' and (10) it can be seen that $\tilde{W}$ is an entanglement witness. The corresponding positive map $\tilde{\Lambda}$ *is identity preserving* because of lemma 2 and the immediate calculation

$$\tilde{W}_A = Tr_B([I \otimes \Lambda](P_+)) + P_{W_A}^\perp = P_{W_A} + P_{W_A}^\perp = I. \tag{12}$$

It also detects entanglement of ϱ as one can easy see that $\langle\Psi|[I \otimes \tilde{\Lambda}](\varrho)|\Psi\rangle = \langle\Psi|Y|\Psi\rangle < 0$ (see (10)). □

Example .- Consider the positive map Λ_r applied in reduction criterion of separability (see [18, 30]). The map satisfies $\Lambda_r : \mathcal{C}^N \to \mathcal{C}^N$, $\Lambda_r(\sigma) \equiv ITr(\sigma) - \sigma$. The corresponding entanglement witness is $W = I - P_+$ so its reduction is proportional to identity $W_A = (dim\mathcal{H}_A - 1)I$. It is immediate then to see that $P_{W_A} = I$ and application of formulas (9), (11) give the map $\tilde{\Lambda}_r = \frac{1}{(dim\mathcal{H}_A-1)}\Lambda_r$ which can be easily seen to be identity preserving.

[5]The positivity of the above map follows from the positivity of Λ.

3.3 Discussion

We have discussed the properties of linear hermitian maps using the maps - operators isomorphism given in [14]. We have shown the characterisation of identity preserving linear hermitian maps in terms of the hermitian operators. Using this we have proven that to characterise completely the set of bipartite separable states one needs only the set of positive maps which are identity preserving. This is important that the proof is constructive i. e. it shows how to construct such a map form the one detecting entanglement of given state.

Further one can consider the "concatenation": instead of all positive IP PM-s maps only some of them can be needed together with the one which are result of embodying of IP MP-s into the higher spaces. There are other interesting questions like, for example, possible application of the above results to the generating set of PM-s from [17, 29] (see sec. 2).

Finally it can be seen that the whole proof from section 3 can be applied for the n-particle case [19]. We get analogous theorem as in section 3. The only difference in the reasoning and final statements are that (i) instead of the $\mathcal{H}_B$ one should put $\mathcal{H}_2 \otimes ... \otimes \mathcal{H}_n$ describing n-1 particles (ii) instead of positive maps one has the maps which are *positive on product states* (see [19]).

The work is supported by Polish Committee for Scientific Research, contract No. 2 P03B 103 16, and by European Union, project EQUIP, contract No. IST-1999-11053.

References

[1] A. Ekert, Phys. Rev. Lett. **67**, 661 (1991).

[2] C. Bennett, G. Brassard, C. Crepeau, R. Jozsa, A. Peres and W. K. Wootters, Phys. Rev. Lett. **70** (1993) 1895.

[3] C. H. Bennett and S. J. Wiesner, Phys. Rev. Lett. **69**, 2881 (1992).

[4] See P. W. Shor, quant-ph/0005003.

[5] R. F. Werner, Phys. Rev. A **40**, 4277 (1989).

[6] P. Horodecki, Phys. Lett. A, **232**, 233 (1997).

[7] V. Vedral and M. Plenio, Phys. Rev. A **57**, 1619 (1998).

[8] R. Horodecki and M. Horodecki, Phys. Rev. A **54**, 1838 (1996).

[9] B. d'Espagnat *Conceptual Foundations of Quantum Mechanics*, Benjamin, Reading, Massachusetts (1976).

[10] K. Blum, *Density Matrix Theory and Applications*, 1981, Plenum Press, New York.

[11] R. Horodecki, P. Horodecki, and M. Horodecki, Phys. Lett. A**230**, 377 (1996).

[12] A. Peres, Phys. Rev. Lett. **76**, 1413 (1996).

[13] M. Horodecki, P. Horodecki and R. Horodecki, Phys. Lett. A**1**, 223 (1996).

[14] A. Jamiołkowski, Rep. Math. Phys. **3**, 275 (1972).

[15] B. Terhal, quant-ph/9810091.

[16] B. Terhal, quant-ph/9911057.

[17] M. Lewenstein, B. Kraus, J. I. Cirac and P. Horodecki, Phys. Rev. A (2000), in print.

[18] M. Horodecki, P. Horodecki, Phys. Rev. A **59**, 4206 (1999).

[19] M. Horodecki, P. Horodecki, R. Horodecki, quant-ph/0006071.

[20] S. L. Woronowicz, Rep. Math. Phys., **10**, 165 (1976).

[21] See P. Horodecki, M. Horodecki and R. Horodecki, Phys. Rev. Lett. **82**, 1056 (1999).

[22] M. Horodecki, R. Horodecki and R. Horodecki, Phys. Rev. A **60**, 1888 (1999).

[23] J. de Pillis, Pacific J. of Math. **23**, (1967) 129. (Berlin,Heidelberg,New York) 1967.

[24] C. H. Bennett, D. DiVincenzo, T. Mor, P. Shor, J. Smolin and B. M. Terhal, Phys. Rev. Lett. **82**, 5385 (1999).

[25] D. DiVincenzo, T. Mor, P. Shor, J. Smolin and B. M. Terhal, quant-ph/9908070.

[26] D. Bruß and A. Peres, Phys. Rev. A **61**, 30301(R) (2000).

[27] P. Horodecki and M. Lewenstein, Phys. Rev. Lett. (2000), in print.

[28] M. D. Choi, Proc. Sympos. Pure Math **38**, 583 (1982).

[29] M. Lewenstein, B. Kraus, P. Horodecki and J. I. Cirac, quant-ph/0005112.

[30] N. Cerf, C. Adami and R. M. Gingrich, Phys. Rev. **60**, 898 (1999).

Decoherence and its Implications in Quantum Computation and Information Transfer
T. Gonis and P.E.A. Turchi (Eds.)
IOS Press, 2001

On capacities of quantum channels and bipartite states

Michał HORODECKI
Institute of Theoretical Physics and Astrophysics, University of Gdańsk, 80–952 Gdańsk, Poland

Abstract

Basing on unified approach to *all* kinds of quantum capacities presented in Ref. [M. Horodecki, P. Horodecki and R. Horodecki, Phys. Rev. Lett. **85**, 433 (2000)] we show that the rate of quantum information transmission is bounded by the maximal attainable rate of any function f satisfying certain modest continuity and normalization conditions. Moreover, we show that, if for any bipartite state the one-way distillable entanglement is no less than the function then one obtains Shannon-like formulas involving f for all the capacities. The coherent information is shown to satisfy the conditions imposed on the function. We also explain, why the quantum Shannon formulas cannot be direct generalizations of the classical ones.

1 Introduction

The challenge for the present quantum information theory is to determine the quantum capacity of noisy channels [1, 2, 3, 4, 5, 6, 7]. The problem is difficult mainly for two reasons. First, according to the present knowledge, unlike in the classical case, there are at least five different types of quantum capacities [1]. This is because quantum information channel can be supplemented by one- or two-way classical channel[1]. Moreover, there are teleportation channels [9, 10], for which a bipartite state is a resource. The second reason is that quantum capacity exhibits a kind of non-additivity [11] that makes them extremely hard to deal with.

As one knows, the cornerstone of classical information theory is the famous Shannon noisy coding theorem, giving the formula for capacity of noisy channel [12]

$$C = \sup_X I(X;Y) \tag{1}$$

where the supremum is taken over all sources X; $I(X;Y) = H(X) + H(Y) - H(X;Y)$ is the Shannon mutual information (with H being the Shannon entropy); Y is the random variable resulting from action of the noise to X. In quantum information theory the potential counterpart of mutual information has been found [2, 3]. It is the

[1]There are also *classical* capacities of quantum channels, see [8]. Here we are concerned with solely *quantum* capacities.

so-called *coherent information* (CI). To define it, consider two coherent informations of the bipartite state ϱ with reductions ϱ_A and ϱ_B:

$$I^X(\varrho) = S(\varrho_X) - S(\varrho), \quad X = A, B, \tag{2}$$

for $S(\varrho^X) - S(\varrho) \geq 0$ and $I^X = 0$ otherwise. Here $S(\varrho) = -\mathrm{Tr}\varrho \log_2 \varrho$ is the von Neumann entropy. Then one defines the coherent information for a channel Λ and a source state σ as

$$I(\sigma, \Lambda) = I^B((I \otimes \Lambda)(|\psi\rangle\langle\psi|)), \tag{3}$$

where ψ is a pure state with reduction σ (the quantity does not depend on the choice of ψ). Now, the following connection between coherent information and a quantum capacity is known [7]

$$Q_\emptyset \leq \lim_{n\to\infty} \frac{1}{n} \sup_{\varrho_n} I(\varrho_n, \Lambda^{\otimes n}) \equiv I^B_\emptyset(\Lambda) \tag{4}$$

where $Q_\emptyset$ is the maximal number of qubits that can be reliably sent down the channel without any supplementary classical channel. Note that if, instead of inequality, there were *equality*, then we would have analogue of Shannon formula.

Unfortunately, despite a huge effort devoted to the problem [4, 6, 7], the equality has not been proven so far. Moreover, remarkably, the similar inequality is not known for other capacities than $Q_\emptyset$, i.e. the ones attainable at the support of backward ($Q_\leftarrow$) or two-way ($Q_\leftrightarrow$) classical communication[2]. There are also capacities of quantum teleportation channels, where the resource is bipartite state rather than channel. In the latter case, transmission requires prior manipulations (called distillation [13, 1]) over the shared pairs, transforming them into pairs in pure maximally entangled states. Then the quantum information can be transmitted by using teleportation [9]. The manipulations include any local actions, and one- or two-way classical communication. Consequently, we have two kinds of one-way distillable entanglement of a state ϱ, $D_\rightarrow$ or $D_\leftarrow$ (since ϱ need not be symmetric, one distinguishes directions of classical communication) and two-way distillable entanglement $D_\leftrightarrow$. In fact, $Q_\leftrightarrow$ also necessarily involves the distillation and teleportation process. The latter processes are so exotic from the point of view of classical information theory, that no analogue of theory of error correcting codes has been worked out for them so far. In contrast, in the case of the capacity $Q_\emptyset$, there exists theory of quantum codes, being generalization of classical error correcting codes theory [14].

In the above context the basic question arises: *Is there possible a consistent approach for all of the capacities*? It turns out that the question has positive answer. In Ref. [15] it has been shown, that one can define so called coherent information (CI) capacity (that depends on kind of supplementary resource). The quantity is the maximal rate of coherent information that can be achieved by large number of uses of channel (or large number of pairs in bipartite entangled state) and given supplementary resource. It was shown that the quantum capacities are bounded from above by the corresponding CI capacities. Moreover, a hypothetical inequality has been conjectured (called hashing

[2] In Ref. [7] it was shown that forward classical communication does not increase capacity, i.e. $Q_\emptyset = Q_\rightarrow$

inequality) that implies *equality* between both kinds of capacities. The inequality is given by the following formula

$$D_{\rightarrow}(\varrho) \geq I^B(\varrho). \tag{5}$$

implies that the answer to the question is "yes". The inequality was tested in various ways, and so far we do not know a single case of violation of the ineqaulity. Thus there is a strong evidence that there is a unified version of Shannon theorem.

In this note we analyse the above results from abstract point of view. We replace coherent information with arbitrary function, and ask for the conditions it should satisfy to produce the same results. We try to minimize the number of conditions, obtaining that a kind of asymptotic continuity and normalization on the maximally entangled state is sufficient. We show in two ways that coherent information satisfies such conditions, which is compatible with the results of Ref. [15] recalled above.

The paper is organized as follows. In Sec. II we provide definitions of quantum capacities, explain the goal of the Shannon theorem. We also explain, why direct generalization of the classical Shannon theorem to quantum domain would give wrong results. In sec. III we present the main results of this paper: we show that for any function f satisfying some conditions, its f-capacity is upper bound for quantum capacity. We also show that inequality $f \leq D_{\rightarrow}$ implies equality of f-capacities and quantum capacities. In Sec. IV we show that coherent information satisfies the conditions imposed on f.

2 Definition of capacities of quantum channel

Let us now define the quantum capacities. As mentioned, we will be concerned with four supplementary resources $C \in \{\rightarrow, \leftarrow, \leftrightarrow, \emptyset\}$. (The last one symbolizes no supplementary resource). If Alice and Bob dispose one use of a channel Λ (directed, by convention, from Alice to Bob) and the supplementary resources symbolized by C, then they can share a bipartite state ϱ. An operation that produced in this way the state ϱ from Λ will be denoted by $\mathcal{E}_C$ so that

$$\varrho = \mathcal{E}_C(\Lambda). \tag{6}$$

If Alice and Bob share initially a bipartite state ϱ_{in}, then we will use notation $\mathcal{D}_C$

$$\varrho_{out} = \mathcal{D}_C(\varrho_{in}) \tag{7}$$

(The letters used in our notation follow from the common associations: usual channel capacity – error correction, teleportation channels – distillation).

Now we define quantum capacities as follows[3]. Define maximally entangled state on the space $\mathcal{H} \otimes \mathcal{H}$ by

$$P_+(\mathcal{H}) = |\psi_+(\mathcal{H})\rangle\langle\psi_+(\mathcal{H})|, \quad \psi_+(\mathcal{H}) = \frac{1}{\sqrt{d}}\sum_{i=1}^{d}|ii\rangle, \tag{8}$$

[3]The used definition of "D" capacities is due to [21] (cf. [1],[20]). In Ref. [19] the definition was shown to coincide with other used definitions. The definition of "Q" capacities is taken from [5]. In Ref. [7] it was shown to be equivalent to other one introduced in [4].

where $|i\rangle$ are basis vectors in $\mathcal{H}$, while $d = \dim \mathcal{H}$. Given a state ϱ, consider sequence of operations $\{\mathcal{D}_C^n\}$ (called protocol) transforming the input state $\varrho^{\otimes n}$ into the state σ_n acting on the Hilbert space $\mathcal{H}_n \otimes \mathcal{H}_n$ and with $\dim \mathcal{H}_n = d_n$, satisfying

$$F(\sigma_n, P_+(\mathcal{H}_n)) \equiv \langle \psi_+(\mathcal{H}_n)|\sigma_n|\psi_+(\mathcal{H}_n)\rangle \to 1. \tag{9}$$

The asymptotic ratio attainable via given protocol is then given by

$$D_{\{\mathcal{D}_C^n\}}(\varrho) = \lim_{n\to\infty} \frac{\log_2 \dim \mathcal{H}_n}{n}.. \tag{10}$$

Then the capacity $D_C(\varrho)$ (call it C-distillable entanglement) is defined by maximum over all possible protocols

$$D_C(\varrho) = \sup D_{\{\mathcal{D}_C^n\}}(\varrho). \tag{11}$$

The usual channel capacities can be defined in the same way. We only need make the following substitutions: $D \to Q$, $\varrho \to \Lambda$ and $\mathcal{D} \to \mathcal{E}$. The protocols $\{\mathcal{E}_C^n\}$ and $\{\mathcal{D}_C^n\}$ that achieve the considered suprema will be called optimal error correction and optimal distillation protocol, respectively.

Capacities of quantum channel have direct physical sense of the *rate of quantum communication* at given resources C. One of the important goals of quantum communication theory is to relate them to abstract quantities, based on entropies. Indeed, the classical Shannon theorem states equality between two functions of the channel:

a) maximal rate of information transmission via the channel (physical quantity)

b) maximal mutual information that can be sent down the channel (abstract quantity)

Since the two quantities turned out to be equal, the mutual information acquired information theoretic sense. In quantum case we would like to have similar equality. The capacities we have defined above. What is the counterpart of the mututal information?

One might consider direct "quantization" of Shannon information theory as follows. The pair of input and output random variables (X, Y) in quantum case is replaced by the state of coumpund system ϱ_{AB} . Then the reductions ϱ_A and ϱ_B play the role of X and Y respectively. Now, the direct counterpart of the mutual information would be the function

$$I_q = S(\varrho_A) + S(\varrho_B) - S(\varrho_{AB}). \tag{12}$$

known as the index of correlation [16, 17]. Let us argue that it will not work in quantum case. We will show that one can reach nonzero mutual information I_q via the channel which certainly cannot convey quantum information reliably. Namely, consider the channel constituted by complete measurement

$$\varrho \to \sum_i P_i \varrho P_i \tag{13}$$

where P_i are one-dimensional projectors. After the action of such channel, any input state ϱ changes into mixture of orthogonal states ψ_i such that $|\psi_i\rangle\langle\psi_i| = P_i$. Then the channel can be simmulated by classical channel in the following way. Alice performs the measurement on the input state, and send classical message on which outcome i was obtained. Then Bob produces in his laboratory the state ψ_i.

Since the channel can be simulated by classical one, it cannot convey quantum information. Indeed, one could clone the bits travelling through the classical channel. If it conveyed quantum information, the latter would be therby cloned, which is impossible, due to fundamental principle of quantum information processing which is *no-cloning theorem* [18]. Thus our channel has quantum capacity zero. However, it is easy to see, that it can convey nonzero amount of mutual information I_q. Indeed, sent a half of maximally entangled state down the channel, one obtains classically correlated state with $I_q = \log d$ where d is the dimension of the Hilbert space of the input. Consequently, the maximum rate of mutual information I_q that can be achieved by use of the channel is not equal to its capacity. Clearly, the quantum counterpart of the mutual information is not its direct generalization. The reasonable counterpart was proposed in Ref. [2], and is given by the formula (2).

3 f-capacities versus quantum capacities

Consider an arbitrary function f of bipartite state. Given one use of channel Λ and a resource C, one can try to produce a bipartite state with maximal possible value of f. Denote this value by $f^1_C(\Lambda)$. Now, one can ask "how much of f" one can produce given n uses of channel, per use, in the asymptotic limit of large n. Thus we are interested in the following quantity

$$f_C(\Lambda) = \limsup_{n\to\infty} \frac{1}{n} f^1_C(\Lambda^{\otimes n}). \tag{14}$$

Of course to get reasonable f_C, the function f should be extensive, i.e. its maximal value on the set of the states of a Hilbert space should scale logarythmically with the dimension of the space (in other words, it should scale linearly with the number of elementary systems i.e. "volume" of the system).

Similarly one can define $f_C(\varrho)$ for bipartite state ϱ.

We will now impose some asymptotic continuity and asymptotic normalization conditions. Assume that a sequence of states ϱ_n defined on the space $\mathcal{H}_n \otimes \mathcal{H}_n$ with $d_n = \dim \mathcal{H}_n \to \infty$ satisfies

$$\lim_{n\to\infty} \langle \psi_+(\mathcal{H}_n)|\varrho_n|\psi_+(\mathcal{H}_n)\rangle \to 1 \tag{15}$$

Then we require that

$$\lim_{n\to\infty} \frac{f(\varrho_n)}{\log d_n} = 1 \tag{16}$$

This means that for large Hilbert space $\mathcal{H}_n \otimes \mathcal{H}_n$

(i) f approaches $\log d_n$ for maximally entangled state

(ii) f is continuous on the maximally entangled state.

This is all we assume about the function f. Now we will show that this suffices for f_C to be an upper bound for Q_C and D_C.

Theorem 1. Quantum capacities are bounded from above by f capacities:

$$Q_C(\Lambda) \leq f_C(\Lambda), \quad D_C(\varrho) \leq f_C(\varrho), \tag{17}$$

for any Λ, ϱ and $C \in \{\rightarrow, \leftarrow, \leftrightarrow, \emptyset\}$.

Theorem 2. If the inequality

$$D_{\rightarrow}(\varrho) \geq f(\varrho) \tag{18}$$

holds for any bipartite state ϱ then the quantum capacities are equal to corresponding f capacities

$$Q_{\leftrightarrow}(\Lambda) = f_{\leftrightarrow}(\Lambda), \quad Q_{\rightarrow}(\Lambda) = f_{\rightarrow}(\Lambda), \tag{19}$$

$$Q_{\emptyset}(\Lambda) = f_{\emptyset}(\Lambda), \tag{20}$$

$$D_{\leftrightarrow}(\varrho) = f_{\leftrightarrow}(\varrho), \quad D_{\rightarrow}(\Lambda) = f_{\rightarrow}(\varrho). \tag{21}$$

Proof of Theorem 1. We will prove the "Q" part of the theorem. The proof for "D" part is similar. Let $\{\mathcal{E}_C^n\}$ be the optimal error correction protocol for Λ. Then we have the following estimates

$$f_C(\Lambda) \geq \frac{1}{n} f(\mathcal{E}_C^n(\Lambda^{\otimes n})) = \frac{\log d_n}{n} \frac{f(\mathcal{E}_C^n(\Lambda^{\otimes n}))}{\log d_n} \rightarrow Q_C(\Lambda), \tag{22}$$

The first inequality follows from the very definition of f_C. The last one is due to the condition (16) and the fact that in optimal error correction protocol we have $F_n \rightarrow 1$, and $\log d_n/n \rightarrow Q_C$.

Proof of Theorem 2. We will also prove only the "Q" part. For $C \in \{\rightarrow, \leftrightarrow\}$, consider the following particular error correcting protocol for the channel $\Gamma = \Lambda^{\otimes n}$. One applies to Γ the operation $\mathcal{E}_C$ that produces the state $\sigma = \mathcal{E}_C(\Lambda^{\otimes n})$ of maximal attainable value of f. Subsequently, one performs optimal distillation protocol for the state σ (the protocol with the resource C). Then we find

$$Q_C(\Lambda) = \frac{1}{n} Q_C(\Lambda^{\otimes n}) \geq \frac{1}{n} D_C(\mathcal{E}_C(\Lambda^{\otimes n})) \geq \frac{1}{n} f(\mathcal{E}_C(\Lambda^{\otimes n})) \rightarrow f_C(\Lambda) \tag{23}$$

where $C \in \{\rightarrow, \leftrightarrow\}$. The equality is a consequence of the very definition of Q_C. The first inequality follows from the fact, that Q_C is supremum over *all* error correction protocols, so it is no less from the rate obtained in the protocol above. The second inequality follows from the assumed inequality, and from the obvious inequality $D_{\leftrightarrow} \geq D_{\rightarrow}$. Finally, the limit is due to the definition of $f(\Lambda)$. The above estimate together with Theorem 1 gives all the desired equalities apart from the one involving $C = \emptyset$. That the latter one is also implied by the inequality (18), it follows immediately from the facts: (a) trivially $f_{\emptyset} \leq f_{\rightarrow}$; (b) $Q_{\rightarrow} = Q_{\emptyset}$ [7]; (c) as just proved, the inequality (18) implies $Q_{\rightarrow} = f_{\rightarrow}$.

As we mentioned in introduction, coherent information is the function that satisfies the condition imposed on the function f. Instead of dealing with coherent information, we will consider simpler quantity, namely conditional entropy, that can be negative:

$$S_{cond}^X(\varrho) = S(\varrho_X) - S(\varrho). \tag{24}$$

(henceforth $X = A$ or B). We will show in two different ways that S^X_{cond} satisfies the condition.

Proof 1. We will refer to two items, into which we have decomposed the condition (16). The first item is satisfied trivially: for the pure state $S(\varrho)$ vanishes, while the maximally entangled state has maximally mixed reductions hence $S(\varrho_X) = \log d_n$. The item (ii) follows from strong continuity property implied by the Fannes inequality for von Neumann entropy. The latter inequality is as follows

$$|S(\varrho) - S(\sigma)| \leq A \log \dim \mathcal{H} D(\varrho, \sigma) + B \tag{25}$$

for all state ϱ, σ acting on the Hilbert space $\mathcal{H}$. Here A, B are constants inependent of the dimension of the Hilbert space. D is the metric induced by the trace norm

$$D_{Tr}(\varrho, \sigma) = \|\varrho - \sigma\|_{Tr} \tag{26}$$

or the Bures metric

$$D_B(\varrho, \sigma) = 2\sqrt{1 - F(\varrho, \sigma)} \tag{27}$$

where F is the Uhlmann-Jozsa fidelity

$$F(\varrho, \sigma) = \mathrm{Tr}\sqrt{\sqrt{\varrho}\sigma\sqrt{\varrho}} \tag{28}$$

The version of the inequality including the latter one was proved in Ref. [6]. Actually, the two metrics are equivalent in the strong sense: if either of them is close to zero, so is the other one in the way independent of the dimension of the Hilbert space. This is expressed by the following inequalities The trace distance D_{Tr} and Bures distance D_B are related by the following inequalities [23]

$$\frac{1}{4} D_{Tr}(\rho, \sigma)^2 = 1 - F(\rho, \sigma) \leq d(\rho, \sigma) \leq \sqrt{1 - F(\rho, \sigma)^2} = \sqrt{\frac{D^2_{Tr}}{2} - \frac{D^4_{Tr}}{16}}. \tag{29}$$

If one of the states is pure, e.g. $\sigma = |\psi\rangle\langle\psi|$ then $F(\sigma, \varrho) = \sqrt{\langle\psi|\varrho|\psi\rangle}$, so that the above metric guarantee continuity in the same sense, as the limit in the condition (16) was taken. From the inequality (25) we will easily get continuity of conditional entropy, if we recall that both metrics are monotonous in the following sense

$$D(\Lambda(\varrho), \Lambda(\sigma) \leq D(\varrho, \sigma) \tag{30}$$

For any trace-preserving completely positive map Λ. In particular, it follows that $D(\varrho_X, \sigma_X) \leq D(\varrho, \sigma)$. Now we have the following estimate for states ϱ and σ acting on the space $\mathcal{H} \otimes \mathcal{H}$

$$\begin{aligned} &|S^X_{cond}(\varrho) - S^X_{cond}(\sigma)| = |S(\varrho_X - S(\varrho) - S(\sigma_X) + S(\varrho)| \leq \\ &|S(\varrho_X) - S(\sigma_X)| + |S(\varrho) - S(\sigma)| \leq A_1 \log \dim \mathcal{H} D(\varrho_X, \sigma_X) + B_1 + \\ &A_2 \log \dim \mathcal{H} D(\varrho, \sigma) + B_2 \leq A \log \dim D(\varrho, \sigma) + B \end{aligned} \tag{31}$$

Proof 2. This proof will be more complicated, but less abstract. We will refer to the following properties of the conditional entropy [22]

- invariance under product unitary transformations.
- convexity

Consider now the state ϱ_n approaching maximally entangled state. Let us apply to it random unitary transformations of the form $U \otimes U^*$ where $*$ is complex conjugation. One knows that the emerging state is of the simple form [24]

$$\varrho(F_n, d_n) = pP_+(\mathcal{H}_n) + (1-p)\frac{1}{d_n^2}I, \quad 0 \le p \le 1, \tag{32}$$

with $F_n = \langle\psi_+(\mathcal{H}_n)|\varrho(F_n, d_n)|\psi_+(\mathcal{H}_n)\rangle = \langle\psi_+(\mathcal{H}_n)|\varrho_n|\psi_+(\mathcal{H}_n)\rangle$, $d_n = \dim \mathcal{H}_n$. Due two above properties of conditional entropy we have

$$S^X_{cond}(\varrho_n) \ge S^X_{cond}(\varrho(F_n, d_n)). \tag{33}$$

Calculating explicitly the right hand side of the inequality we obtain that

$$\frac{S^X_{cond}(\varrho(F_n, d_n))}{\log d_n} \to 1 \tag{34}$$

if $F_n \to 1$. Recalling that $S^X_{cond} \le S(\varrho_X) \le \log \dim \mathcal{H}_n$, we obtain the required convergence.

The work is supported by Polish Committee for Scientific Research, contract No. 2 P03B 103 16, and by European Union, project EQUIP, contract No. IST-1999-11053.

References

[1] C. H. Bennett, D. P. DiVincenzo, J. Smolin and W. K. Wootters, Phys. Rev. A **54**, 3814 (1997).

[2] B. Schumacher, Phys. Rev. A **54**, 2614 (1996); B. Schumacher and M. A. Nielsen, Phys. Rev. A **54**, 2629 (1996).

[3] S. Lloyd, Phys. Rev. A **56**, 1613 (1997).

[4] H. Barnum, M. Nielsen and B. Schumacher Phys. Rev. A **57**, 4153 (1998).

[5] C. H. Bennett, D. P. DiVincenzo and J. Smolin, Phys. Rev. Lett. **78**, 3217 (1997).

[6] H. Barnum, J. Smolin, and B. Terhal Phys. Rev. A **58**, 3496 (1998).

[7] H. Barnum, E. Knill and M. Nielsen, quant-ph/9809010.

[8] B. Schumacher and M. Westmoreland Phys. Rev. A **56**, 131 (1997); S. Holevo, IEEE Trans. IT **44**, 269 (1998); Bennett *et al.* Phys. Rev. Lett. **83**, 3081 (1999).

[9] C. Bennett, G. Brassard, C. Crepeau, R. Jozsa, A. Peres and W. K. Wootters, Phys. Rev. Lett. **70** (1993) 1895.

[10] S. Popescu, Phys. Rev. Lett. **72**, 797 (1994).

[11] P. Shor and J. Smolin, quant-ph/9706061; D. P. DiVincenzo, P. Shor and J. Smolin Phys. Rev. A **57**, 830 (1998).

[12] T. M. Cover and J. A. Thomas, *Elements of Information Theory*, John Wiley and Sons, New York, 1991.

[13] C. H. Bennett, G. Brassard, S. Popescu, B. Schumacher, J. Smolin and W. K. Wootters, Phys. Rev. Lett. **76**, 722 (1996).

[14] P. Shor, Phys. Rev. A, **52**, 2439 (1995); A. Steane, Phys. Rev. Lett. **77**, 793 (1996); Th. Beth and M. Grassl, Fortschr. Phys. **46**, 459 (1998) and references therein. K. Kraus, *States, Effects and Operations: Fundamental Notions of Quantum Theory*, Wiley, New York, 1991.

[15] M. Horodecki, P. Horodecki and R. Horodecki, Phys. Rev. Lett. **85**, 433 (2000).

[16] S.M. Barnett and S.J. Phoenix, Phys. Rev. A 44 (1991) 535.

[17] R. Horodecki, Phys. Lett. A 187 (1994) 145.

[18] D. Dieks: Phys. Lett. A **92**, 271 (1982); W. K. Wooters and W. H. Żurek, Nature (London) **299**, 802 (1982)

[19] E. Rains, Phys. Rev. A **60**, 173 (1999).

[20] M. Keyl, R. F. Werner, quant-ph/9910124.

[21] E. M. Rains, Phys. Rev. A **60**, 179 (1999).

[22] A. Wehrl, Rev. Mod. Phys. **50**, 221 (1978).

[23] C. A. Fuchs and J. van de Graaf, IEEE T. Inform. Theory 45: (4) 1216 (1999), quant-ph/9712042.

[24] M. Horodecki and P. Horodecki, Phys. Rev. A, **59** 4206 (1999).

Decoherence and its Implications in Quantum Computation and Information Transfer
T. Gonis and P.E.A. Turchi (Eds.)
IOS Press, 2001

SQUID Qubits and Quantum Gates

Siyuan Han[(a)] and R. Rouse[(b)]
(a) Department of Physics and Astronomy, University of Kansas, Lawrence, KS 66045
(b) Sun Microsystems, Sunnyvale, CA 94303

Abstract

As a quantum two-state system with macroscopically distinct states the superconducting quantum interference device (SQUID) is promising for a solid state implementation of quantum computing. We show that the level structure of a SQUID can be engineered to have properties desirable for a qubit and its dynamics can be precisely controlled by the application of microwave pulses. Quantum controlled-NOT (QCN) gates can also be readily constructed from SQUIDs. An example of the SQUID QCN gate is presented which has a four-entangled-state structure similar to that of a Be^+ ion in which QCN gate operation has been demonstrated. The effective damping in a SQUID qubit was measured using double resonance photon-assisted tunneling spectroscopy. Damping from various sources is analyzed and their implication to decoherence in SQUID qubit is discussed.

1. *Introduction*

Superconducting quantum interference device (SQUID) has been demonstrated to be a well behaved macroscopic quantum system. Many quantum phenomena which are common in the microscopic world but rarely occur in macroscopic systems have been observed recently in SQUIDs. Examples include tunneling through energy barriers [1,2], energy level quantization and resonant tunneling between quantized levels [3–6], photon-assisted tunneling and photon-induced transition [7, 8], population inversion [9], and coherent superposition of quantum states [10]. The two macroscopically distinct fluxoid states of a SQUID is naturally well represented by a quantum two-level system. The quantum dynamics of a SQUID can be precisely controlled by applying an external magnetic flux and/or microwave pulses. Since SQUID is thin film device that can be fabricated by the very large scale integrated circuits (VLSC) technology, which is the foundation of present day digital computing and information industry, it is a very attractive candidate for the physical realization of quantum computer.

Just like a bit is the most basic element of any classical computer a quantum bit (qubit) is the most fundamental building block of a quantum computer (QC) [11]. It has been shown that the use of a single type of quantum logic gate, the quantum controlled-NOT (QCN) gate, is sufficient to build a quantum computer [12–15]. Although quantum computing with solid state qubits has many advantages, which include flexibility in design and compatible to VLSC fabrication technology, it also has many drawbacks. Among them, damping induced decoherence is the most critical one because solid state systems usually have rather strong interactions with their surrounding environment and therefore are much more difficult to isolate. It is therefore critical to have a in depth knowledge about damping in SQUID to decide whether the system is suitable for QC application. In this work we report measurement of effective damping in an rf-SQUID qubit and discuss its implication to SQUID quantum computing.

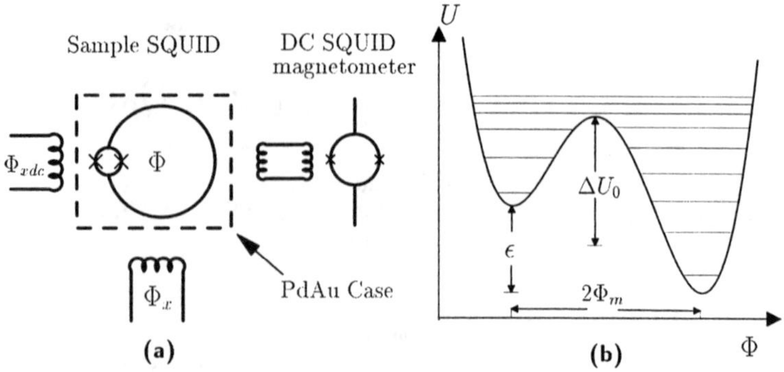

Figure 1: a) Schematic of a variable barrier SQUID with a magnetometer and two flux-applying coils. (b) The potential energy of the SQUID along the x direction and the associated energy levels.

2. *SQUID Qubits*

The simplest SQUID, also called rf-SQUID, is made of a superconducting loop of inductance L interrupted by a Josephson element of critical current I_c. The SQUID, its potential energy and energy levels are shown schematically in Fig. 1. For qubit application damping in the Josephson element should be as low as possible. This requirement rules out the use of many types of Josephson weak link including microbridge, point contact, and SNS junctions from the consideration because they are usually heavily overdamped. For this reason, only high quality tunnel junctions of $s-$wave superconductors, which are extremely underdamped, can meet the requirement. The Hamiltonian of an isolated Josephson tunnel junction based rf-SQUID is

$$H(\Phi, \Phi_x) = \frac{p_\Phi^2}{2C} + U(\Phi, \Phi_x), \tag{1}$$

where the total magnetic flux in the SQUID Φ is the general coordinate of the system and the macroscopic variable of interest. Φ_x is the externally applied flux and $p_\Phi = -i\hbar\partial/\partial\Phi$ is the charge on the junction capacitance C. With both Φ and Φ_x referenced to $\Phi_0/2 \equiv h/4e$ the potential energy is:

$$U = \frac{(\Phi - \Phi_x)^2}{2L} + E_J \cos\left(2\pi\frac{\Phi}{\Phi_0}\right), \tag{2}$$

where $E_J \equiv I_c\Phi_0/2\pi$. The shape of the potential depends on the parameter $\beta_L = 2\pi L I_c/\Phi_0$ and Φ_x. For $\beta_L \gtrsim 1$ and $\phi_x \equiv \Phi_x/\Phi_0 = 0$ the potential is a symmetric double-well separated by a barrier ΔU. The energy difference between the two wells ε is proportional to ϕ_x for $\phi_x \ll 1$. By replacing the single Josephson junction with a dc SQUID of inductance $l \ll L$ the parameter β_L can be varied *in situ* by applying a flux Φ_{xdc} to the dc SQUID loop. The detail of this variable β_L SQUID has been described elsewhere [16].

The energy levels and wavefunctions of the SQUID can be obtained by numerically diagonalizing an N by N matrix using the simple harmonic oscillator (SHO) basis. N is chosen such that further increase of N produces negligible change in the eigenenergies of the relevant levels. Except at $\varepsilon = 0$, the levels (indexed by $\{f, i\}$) with energies $E_{f,i} \ll \Delta U$ are well localized in either the $f = 0$ or $f = 1$ fluxoid state. When ε is much less than the typical intrawell level spacing $\hbar\omega_0$ and the condition $k_B T \ll \varepsilon$ is satisfied the SQUID is well described by a quantum two-level system [17–19] . Naturally, the lowest states in each fluxoid well, $\{0, 0\}$ and $\{1, 0\}$, form the computational basis of the SQUID qubit. For the sake of simplicity these two states are denoted as $|0\rangle$ and $|1\rangle$. The energy difference between the two qubit states is $\Delta E_{01} \equiv E_1 - E_0 \simeq \varepsilon$ which can be adjusted in situ by varying the applied static flux Φ_x. When a *resonant* microwave field of frequency $\omega = \Delta E_{01}/\hbar$ is applied to the SQUID it generates an ac flux $\phi_{rf} \cos\omega t = \frac{1}{2}\phi_{rf}(e^{i\omega t} + e^{-i\omega t})$ in the SQUID which produces Rabi oscillations among the two qubit states. The frequency of the Rabi oscillation is given by

$$\Omega = \frac{\Phi_0^2}{2L}\phi_{rf} \left|\langle 0|\phi|1\rangle\right| /\hbar, \tag{3}$$

which is a linear function of the ac flux amplitude ϕ_{rf} and the matrix element $\langle 0|\phi|1\rangle$. At zero damping, the dynamics of the SQUID qubit is entirely determined by its time-dependent wavevector

$$|\psi\rangle = C_0(t)|0\rangle + C_1(t)|1\rangle, \tag{4}$$

where $C_0(t)$ and $C_1(t)$ are in general complex quantities which satisfy the normalization condition $|C_0|^2 + |C_1|^2 = 1$. $C_{i=1,2}(t)$ contain all information about the time evolution of the qubit. With the rotating wave approximation (RWA) the Hamiltonian of the two-state qubit becomes

$$H_{RWA} = \frac{\hbar}{2}\begin{pmatrix} 0 & \Omega^* \\ \Omega & 0 \end{pmatrix}. \tag{5}$$

A quick examination of the Hamiltonian (5) shows that the dynamics of the qubit is entirely determined by Ω. It is well known that when such a quantum two-level system is subjected to a resonant pulse excitation its final state depends only on its initial state and the pulse area $A = \int_0 \tau\Omega(t)dt$, where the pulse is on during time interval $t = 0$ to τ. Therefore, the qubit can be controlled at our will using resonant microwave pulses. As demonstrated in Fig. 2, by changing SQUID parameters one can vary the energy level structure of the SQUID to obtain ω and Ω suitable for specific qubit operations.

3. *Two-Qubit SQUID Quantum Controlled-NOT Gates*

A quantum controlled-NOT (QCN) gate is the basic information processing unit of a quantum computer. A QCN provides the transformation

$$|\,\epsilon_1\rangle\,|\,\epsilon_2\rangle \rightarrow |\,\epsilon_1\rangle\,|\,\epsilon_1 \oplus \epsilon_2\rangle, \tag{6}$$

where $\epsilon_1 = (0 \text{ or } 1)$ and $\epsilon_2 = (\uparrow \text{ or } \downarrow)$ and $\oplus$ is addition modulo 2. In this notation ϵ_1 is the control bit and ϵ_2 is the target bit. If $\epsilon_1 = 0$, the target bit remains unchanged. If $\epsilon_1 = 1$, the target bit flips.

A spectroscopic experiment on any four-level quantum system, where the level spacings are unequal, shows this type of logical structure if we make the appropriate labeling of the levels. In fact QCN logic operations have been demonstrated in trapped $^9Be^+$ ions using a 4-level scheme [20] . In this pioneering work, the two quantum bits are stored in the internal

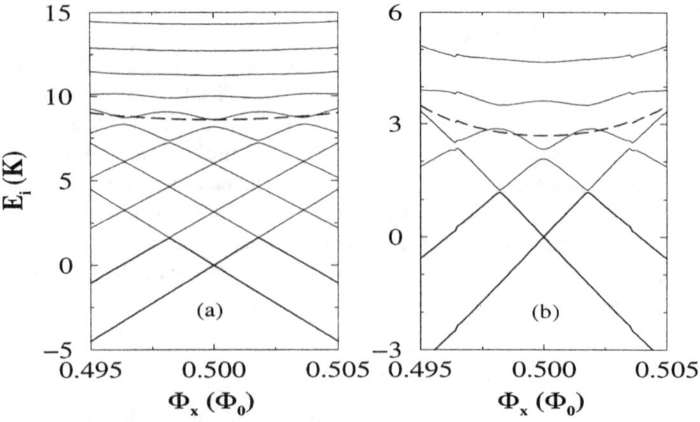

Figure 2: Energy level diagrams of two rf-SQUIDs with $L = 60$ pH, $C = 40$ fF, and (a) $\beta_L = 1.25$, and (b) $\beta_L = 1.15$ vs. the applied flux Φ_x. The two lowest levels (heavy solid lines) form the two basis states of the qubits. The dashed line is the top of the barrier between the two potential wells.

and external degrees of freedom of a single trapped ion, which is laser cooled to the zero-point energy. In the implementation of a QCN logical gate in a trapped ion, the target qubit is spanned by the two $^2S_{1/2}$ hyperfine ground state of a single Be^+ ion, denoted by the equivalent spin-1/2 states $|\uparrow\rangle$ and $|\downarrow\rangle$ states. The control qubit $|\epsilon_1\rangle$, $\epsilon = 0, 1$, is spanned by the first two quantized harmonic oscillator states of the center-of-mass motion of the trapped ion. The energy level spacing of the harmonic oscillator is smaller than that of the internal state spacing in the case of trapped ion. The transformation , Eq. (6), has been experimentally realized for a single trapped ion [20] . Transitions between levels were produced using two laser beams to realize stimulated Raman transitions. In the first experiment [20] the CN gate operation between these states was realized by sequentially applying three pulses of Raman laser beams to the ion. In the second experiment, a simplified QCN gate scheme between an ion's internal and motional states has been proposed which uses only a single laser pulse and eliminates the requirement of the auxiliary internal electronic level [21] .

It is interesting to notice that with a proper choice of SQUID parameters a variable β_L (also called tunable barrier) SQUID can have a level structure similar to the trapped ions discussed above. Since a variable β_L SQUID contains two Josephson junctions it has two degrees of freedom associated with the superconducting gauge invariant phase difference of the two junctions. The potential energy is therefore 2-dimensional [22] and the Hamiltonian, in the case of two identical junctions, can be written as

$$H(x,\, y) = \frac{p_x^2}{2m_x} + \frac{1}{2}m_x\omega_x^2 x^2 + \left(\frac{\vartheta}{2\pi}\right)^2 \beta_L \cos[2\pi(x+x_0)]\cos[\pi(y+y_0)] + \frac{p_y^2}{2m_y} + \frac{1}{2}m_y\omega_y^2 y^2, \tag{7}$$

where, $x \equiv (\Phi - \Phi_x)\Phi_0$ and $y \equiv (\Phi_{dc} - \Phi_{xdc})\Phi_0$ are the normalized total fluxes in the large (rf) and small (dc) SQUID loops, respectively. $x_0 \equiv \Phi_x/\Phi_0$ and $y_0 \equiv \Phi_{xdc}/\Phi_0$ are the normalized

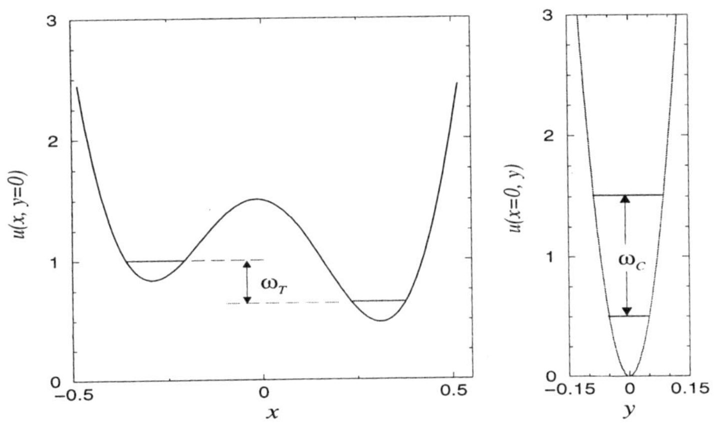

Figure 3: Normalized potential energy u of a SQUID QCN gate with $\beta_L = 2.0$, $\gamma = 10.0$, $x_0 = 0.515$, and $y_0 = 0$. Left: the $y = 0$ section of the potential and the two target qubit states $|\downarrow\rangle$ and $|\uparrow\rangle$. Right: the $x = 0$ section of the potential and the two control qubit states $|\, 0\rangle$ and $|\, 1\rangle$. For clarity, curves are shifted vertically.

external fluxes applied to the two loops. Again, all fluxes are measured with respect to $\Phi_0/2$. The anisotropic mass of the 'flux particle' is given by $m_x \equiv C\Phi_0^2$ and $m_y = m_x/4$, where C is the total capacitance of the two junctions in parallel. The harmonic oscillation frequencies in the x and y directions are $\omega_x \equiv 1/\sqrt{LC}$ and $\omega_y \equiv \sqrt{4\gamma}\omega_x$ with $\gamma \equiv L/l$. In Eq. (7) $\vartheta \equiv \sqrt{2\pi R_Q/Z_0}$, where $R_Q \equiv h/4e^2$ is the resistance quantum and $Z_0 \equiv \sqrt{L/C}$ is the characteristic impedance of the rf-SQUID. If the parameter $\beta_L \equiv 2\pi L I_C/\Phi_0$ is greater than unity then the potential has two extrema in the x direction that corresponding to clockwise and counterclockwise circulating supercurrent in the large loop. When the system is in the lowest level in the right (left) well $\Phi \approx \pm\Phi_m$, where $2\Phi_m$ is the distance between the two fluxoid wells in the x direction. In analogy to spin notation these two levels are labelled as the up ($|\uparrow\rangle$) and down ($|\downarrow\rangle$) states of the target qubit. The energy bias ($\approx \varepsilon$) between the up and down states can be continuously varied by changing magnetic flux applied to the large loop. For example, changing x_0 by 10^{-3} will change ε by about 1 K ($\sim$ 20 GHz) for a typical SQUID [23]. The potential shape in the y direction and the strength of coupling between the x and y degrees of freedom is determined by the value of γ. Roughly speaking, for γ not much greater than unity potential in the y direction could also have two potential wells while for $\gamma \gg 1$ the y direction potential becomes increasingly like that of a simple harmonic oscillator and the coupling between the x and y degrees of freedom decreases (see Fig. 3). For QCN application we choose $\gamma \gg 1$ so that the control qubit is formed by the two lowest states in the y direction which are denoted as the $|0\rangle$ and $|1\rangle$. Therefore, we have all of the four entangled states, $|0\rangle\,|\uparrow\rangle$, $|0\rangle\,|\downarrow\rangle$, $|1\rangle\,|\uparrow\rangle$, and $|1\rangle\,|\downarrow\rangle$, which form the basis for the required CN gate operations. This level structure resembles that of the $^9Be^+$ ions used to successfully demonstrate the QCN gate operation. For example, resonant microwave pulse with frequency $\omega = \omega_T \equiv (E_\downarrow - E_\uparrow)/\hbar \approx \varepsilon/\hbar$ or $\omega = \omega_C \equiv (E_1 - E_0)/\hbar \approx \omega_y$ can be used to create arbitrary superposition of the $|\uparrow\rangle$ and $|\downarrow\rangle$ states of the target qubit or the $|0\rangle$ and $|1\rangle$ states of the control qubit. In addition, "blue" ($\omega = \omega_C + \omega_T$) and "red" ($\omega = \omega_C - \omega_T$) microwaves can be used to obtain required QCN gate operations. A unique property of this type of SQUID QCN is that the matrix element $|\langle 0|y|1\rangle| \gg |\langle\downarrow|y|\uparrow\rangle|$ and $\omega_C \gg \omega_T$ so that a very weak microwave field is sufficient to manipulate the control qubit with negligible crosstalk effect on the target qubit. The highly asymmetric nature of this SQUID QCN also reduce the backaction of the target qubit on the control qubit.

4. *Damping-Induced Decherence and Its Effects*

The damping in SQUID qubits can be characterized by the quantum damping parameter $\alpha \equiv (2\phi_m)^2(R_Q/R)$, where $\phi_m \equiv \Phi_m/\Phi_0$ and R is the effective shunt resistance of the junction. Damped coherent oscillation can occur only if $\alpha < \frac{1}{2}$ [17,19]. In this regime, the decoherence time of a two-level system is $\tau_d = \alpha\pi k_B T$, which is proportional to R^{-1}. On the other hand, whether the energy spectrum is quantized or continuous depends on the value of the classical damping parameter $Q_c = RC\omega_0 \propto \alpha^{-1}$, where ω_0 is the small oscillation frequency of the well. In systems with $Q_c \gg 1$ the most significant effect of damping on the excited states is to broaden the levels [4, 24]. Therefore a measurement of the lifetime of excited states can provide important information on the strength of dissipation, and thus decoherence, in SQUID qubits.

4.1 *Measurement of effective damping*

The level broadening due to dissipation (energy relaxation) can be approximated by calculating the damping-induced decay of excited states in a well. The rate of this intrawell spontaneous decay from a level i to a lower level i' is [25]

$$\mathcal{W}^{sp}_{i\to i'} = \frac{2\pi\delta E_{i,i'}}{\hbar}\frac{R_Q}{R}|\phi_{i,i'}|^2\left[1+\coth\left(\frac{\delta E_{i,i'}}{2k_BT}\right)\right], \tag{8}$$

where $\delta E_{i,i'} \equiv E_i - E_{i'}$ and $\phi_{i,i'} \equiv \langle i|\phi|i'\rangle$ is the matrix element between the levels. Since the wavefunctions of the localized levels are close to those of SHO, the dominate decay channel is always $i \to i-1$. This gives, for $k_BT \ll \delta E_{i,i'}$, a level width $\gamma_i \simeq (E_i - E_0)/Q_c$, a result which becomes exact for a SHO [26]. Equation (8) shows that for $k_BT \ll \delta E_{i,i'}$ the width of energy levels depend linearly on R^{-1} and is insensitive to T.

The energy level spectrum can be investigated by measuring the tunneling rate as a function of energy bias ε between the wells. The tunneling rate from a level $\{0,i\}$ in the upper well to the lower well is given by [3]

$$\Gamma_i = \sum_j \frac{\Delta^2_{i,j}}{4\hbar}\frac{\gamma_{i,j}}{(E_i - E_j)^2 + \gamma^2_{i,j}/4}, \tag{9}$$

Here, $\Delta_{i,j}$ is the tunneling matrix element between levels $\{0,i\}$ and $\{1,j\}$ and $\gamma_{i,j} = \gamma_i + \gamma_j$ is the combined linewidth. At low temperatures only the lowest level of the initial fluxoid state ($f = 0$) is occupied hence the total tunneling rate from the upper well is $\Gamma \equiv \sum_i \Gamma_i \simeq \Gamma_0$, i.e., most of the tunnelings from the $f = 0$ to $f = 1$ state are from level $\{0,0\}$. Equation (9) shows that for $\gamma_{i,j} \ll |\delta E_{i,j}|$ (i.e., $Q_c \gg 1$) resonant tunneling occurs at values of ε where $\delta E_{i,j} \equiv E_i - E_j = 0$, i.e., the tunneling rate is greatly enhanced as levels in the opposite wells are aligned. In this regime, the shape of spectral line (tunneling rate vs. ε) is Lorentzian. The full width at half maximum of the Lorentzian peak is $\gamma_{i,j}$.

Experimentally it is much more efficient to measure the tunneling probability distribution P as a function of ε [27] . $P(\varepsilon)$ can be obtained by linearly ramping Φ_x, increasing ε, until an interwell fluxoid tunneling occurs. The resulting abrupt change in Φ is sensed by a level detector on the output of a dc SQUID magnetometer weakly coupled to the sample. This event causes the value of Φ_x, where the tunneling occurs, to be recorded. The ramp direction is then reversed, leading to a tunneling which returns the system to its original state. This cycle is repeated about 10^4 times to produce the histogram. The normalized histogram is a good approximation to $P(\varepsilon)$ which is related to the tunneling rate $\Gamma(\varepsilon)$ by

$$P(\varepsilon) = \frac{\Gamma}{d\varepsilon/dt} \exp\left[-\frac{1}{d\varepsilon/dt}\int \varepsilon\Gamma(\varepsilon')d\varepsilon'\right]. \qquad (10)$$

The inverse conversion from the transition probability $P(\varepsilon)$ to transition rate $\Gamma(\varepsilon)$ is given by

$$\Gamma(\varepsilon) = [P(\varepsilon)d\varepsilon/dt] / \left[1 - \int_0 \varepsilon P(\varepsilon')d\varepsilon'\right]. \qquad (11)$$

The experiment is preformed in a dilution refrigerator with a helium filled copper sample cell mounted on a temperature regulated platform. Two cylindrical mu-metal shields at room temperature and a Nb enclosure inside the copper cell provide satisfactory shielding against the environmental magnetic field. Both Φ_x and Φ_{xdc} field coils were weakly coupled to the sample and were carefully filtered. The parameters of the SQUID potential were determined from independent measurements which yield $L = 231\pm3$ pH, $C = 77\pm2$ fF, $L/l = 25\pm5$, and $\beta_{L0} = 2.010 \pm 0.005$. The two junctions have equal I_c to within 1% [28] . Coupling between the magnetometer, which has an integrated thin film flux transformer, and the sample was achieved by placing the magnetometer chip orthogonal to the sample plane. This coupling can be varied by changing the spacing between the two chips. To shield the SQUID from microwaves generated by the magnetometer the SQUID chip was enclosed by a 1.6 μm AuPd foil of sheet resistance $R_\square = 0.18\,\Omega$. Fig. 4 shows the result of the lifetime measurement. The data (triangles) is obtained with a SQUID-foil separation of 150 μm resulting in an intrinsic linewidth much smaller than the level spacing (see the next section). The multi-peak tunneling probability $P(\varepsilon)$ clearly shows that the system's energy levels are quantized.

By comparing the calculated and the measured $\Gamma(\varepsilon)$ in Fig. 4 it can be seen that the observed widths of the resonant tunneling peaks are much wider than that expected from Eq. (8) and the shape of the measured $\Gamma(\varepsilon)$ is closer to Gaussian than Lorentzian. This is caused by Gaussian noise which is primarily due to the effect of Nyquist noise in the magnetometer on the level detection circuit ($\sigma'_{\phi_x} \approx 2 \times 10^{-4}$) and to a much lesser extent to the low frequency noise in Φ_x generated by the foil ($\sigma_{\phi_x} < 1 \times 10^{-4}$) rather than the dissipation-induced level width discussed above [28] . To find the intrinsic Lorentzian linewidth, we have deconvoluted one of the peaks, obtained at $T = 30$ mK in a dilution refrigerator with a very weak coupling to the AuPd foil, into its Gaussian and Lorentzian components. The original data along the fits are shown in Fig. 4. Clearly, the Gaussian component is by far the broader one. Unfortunately, it was only possible to perform a meaningful deconvolution for the strongest peak, i.e. the doubly resonant peak which occurs only at specific values of β_L and ε where a weak radiation field with frequency $\omega_r = (E_{0,i} - E_{0,0})/\hbar$ and the resonant tunneling condition $E_{0,i} = E_{1,j}$ are both satisfied. The resulting Lorentzian width of $\gamma_{12}/k_B = 0.030 \pm 0.005$ K implies $Q_c \approx 10^3$, or $R = (5 \pm 0.8)R_Q \simeq 32$ kΩ and $\alpha \approx 0.03 \ll 1/2$ which is well into

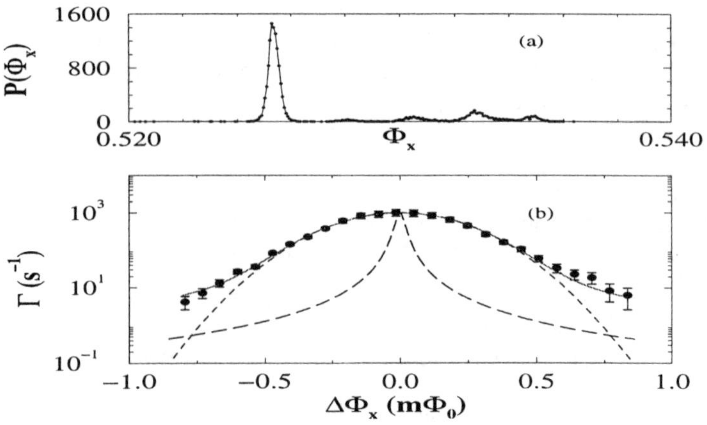

Figure 4: (a) Measured transition probability of the sample SQUID with $\beta_L = 1.70$. The left most peak is the result of double resonant photon-assisted tunneling between the two fluxoid states described in the text. (b) The data (solid dots with error bars) is deconvoluted into its Gaussian (short dashed curve) and Lorentzian (long dashed curve) components. Agreement between the data and convolution of the Gaussian and Lorentzian (solid line) is very good.

the regime where macroscopic quantum coherence could occur. This result has been confirmed recently through a careful quantitative analysis of photon induced resonant transitions between the two fluxoid states in the SQUID [8]. In fact, energy level splitting as a result of coherent superposition of macroscopically distinct fluxoid states in a SQUID very similar to our sample has been reported recently [10]. For the two qubit states, the measured effective damping would imply a decoherence time of $\tau_d = \pi\alpha k_B T/\hbar \approx 8$ ns at $T = 10$ mK.

4.2 *The sources of damping*

It is obvious that the measured effective damping resistance is much smaller than the quasiparticle tunneling resistance R_{qp} which was greater than 1 MΩ at $T < 1.2$ K for our low critical current Nb tunnel junctions. Other sources of damping in the sample include inductive coupling to other lossy environmental elements (e.g. the AuPd foil, the magnetometer, and the various filters and coils), loss due to magnetic dipole radiation, and loss due to finite surface resistance of the Nb film at microwave frequencies.

The damping effect of each of the inductively coupled environmental elements, including the magnetometer, the AuPd foil, and the coils and filters, to the SQUID sample can be modeled by a resistance R' in series with an inductance L'. The inductance L' is coupled to the sample through a mutual inductance M, as shown in Fig. 1. A simple circuit analysis shows that at frequency ω, this can be transformed to an equivalent shunt resistance R_{eff} across the junction with

$$R(\omega) = \left(\frac{L}{M}\right)^2 R' \left(1 + \frac{\omega^2}{\omega_D^2}\right), \tag{12}$$

where L is the inductance of the SQUID and the cutoff frequency $\omega_D \equiv R'/L'$. Based on this analysis, the equivalent damping resistances R from the Φ_x and Φ_{xdc} coils and filters are estimated to be much greater than 10 MΩ and thus have had very little effect on the observed linewidth.

Damping due to the AuPd foil can be calculated by equating the power dissipated by the eddy currents in the foil due to currents in the sample inductance, $P_{\text{foil}} = \frac{1}{2}\int_{\text{foil}} J(r)^2 R_{\square} ds$, with that dissipated in an equivalent shunt resistor across the junction. The eddy current density, $J(r)$, is related to the vector potential $A(r)$ generated by a current I_s in the sample by $R_{\square} J(r) = Re\left[-i\omega A(r)\right]$. For sufficiently low frequencies, the eddy current contribution to $A(r)$ can be neglected. Thus $A(r) \propto I_s$ and is frequency independent. The power dissipated in an equivalent shunt resistor R in the sample SQUID is $P_{\text{eff}} = \frac{1}{2} I_s^2 L^2 \omega^2 / R$. At low frequency, both powers scale as ω^2, so the damping is ohmic. Since the magnitude of the vector potential depends on the SQUID-foil distance d, the damping can be tuned by varying d. The damping can also be varied by changing the material or the thickness of the foil. The foil used has a sheet resistance of $R_{\square} \simeq 0.18\ \Omega$ and $\omega_D/2\pi \approx 0.5$ GHz. At a distance of 150 μm from the SQUID chip $R(0) \approx 6$ kΩ which result in an effective damping resistance of greater than $3600 R_Q$ at the interwell level spacing frequency of about 30 GHz. If the above analysis is valid it shows that damping due to the AuPd foil is also negligible.

The magnetometer used in our experiment is a dc SQUID with a total shunt resistance of $R' = 3.3\ \Omega$ and loop inductance of $L' = 26$ pH which gave a cutoff frequency $\omega_D/2\pi \simeq 20$ GHz. The mutual inductance between the magnetometer and the sample SQUID, taking into account the superconducting dc flux transformer, was $M \approx 0.4$ pH. Therefore, the effective damping due to coupling to the magnetometer, at the sample SQUID level spacing frequency $\omega_0/2\pi \approx 30$ GHz, is $R_{\text{eff}} \approx 1$ M$\Omega \approx 170 R_Q$. It is obvious that in this case coupling to the magnetometer was not the dominant source of damping.

When a SQUID loop has an oscillating current of frequency ω it produces energy loss in the form of radiation. In an rf SQUID magnetic dipole radiation is the dominant mechanism. The quality factor of this dipole radiation is $Q = \omega L/R_r$, where $R_r \simeq 3 \times 10^5 (b/\lambda)^4\ \Omega$ is the radiation resistance of a magnetic dipole of radius b and λ is the wavelength of radiation. Equate this quality factor to $RC\omega$ the effective shunt resistance due to radiation energy loss can then be expressed as $R = Z_0^2/R_r$. From the sample geometry ($b \approx 50\ \mu$m) and wavelength ($\lambda \approx 1$ cm) an estimate of $R \approx 1.6 \times 10^7\ \Omega \approx 2500 R_Q$ is obtained which shows that the effect of radiation loss on measured linewidth is negligible.

Finally, the effective damping due to Nb film's surface resistance can be modeled in a similar fashion. If the surface resistance of the Nb film that forms the SQUID is R_s the effective shunt resistance R across the junction due to R_s is given by $R = Z_0^2/R_s$. It is well known that surface resistance of superconducting films is not only dependent of intrinsic superconducting parameters (e.g., energy gap, critical current, and critical temperature) but also greatly influenced by extrinsic factors including surface morphology, grain boundaries, and surface contaminations. It has been observed that the surface (sheet) resistance of Nb films prepared by sputtering bottoms out at about $R_{\square} \approx 1$ mΩ below $T \approx 1.5$ K [29] . This yields an estimated total surface resistance $R_s \sim 84$ mΩ and an effective shunt resistance $R \sim 5.5 R_Q$, which agrees with the measured effective shunt resistance pretty well. The analysis shows that the measured linewidth of the excited state in SQUID was most likely dominated by energy loss due to surface resistance of the Nb film. For this reason, efforts must be made to significantly reduce surface resistance of Nb films used to fabricate SQUIDs.

4.3 *Implication to SQUID qubits and QCN gates*

It has been shown that in order to have coherent oscillation in a dissipative quantum two-level system the quantum damping parameter α must be less than $\frac{1}{2}$ and $k_BT < \hbar\Delta/\alpha\pi$, where Δ is the frequency of the underdamped coherent oscillation. The decoherence time is $\tau_d = \alpha\pi k_B T$. Since α is inversely proportional to the effective damping R it is thus crucial to reduce damping as much as possible. The measured effective shunt resistance $R \approx 5R_Q$ gave $\alpha \approx 0.03$ and $\tau_d \approx 8$ ns at $T = 10$ mK. Obviously this decoherence time is too short for any realistic qubit operation. However, this problem is not fundamental since surface resistance of sputtered Nb films are much higher than that of bulk Nb and therefore can be significantly reduced by improving the quality of Nb film.

On the other hand, energy dissipation due to radiation decreases rapidly as the size of the SQUIDs and the frequency of radiation are reduced ($R^{-1} \propto b^4/\lambda^4$). For example, reducing the radius of SQUID loop b by a factor of five to 10 μm and taking $\omega/2\pi = 10$ GHz will decrease the effective damping (increase R) by a factor of 5×10^4. Therefore, in general the effect of radiative energy loss on decoherence is negligible in SQUID qubits.

Although quasiparticle resistance of low J_c (critical current density) Josephson tunnel junctions made of conventional superconductors at $T \ll T_c$ is high enough to produce decoherence time $\tau_d \gg 1$ μs it is not clear whether high J_c junctions can deliver the same performance. The use of high J_c junction SQUIDs is necessary to bring the qubits' size down. As the size of the SQUID qubits is reduced so is their inductance L. This in turn requires a reduction in junction capacitance C and a increase in I_c to keep Z_0 (i.e., L/C) and β_L (i.e., LI_c) constant so that the level structure of the qubits will not change. Hence, the critical current density required scales linearly with the areal density of qubits. For this reason, dissipative quasiparticle tunneling in high J_c junctions deserves further investigation.

The environment induced dissipation in SQUID qubits that is due to inductive coupling to external dissipative elements can be kept low enough by reducing the mutual inductance between the qubits and the external elements. A common source of dissipation is due to the coupling between a qubit and a magnetometer used for readout. As demonstrated before, using a weakly coupled magnetometer can minimize this type of dissipation. However, it should be pointed out that a conventional overdamped dc SQUID magnetometer is also an active source of microwave irradiation. The irradiation can be a significant mechanism of decoherence for the SQUID qubits [8].

Finally, an rf-SQUID is really a multi-level quantum system. The ϕ matrix elements between nearest intrawell levels are exponentially larger than their interwell counterpart that determines the Rabi frequency. Therefore, probability leakage to levels other than the two computational basis is unavoidable. This can be a serious problem for quantum logic operation using SQUID qubits and must be understood through theoretical and experimental investigations.

5. *Summary*

In summary, we have shown that an rf-SQUID is a quantum system which is attractive for qubit application. The rf-SQUIDs can be engineered to have properties required for qubits and the quantum dynamics of the rf-SQUIDs can be precisely controlled by microwave pulses. In addition, we have shown that a simple variation of the rf-SQUID—the variable β_L SQUID—has a similar energy level structure to the Be^+ ion qubits so that quantum controlled-NOT gate operation can be readily implemented. However, like all other solid state qubits SQUIDs suffer from the perils of the damping-induced decoherence. To determine the effective damping in SQUIDs under real world condition we measured the lifetime of an excited level in an rf-SQUID qubit. The Lorentzian linewidth obtained from a deconvolution procedure implies a quantum damping parameter of $\alpha \approx 0.03$, indicating that coherent oscillation could occur in this SQUID but the inferred decoherence time of $\tau_d \approx \alpha\pi k_B T/\hbar \approx 8$ ns at $T = 10$ mK is not long enough for serious qubit operation. A quantitative analysis shows that for SQUID qubits dissipation due to superconducting Nb film's microwave surface resistance could be the dominant mechanism of decoherence. However, this problem can be addressed by optimizing the qubit design and improving the superconducting Josephson tunnel junction fabrication technology.

The authors are grateful to Prof. Shih-I Chu and J. E. Lukens for helpful discussions. The work at the University of Kansas is supported in part by the AFOSR Grant No. F49620-99-1-0205, the NSF Grant No. DMR9876874, and the State of Kansas Grant No. S99041.

6. *References*

[1] A. O. Caldeira and A. J. Leggett, Phys. Rev. Lett. **46**, 211 (1981).

[2] D. B. Schwartz, B. Sen, C. N. Archie, and J. E. Lukens, Phys. Rev. Lett. **55**, 1547 (1985).

[3] A. Garg, Phys. Rev. B **51**, 15161 (1995).

[4] J. M. Schmidt, A. N. Cleland, and J. Clarke, Phys. Rev. B **43**, 229 (1991).

[5] R. Rouse, S. Han, and J. E. Lukens, Phys. Rev. Lett. **75**, 1614 (1995).

[6] B. Ruggiero and P. Silvestrini, IEEE Trans. Appl. Supercon. **5**, 3081 (1995).

[7] S. Chakravarty and S. Kivelson, Phys. Rev. Lett. **50**, 1811 (1983).

[8] S. Han, R. Rouse, and J. E. Lukens, Phys. Rev. Lett. **84**, 1300 (2000).

[9] S. Han, R. Rouse, and J. E. Lukens, Phys. Rev. Lett. **76**, 3404 (1996).

[10] J. R. Friedman, V. Patel, W. Chen, S. K. Tolpygo, and J. E. Lukens, Nature (2000), to be published.

[11] D. P. DiVincenzo, Science **270**, 255 (1995), and references therein.

[12] S. Lloyd, Phys. Rev. Lett. **75**, 346 (1995).

[13] J. F. Poyatos and J. I. Cirac, Phys. Rev. Lett. **78**, 390 (1997).

[14] A. Barenco, D. Deutsch, and A. Ekert, Phys. Rev. Lett. **74**, 4083 (1995).

[15] T. Sleator and H. Weinfurter, Phys. Rev. Lett. **74**, 4087 (1995).

[16] S. Han, J. Lapointe, and J. E. Lukens, in *Single-Electron Tunneling and Mesoscopic Devices*, edited by H. Koch and Lübbig (Springer-Verlag, Berlin Heidelberg, 1991), Vol. 31, pp. 219–222.

[17] A. J. Leggett, S. Chakravarty, A. T. Dorsey, M. P. A. Fisher, A. Garg, and W. Zwerger, Rev. Mod. Phys. **59**, 1 (1987).

[18] W. Zwerger, Z. Phys. B **53**, (1983).

[19] U. Weiss, H. Grabert, and S. Linkwitz, J. Low Temp. Phys. **68**, 213 (1987).

[20] C. Monroe, D. M. Meekhof, B. E. King, W. M. Itano, and D. J. Wineland, Phys. Rev. Lett. **75**, 4714 (1995).

[21] C. Monroe, D. Leibfried, B. E. King, D. M. Meekhof, W. M. Itano, and D. J. Wineland, Phys. Rev. A **55**, R2489 (1997).

[22] S. Han, J. Lapointe, and J. E. Lukens, Phys. Rev. Lett. **63**, 1712 (1989).

[23] R. Rouse, S. Han, and J. E. Lukens, Phys. Rev. Lett. **75**, 1614 (1995).

[24] W. Bialek, S. Chakravarty, and S. Kivelson, Phys. Rev. B **35**, 120 (1987).

[25] A. I. Larkin and Y. N. Ovchinnikov, Sov. Phys. JETP **64**, 185 (1986).

[26] L. R. Senitzky, Phys. Rev. **119**, 670 (1960).

[27] J. Kurkijarvi, Phys. Rev. B **6**, 832 (1972).

[28] R. P. Rouse, Ph.D. thesis, SUNY at Stony Brook, 1996.

[29] B. Bi, K. Wan, W. Zhang, S. Han, and J. E. Lukens, IEEE Trans. Appl. Supercon. **1**, 145 (1991).

Decoherence and its Implications in Quantum Computation and Information Transfer
T. Gonis and P.E.A. Turchi (Eds.)
IOS Press, 2001

QUANTUM IMPLICATIONS of QUANTUM AUTOMATA

Jozef Gruska[1]
Faculty of Informatics, Masaryk University, Botanická 68a, 602 00 Brno, Czech Republic

Abstract On one side the aim of the paper is to provide a basic insight, especially for non-computer scientists, into the aims, motivations, methods, results and the main open problems in the area of quantum finite automata (QFA) and quantum almost finite automata (QAFA) On the other side, the aim of the paper is to discuss and analyze new ideas and further potentials of this area, that has so far developed mainly as a natural quantumization of the classical theory of FA and AFA, with respect to the main goals and difficult problems of the research in QIPC.

1 Introduction

Classical theory of automata is an area of research that has brought a variety of deep concepts, methods and results of the great importance not only for theoretical computer science, but for all computing. This is due to the following:

- The concepts of finite automata and finite-state machines are very robust and there is a variety of important and very different models of automata and of descriptive systems (regular expressions, finite-state grammars, formulas of monadic logic, ...), that have the same (descriptional) power.

- Finite automata can be seen as the simplest model of automata that is sufficiently robust and has many interesting and important properties.

- Finite automata and related regular expressions have turned to be very closely related to a variety of algebraic structures. This has allowed to use deep algebraic techniques to deal with problems of finite automata and regular expressions. Algebraic theory of automata is now much developed.

- Finite automata can be seen as components of other more powerful models of automata.

- From the application point of view, all machines can be seen as finite state machines and whenever really powerful methods have to be developed for linear time solution of some important problems they often result to methods implementable by finite automata. From this point of view, finite state machines play in computation a role analogous to linear systems in applied mathematics.

- The concept of universal automata (for example Turing machines) is also very robust and there is also a variety of very different computation models and formal systems that have the same computational power. This has been used to introduce, as very robust, the very basic concept of *computability*.

[1]The paper has been written during the author's stay with the University of Karlsruhe, Department of Informatics, in summer 2000. Support of the grants GAČR 201/98/0369, CEZ:J07/98:143300001 and VEGA 1/7654/20 is to be acknowledged.

- The main models of the universal machines are also robust with respect to computation in deterministic (or randomized) polynomial time. This has been used to define as very robust computational complexity classes **P** (and **BPP**) for polynomial time computation on deterministic (probabilistic) machines and the corresponding concepts of *feasibility*.
- From the theoretical point of view, all automata models can be seen as restricted variants of the universal models.

Models of universal computers played the key role when defining such concepts as effective computability and also efficient computation and in defining Kolmogorov complexity. On the other hand models of quantum automata are actually the main model of real-time computations. Theory of quantum automata has so far developed mainly on the basis of the same motivations, insights, paradigms and aims, as the classical automata theory (and this can be seen as one of the reasons why entanglement played no role in it so far.). For several reasons – one of them is a (puzzling for many) preoccupation of this theory with **language recognition** – it seems that the aims of this theory as well as its main methods and results are often quite unknown and even puzzling for people outside of (theoretical) computing. This seems to be especially the case for finite automata and therefore this is the model we concentrate on in this paper.

1.1 Aims

The research in the area of quantum finite automata has several closely related goals.

- To get an insight into the power of different quantum computing models and modes using automata/language theoretic methods and concepts.
- To get an insight and some qualitative and quantitative understanding of the power of various quantum resources (superposition, parallelism, entanglement, measurement)?
- To explore potential of the most simple models of quantum computing (the very basic quantum processors are expected to perform quantum actions on classical inputs).
- To investigate, using the most simple models of quantum processors, how different computational features and modes influence the overall power in the case of quantum devices.
- To determine the simplest models of automata for which we can prove that their quantum versions are more powerful, in a reasonable sense, than their classical (randomized) counterparts.
- To explore, using at first the simplest models of quantum computation, how much of quantumness is needed and how pure it has to be in order to have automata (machines or algorithms) more powerful than classical ones.
- To develop design and optimization methodologies at first first for very simple models of computation.
- To develop lower bound techniques for showing limitations at first for the simplest models of quantum processors.
- To discover, in a transparent (and elegant) form, limitations and potentials of quantum computations and communications.

For a detailed presentation of the basic concepts, results and methods in this area, see [16, 17].

1.2 Language recognition

Since one of the most puzzling feature (for many) of the automata theory is its preoccupation with language recognition, let us first discuss this problem in some details.

A (formal) language L is just a subset of the set of all words, denoted Σ^*, over an alphabet Σ. A **language recognition problem** for an alphabet Σ and a language $L \subseteq \Sigma^*$ is the problem to decide, given an arbitrary string (world) $x \in \Sigma^*$, whether $x \in L$.

The importance of the language recognition problem is due to the fact that (formal) languages present a convenient formalization of the intuitive concept of the decision problems..

For example the problem, P_1, to decide whether a given graph G is 3-colorable (that is whether its nodes can be coloured by three colours in such a way that no two nodes connected by an edge have the same colour), or problem, P_2, to decide, given two graphs, G_1 and G_2, whether they are isomorphic, or, finally, the problem, P_3, to decide whether a given integer n is prime. In all such cases we can associate to the problem P_i, using a proper encoding of inputs (say of a graph G for problem P_1, or of two graphs G_1 and G_2 for the problem P_2, and of an integer n for the problem P_3, into a fixed (mostly binary) alphabet Σ, a language L_{P_i} of all strings (words) that represent those inputs for which the corresponding decision problem has the positive answer. That is L_{P_1} is the language of those words that are encodings of 3-colorable graphs, ...

A procedure to check, given $x \in \Sigma^*$, whether $x \in L$, usually consists of two parts. The first part is to decide whether a given x is really an encoding of the corresponding problem (what is usually not so difficult) and then to decide (what can be very difficult), whether x is an encoding of an input (problem) with the positive answer.

1.3 Classical models of finite automata

Two basic approaches to the classical finite state machines are to see them either as **transducers** or as **acceptors**. In both cases, such a machine M is specified by a finite set Q of states, an initial state q_0, a finite alphabet Σ), and a transition mapping δ. In the case of transducers we have also an output alphabet Δ and $\delta : Q \times \Sigma \rightarrow Q \times \Delta$. Such a transducer is called a *Mealy machine* and in the case the output does not depend on the input we have a *Moore machine.*

In the case of an acceptor A there is no output alphabet, but a set Q_a of accepting states is considered. In the very basic model, see Figure 1a, of the so-called *deterministic finite automaton* (DFA), the behaviour of the automaton is specified by the transition function $\delta : Q \times \Sigma \rightarrow Q$, which is then extended to $\delta : Q \times \Sigma^* \rightarrow Q$ by defining, $\delta(q, wa) = \delta(\delta(q, w), q)$ for any $w \in \Sigma^*$ and $a \in \Sigma$. An input word w is said to be accepted by the automaton A if $\delta(q_0, w) \in Q_a$. The set of all accepting words is said to be the language accepted by the automaton A.

A different way to see DFA is depicted in Figure 1b. An input word is written on an input tape with endmarkers. The control unit is connected with the input tape by a reading head that is initially positioned on the first symbol of the input and at each step the head reads an input symbol, changes its state according to the given transition mapping and the head then moves to the right (in such a case we sometimes speak about one-way FA (1FA)). A natural generalization is a two-way FA (2FA), see Figure 1c, at which the head can move in both directions (and its transition mapping has the form $\delta : Q \times \Sigma \rightarrow Q \times \{\leftarrow, \downarrow, \rightarrow\}$.

Another generalization of the concept of DFA is that of the nondeterministic FA (NFA), at which the transition function is defined as $\delta : Q \times \Sigma \rightarrow 2^Q$, that is several next states can be possible at each transition. Such an automaton accepts an input word (string) if there is

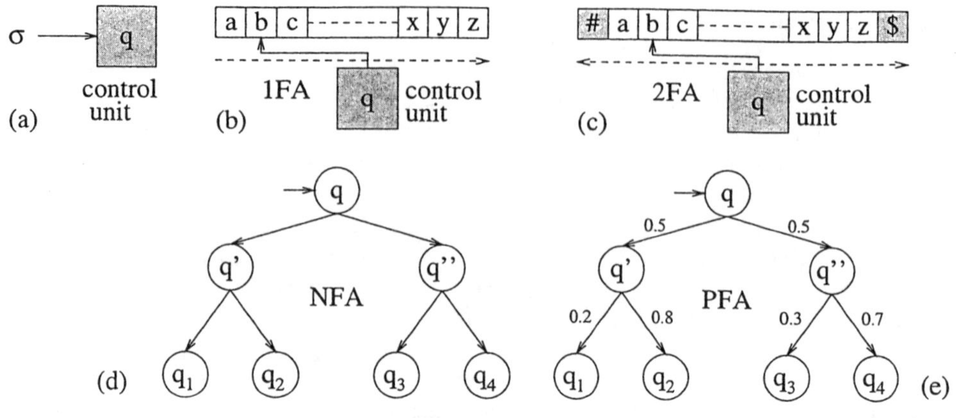

Figure 1: Models of classical automata

a way to get on that input to an accepting state. In other words, the transition function may associate to any configuration a set of possible next configurations, see Figure 1d. The next natural step is to consider probabilistic FA (PFA), at which for each two states q and q' and an input a a probability is given that the automaton moves under the input a from the state q to the state q' (see Figure 1e for the case that all input symbols are identical and they are therefore not written in the Figure).

One of the main result of the classical theory of automata is that all above models of finite automata are equivalent in the sense that they accept exactly the class of regular languages -the set of languages containing all finite languages and closed under operation of concatenation and iteration. (For two language L_1 and L_2 their *concatenation* is the language $L_1L_2 = \{w \mid w = w_1w_2, w_1 \in L_1, w_2 \in L_2.\}$ For a language L and an integer $i \geq 0$ let $L^0 = \{\varepsilon\}$, where ε denotes the empty word and $L^{i+1} = l^I L$. An iteration L^* of a language L is the language $L^* = \bigcup_{i=0}^{\infty} L^i$. For more on the classical models of finite automata see [15].

Very important are for classical finite automata results showing that, up to an isomorphism, to each finite deterministic automaton there exists a single minimal automaton (with respect to the number of states), and that there is a simple and efficient procedure to find such a minimal automaton.

1.4 Two ways to go from the classical to quantum automata

One way to go from the classical to the (restricted) quantum finite automata, and historically the first one, was through **reversible DFA**. They are such DFA that for each state q and each input symbol a there is at most one state q' such that the automaton gets from the state q' to the state q on the input a. (If we draw a flow-diagram for a DFA, then it is reversible if it remains deterministic by reversing the order of edges.)

Reversible FA are weaker than DFA. For example there is no RFA accepting the language 0^*1^*. The basic characterizations of the languages accepted by reversible FA was provided in [27].

The second way to come to QFA is as a natural generalization of PFA. Instead of assigning a probability to each transition from a state (configuration) to a state (configuration), probability amplitudes are assigned. This step is easy. However, the probability amplitudes have to be assigned in such a way that the resulting evolution is unitary.

One of the main sources of difficulties with going to quantum finite automata from the classical ones is connected with the fact that it is very nontrivial to combine probabilistic features with the reversibility features, that is to achieve that the corresponding evolution is indeed unitary.

There is also a concept of the probabilistic reversible FA (PRFA), due to [5]. They are such PFA that for every state q and each input symbol a there is at most one state from which the automaton can get to the state q under the input a with non-zero probability.

1.5 Ways to see the power of finite automata

There are two basic ways to evaluate and compare the power of various models of automata. Each of them has its own merit.

The main one is with respect to the set of languages acceptable by automata of a certain type. This approach considers automata models that accept the same class of languages as having the same recognition power.

The second approach is mainly related to a comparison of two types of automata that accept the same family of languages, or two types of automata, one of which accept more languages than the other one. In this (descriptional complexity) approach automata models are compared with respect to their *descriptional power* — or with respect to their *succinctness* — how large they need to be (with respect to the number of states) to recognize the same languages. A well known example is that for each regular language R and each DFA $\mathcal{A}_R$ recognizing the language R there is a NFA accepting the same language and having not more states than $\mathcal{A}_R$. In addition, for every integer n there is a regular language R_n in the alphabet of two symbols such that there is a NFA with n states recognizing R_n, but no DFA recognizing R_n has less than 2^n states. This mean that NFA that have the same recognition power as DFA, are never larger and sometimes they can be exponentially smaller than the smallest DFA recognizing the same language.

2 Potentials and Limitations of Quantum Automata

Several approaches have been developed how to define models of quantum finite automata and a variety of results have been obtained for these models. Let us discuss briefly the main approaches and outcomes.

2.1 Abstract approaches

On both classical and quantum level there are very abstract approaches to the concept of (finite state) sequential machines.

In the classical case we have the so called *transition systems* $S = \langle Q, q_0, R\rangle$, where R is a relation on Q. A sequence of states $q_0, q_1, q_2, \ldots$ repesent a computation if $(q_i, q_{i+1}) \in R$ for any $i \geq 0$. In the case R is a function we speak about a sequential (deterministic) machines. These concepts have been intensively investigated in theoretical informatics and in mathematics.

On the quantum level a certain analogy is the concept of the *quantum transition machine* (QTRM) $\mathcal{M} = \langle H, |q_0\rangle, U\rangle$, where $\mathcal{H}$ is a Hilbert space, $|q_0\rangle$ is the (initial) state and U is a unitary transformation. Closely related is the concept of *quantum sequential machine* (QSM) $\mathcal{M} = \langle Q, q_0, \delta\rangle$, where Q is the set of states, $q_0 \in Q$ and $\delta : Q \times Q \to \mathbf{C}$ is a transition mapping such that the following well-formedness condition

$$\sum_{q'} \delta(q,q')\delta^*(q_1,q') = \begin{cases} 1 & \text{if i=j;} \\ 0 & \text{otherwise,} \end{cases}$$

is satisfied that guarantees unitarity of the corresponding evolution operator [18]

$$U : q \rightarrow \sum_{q'} \delta(q,q')q'.$$

The above concepts of quantum machines can naturally be extended in two ways. We can consider a set Q_t of terminating states in the case of QSM and a subspace of terminating states H_t in the case of QTRM. A computation is considered as terminating if it gets into a terminating state. The second way of generalization is to consider transition mapping as being dependent also on input symbols. This can then naturally be combined also with the case of having terminating states.

There is a close relation between these two concepts of the machine. Indeed, to each quantum sequential machine we can construct an equivalent quantum transition machine by considering as the corresponding Hilbert space $l_2(Q)$ and as the unitary operator the evolution operator defined above. Conversely, suppose that $\mathcal{M} = \langle Q, q_o, U\rangle$ is a quantum transition machine. Let $\mathcal{B}$ be an orthonormal basis for H that includes the state q_0 and let us define $\delta(q,q') = \langle Uq|q'\rangle$. (That is a quantum transition machine can be seen as quantum sequential machine at which computational basis is unspecified).

2.2 Basic components of the quantum finite automata models

Several nonequivalent models of quantum finite automata have been introduced. Most of them have the following general structure:

An **input** has the form $\#w_1 \ldots w_n\$$ or, shortly, $\#w\$$, where $w \in \Sigma^*$, $|w| = n$, Σ is an input alphabet and $\#, \$$ are endmarkers.

The set of **states** $Q = Q_a \cup Q_r \cup Q_n$ is composed of the mutually disjoint subsets of **accepting states** Q_a , **rejecting states** Q_r, and **nonhalting states** Q_n.

A **configuration** is a pair (q, i) — a state and a position on the input tape. The **set of configurations** has the form $C(Q,w) = \{(q,i) \mid q \in Q, 0 \leq i \leq |w|+1\}$ This is used to introduce the corresponding **Hilbert space** $l(C(Q,w))$.

Transition mappings, for transitions between configurations, δ are defined as follows

$$\delta(|q,i\rangle) = \sum_{q' \in Q, 0 \leq j \leq n+1} \alpha_{q',j}|(q',j)\rangle.$$

2.3 Basic models of quantum finite automata

The very basic model is that of one-way quantum finite automata (1QFA). At this model a state is a superposition of the basis states that correspond to "the heads positioned on the same square of the tape" and at each evolution step "all heads move" one square right. A formal description of 1QFA follows.

Definition 2.1 *A* **one-way (real-time) quantum finite automaton** $\mathcal{A}$ *is given by:* Σ *— the input alphabet;* Q *— the set of the states;* q_0 *– the initial state;* $Q_a \subseteq Q$, $Q_r \subseteq Q$ *— sets of accepting and rejecting states and the transition function*

$$\delta : Q \times \Gamma \times Q \rightarrow C_{[0,1]},$$

where $\Gamma = \Sigma \cup \{\#, \$\}$, $\#, \$$ *are endmarkers and* $\mathbf{C}_{[0,1]}$ *is the set of complex numbers absolute value of which is not larger than* 1.

The evolution (computation) of $\mathcal{A}$ is performed on the Hilbert space $l_2(Q)$, with the basis states $\{|q\rangle \mid q \in Q\}$, using the unitary operators $V_\sigma, \sigma \in \Gamma$, defined by

$$V_\sigma|q\rangle = \sum_{q' \in Q} \delta(q, \sigma, q')|q'\rangle.$$

For an input $w = w_1 w_2 \dots w_n$ the whole evolution is then given by the mappings

$$V_\$ V_{w_n} \dots V_{w_2} V_{w_1} V_\# |q_0\rangle.$$

The second very basic model is that of the two-way quantum finite automaton (2QFA) at which a state is a superposition of basis states that can correspond to different heads positioned on different squares of the input type. Formally, 2QFA are defined as follows [29].

A **two-way quantum finite automaton** $\mathcal{A}$ is specified again by an alphabet Σ, a finite set of states Q, an initial state q_0, sets $Q_a \subseteq Q$ and $Q_r \subseteq Q$, such that $Q_a \cap Q_r = \emptyset$, and the transition function

$$\delta : Q \times \Gamma \times Q \times \{\leftarrow, \downarrow, \rightarrow\} \longrightarrow \mathbf{C}_{[0,1]},$$

where $\Gamma = \Sigma \cup \{\#, \$\}$ is the tape alphabet of $\mathcal{A}$ and $\#$ and $\$$ are endmarkers not in Σ, which satisfies the following conditions (of well-formedness) for any $q_1, q_2 \in Q$, $\sigma, \sigma_1, \sigma_2 \in \Gamma$, $d \in \{\leftarrow, \downarrow, \rightarrow\}$:

1. **Local probability and orthogonality condition.**
   $\sum_{q',d} \delta^*(q_1, \sigma, q', d)\delta(q_2, \sigma, q', d) = \begin{cases} 1, & \text{if } q_1 = q_2; \\ 0, & \text{otherwise.} \end{cases}$

2. **Separability condition I.**
   $\sum_{q'} \delta^*(q_1, \sigma_1, q', \rightarrow)\delta(q_2, \sigma_2, q', \downarrow) + \sum_{q'} \delta^*(q_1, \sigma_1, q', \downarrow)\delta(q_2, \sigma_2, q', \leftarrow) = 0.$

3. **Separability condition II.** $\sum_{q'} \delta^*(q_1, \sigma_1, q', \rightarrow)\delta(q_2, \sigma_2, q', \leftarrow) = 0.$

The above conditions are not easy to verify. Fortunately there is a simpler concept of 2QFA that is equally powerful.

Definition 2.2 *A 2QFA* $\mathcal{A} = \langle \Sigma, Q, q_0, Q_a, Q_r, \delta \rangle$ *is* **simple**, *or* **unidirectional**, *if for each* $\sigma \in \Gamma$ *there is a unitary operator* V_σ *defined on the Hilbert space* $l_2(Q)$ *and, in addition, a function* $D : Q \rightarrow \{\leftarrow, \downarrow, \rightarrow\}$ *such that for each* $q \in Q, \sigma \in \Gamma$,

$$\delta(q, \sigma, q', d) = \begin{cases} \langle q'|V_\sigma|q\rangle, & \text{if } D(q') = d; \\ 0, & \text{otherwise,} \end{cases} \tag{1}$$

It is straightforward to verify that if we rewrite the well-formedness conditions using the relation (1) , then we get that a simple 2QFA $\mathcal{A}$ satisfies the well-formedness condition if and only if

$$\sum_{q'} \langle q'|V_\sigma|q_1\rangle^* \langle q'|V_\sigma|q_2\rangle = \begin{cases} 1, & \text{if } q_1 = q_2; \\ 0, & \text{otherwise,} \end{cases}$$

for each $\sigma \in \Gamma$, which holds if and only if every operator V_σ is unitary.

The last main model is that of the **generalized 1QFA** (g1qFA) that is a special case of 2QFA at which no head can move left (heads can move right or stay stationary, or both, of course).

2.4 Measurement issues in quantum automata

In order to study in depth the power of various models of QFA so far only projection measurements were considered explicitly in a way that mimic how we deal with acceptance in the classical automata. The set of states Q is decomposed into the set of accepting states Q_a, the set of rejecting states Q_r, and the set Q_n of nonterminating states. The observable $\mathcal{O}$ used then corresponds to the decomposition of the corresponding Hilbert space of all configurations to the subspaces $E_a(E_r)[E_n]$ spanned by configurations with accepting (rejecting) [nonterminating] states. Two measurement modes intensively explored for 1QFA are:

- **Measurement many mode** [20], at which the projection measurement, with respect to the observable $\mathcal{O}$, is performed after each evolution step. 1QFA with the above measurement mode are denoted as MM-1QFA.
- **Measurement once mode**, [23, 11], at which measurement with respect to the observable $\mathcal{O}$ is performed only once, at the end of each computation. 1QFA with such a measurement mode are denoted as MO-1QFA.

As discussed later these two measurement modes lead to automata with different power.

In the case of 2QFA and $g1QFA$ only MM-mode of measurement has been considered so far.

The above results should be seen as the very first ones. A much more in depth investigation of the power of finite automata with respect to the types of measurement used and modes how often measurements are performed is needed.

Also the problem of quantum automata working with mixed states and with POVM or other types of measurement needs to be explored.

2.5 Acceptance probabilities for quantum finite automata

There are various ways how to define languages acceptable by quantum finite automata – and the same is true for PFA.

In the following we overview basic approaches for 1QFA. An approach to acceptance for the case of other models is similar.

A 1QFA $\mathcal{A}$ accepts (rejects) a word w of length n with probability p if p is the sum of probabilities that w is accepted (rejected) after i symbols of w are scanned for $i = 1, \ldots, n$.

A 1QFA $\mathcal{A}$ accepts a language L with probability $\frac{1}{2} + \varepsilon, \varepsilon > 0$, if $\mathcal{A}$ accepts (rejects) any $x \in L$ ($x \notin L$) with probability at least $\frac{1}{2} + \varepsilon$.

If there is an ε such that $\mathcal{A}$ accepts L with probability $\frac{1}{2} + \varepsilon$, then $\mathcal{A}$ is said to accept L with **bounded error probability**.

A language L is accepted by $\mathcal{A}$ with **unbounded error probability** if $x \in L$ ($x \notin L$) is accepted (rejected) with probability at least $\frac{1}{2}$.

The acceptance with bounded error probability is considered to be the main and the most realistic one because it is robust with respect to small errors. Acceptance with respect to unbounded errors can lead to unrealistic conclusions.

A language L is said to be **Monte Carlo** accepted by a QFA $\mathcal{A}$ if every $x \in L$ ($x \notin L$) ic accepted (rejected) with probability 1 $(1 - \varepsilon)$.

Finally, a language L is said to be **Las Vegas** accepted by a QFA, if every $x \in L$ ($x \notin L$) is accepted (rejected) with probability 1.

2.6 Examples

As the first example [5], we describe a 1QFA accepting the language $L = \{0^i1^j \mid i \geq 0.j > 0\}$ with probability $p = 0.68$ (such that $p = 1 - p^3$).

States: $Q = \{q_0, q_1, q_2, q_a, q_r\}, Q_a = \{q_a\}, Q_r = \{q_r\}$. **Transitions:**

$$V_{\#}|q_0\rangle = \sqrt{1-p}|q_1\rangle + \sqrt{p}|q_2\rangle,$$

$$V_0|q_1\rangle = (1-p)|q_1\rangle + \sqrt{p(1-p)}|q_2\rangle + \sqrt{p}|q_r\rangle,$$
$$V_0|q_2\rangle = \sqrt{p(1-p)}|q_1\rangle + p|q_2\rangle - \sqrt{1-p}|q_r\rangle,$$
$$V_1|q_1\rangle = |q_r\rangle, V_1|q_2\rangle = |q_2\rangle, \quad V_{\$}|q_1\rangle = |q_r\rangle, V_{\$}|q_2\rangle = |q_a\rangle.$$

The remaining transitions are defined arbitrarily to satisfy the unitarity condition for mappings V_σ.

As the second example let us mention a g1QFA that can accept, with respect to unbounded error probability, the language $L_1 = \{w \mid w \in \{0,1\}^*, |w|_0 = |w|_1\}$, where $|w|_a$ denotes the number of symbols a in the world w.

As the next example let us mention that, as shown by [29], 2QFA can accept the language $\{0^i1^i \mid i \geq 1\}$ with respect to the bound-error probability.

Languages of the last two examples cannot be accepted by DFA.

Finally, 2QCFA can recognize, with rational amplitudes, the following language of palindroms

$$L = \{w \mid W = W^R, w \in \{0,1\}^*\}$$

that cannot be accepted by 2PFA.

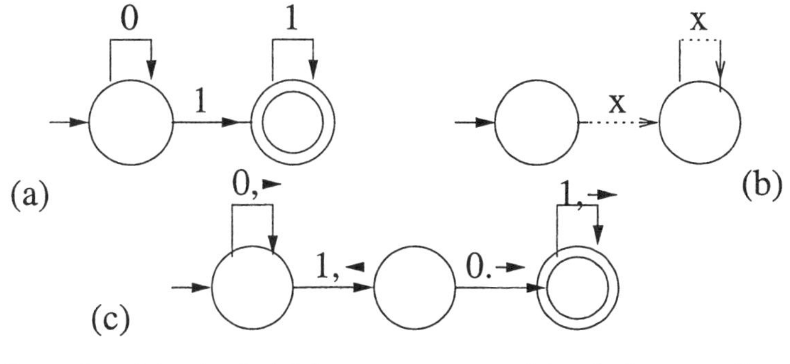

Figure 2: Examples of the classical finite automata recognizing the language $L = \{0^i1^j \mid i \geq 0.j > 0\}$, figures (a) and (c), and a "forbidden construction" (figure (c)). Figure (a) contains a DFA that is not reversible, and figure (c) a reversible 2FA.

2.7 Acceptance power and closure properties

Experiences with the study of classical automata show that it is of large importance to determine fully, by various means, classes of languages accepted by automata of various types. This helps often better to see what can be done with automata of a certain type. Of similar importance is to know under which operations is closed the class of languages accepted by automata of a certain type. Since the proofs showing the acceptance under particular operations are usually constructive, they actually provides a certain methodology how to construct more complex automata form simpler automata of the same type.

The most detailed characterization is known so far for languages accepted by RFA. Pin [27], formulated five equivalent conditions, in term of various algebraic concepts, for a regular language to be accepted by a RFA.

Quite well characterized are also languages accepted by MO-1QFA, with respect to the bounded error acceptance. They form exactly the class of languages accepted by the so-called *group automata*. This family of languages is strictly smaller than that of regular languages. In addition, this family of languages is closed under Boolean operations, inverse homomorphism and word quotients, but not under homomorphism [11]. On the other hand, with respect to the unbounder error acceptance M0-1QFA can accept also non-regular languages, as it was already mentioned above.

MM-1QFA can accept only regular languages but not all of them. For example not the language $L = \{0,1\}^*0$. The family of languages accepted by MM-1QFA is closed under complement, inverse homomorphism and word quotients, but not under homomorphism and not under binary Boolean operations (see [5] and [28]). A characterization is known for languages accepted by MM-1QFA with probability larger than $\frac{7}{9}$. They are exactly languages accepted by RFA.

2.8 Probability acceptance hierarchies

In the case of PFA it holds that if a language is accepted with a probability $p > \frac{1}{2}$, then it can be accepted with probability arbitrarily close to 1. The same is true for languages accepted by MO-1QFA, but not for MM-1QFA (at least not always for languages accepted with probability smaller than $\frac{7}{9}$).

For example, the language from Example 1, which can be accepted by a MM-1QFA with probability 0.68 cannot be accepted by a MM-1QFA with probability greater than $\frac{7}{9}$. In addition, an infinite hierarchy of languages $\{L_i\}_{i=1}^{\infty}$ has been designed such that each next one can be accepted only by probability smaller than previous one and these probability converges from above to $\frac{1}{2}$. This has been shown by [3] for the case that each language L_n is over an alphabet of n symbols and later by [19] for the case that each language L_n is over the two letter alphabet.

In the first case they considered the language $L_n = a_1^* \ldots a_n^*$ and in the second case languages L_n defined as follows $L_1 = a^*$, $L_i = L_{i-1}b^*$ if i is even and $L_i = L_{i-1}a^*$ if i is odd. Such a language L_n can be recognized with probability p such that $p^{\frac{n+1}{n-1}} + p = 1$ by the automaton from Example 1. For these sequences of languages it holds that the language L_n can be recognized with probability $\frac{1}{2} + \frac{1}{4n}$, but not with probability greater than $\frac{1}{2} + \frac{3}{\sqrt{n-1}}$.

Lower bounds for accepting probabilities have been shown using the idea presented in the following result [5].

Theorem 2.3 *Let L be a regular language and $\mathcal{A}$ a minimal DFA for L with the transition function δ and a set F of accepting states. Let there be in $\mathcal{A}$ states q_1, q_2 and an input word w such that (see Figure 2b: (a) $q_1 \neq q_2$; (b) $\delta(q_1, w) = \delta(q_2, w) = q_2$; (c) q_2 is neither "all-accepting" nor "all-re jecting" state. Then L cannot be accepted by a MM-1QFA with probability at least $\frac{7}{9} + \varepsilon$ for any $\varepsilon > 0$.*

2.9 Succinctness results

Though 1QFA are clearly weaker models than 1FA, with respect to the recognition power, the situation is quite different with respect to their descriptional power.

It has been shown in [5], that for some languages 1QFA can be exponentially smaller than 1FA recognizing the same language. This statement is true, for example for the languages from the following examples [5].

- For any prime p any DFA recognizing the language $L_p = \{a^i \mid i \text{ is divided by } p\}$ has to have at least p states, but there is a MM-1QFA with $\mathcal{O}(\lg n)$ states recognizing L_p.
- For any integer n, each DFA recognizing the language $L_n = \{0^n\}$, containing the single string, has to have n states, but there is a MM-1QFA with $\mathcal{O}(\lg n)$ states recognizing L_n.

However, this is not the whole story. For some languages 1QFA have to be exponentially larger than any 1FA accepting the same language. This is true, for example, for the language [6, 26] discussed next.

- For each integer n there is a DFA of size $\mathcal{O}(n)$ recognizing the language $L_n = \{w0 \mid w \in \{0,1\}^*, |w| \leq n\}$, but each MM-1QFA recognizing L_n with probability greater than $\frac{1}{2}$ has to have $2^{\Omega(n)}$ states.

We can conclude that in some cases, it seems that due to quantum parallelism, 1QFA can be much smaller than their classical counterparts and in some cases, it seems that due to the requirement on unitarity (reversibility), this is just the opposite way. Also the following result indicates that. By [5] there is for any integer n a language that can be accepted by a DFA with $\mathcal{O}(n)$ states but each RFA accepting that language has to have $\Omega(2^n)$ states.

Of interest are specially methods to show lower bounds. Some are brought from those developed for probabilistic automata, but some are more inherently quantum. For example, the method based on results concerning dense coding [6] as well as the method based on quantum entropy [26].

For a more detailed treatment of the problem of succinctness in quantum computing see [17].

2.10 Decidability results

Two most basic algorithmic decision problems related to the (classical) automata are the *emptiness problem* - to decide whether a given automaton, considered as an acceptor, accepts a non-empty language – and the *equivalence problem* – to decide given two automata of a certain type whether they accept the same language.

There is simple relation between these two problems. Decidability of the equivalence problem implies decidability of the emptiness problem and, consequently, undecidability of the emptiness problem implies undecidability of the equivalence problem.

For all versions of the classical finite automata considered above these problems are decidable. For MO-1QFA the equivalence problem is decidable [11]. It is not known whether this is true for MM-1QFA. On the other hand, the emptiness problem is undecidable already for g1QFA, see [?], and for a one-way two tape generalization of 1QFA [8].

Another basic algorithmic decision problem is the *membership problem*: To decide whether a given language from a a class of languages ic accepted by a QFA of a certain type. For example, as shown by [27], membership problem is decidable for RFA with respect to the family of regular languages. Decidability of the membership problem for MO-1QFA was also shown [11].

2.11 Price to pay for reversibility

In his seminal paper Bennett [7] has shown that on the level of Turing machines and time as the resource ther is not too much (only by a constant factor) to pay for transfering a Turing machine to a reversible one.

The situation seems to be different on the level of finite automata. Since not every language is accepted by a 1QFA more powerful models are needed to achieve reversibility in some cases.

Watrous [29] has shown a way how to design for each regular language a 2RFA accepting that language. This construction leads in some cases to an exponential increase of states.

2.12 Pure and mixed quantum finite automata

The models of QFA considered so far have all been natural (pure) quantum versions of the classical models of automata. Of a different type is the model introduced in [4], and called *two-way finite automata with quantum and classical states* (2QCFA). This model is also more powerful than classical (probabilistic) 2FA and at the same time it seems to be more realistic, and really more "finite" than 2QFA because 2QFA actually need quantum memory of size $\mathcal{O}(n)$ to process an input of the size n. 2QCFA can be seen as an intermediate model between 1QFA and 2QFA.

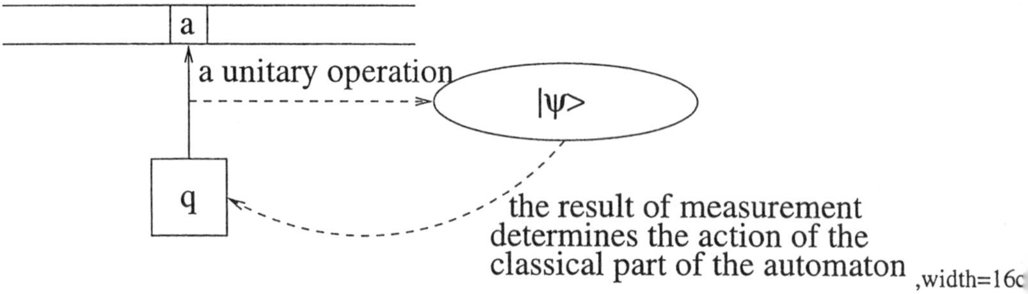

Figure 3: A model of 2QCFA

A 2QCFA $\mathcal{A}$ is defined similarly as a classical 2FA, but, in addition, $\mathcal{A}$ has a fixed size quantum register (which can be in a mixed state) upon which the automaton can perform either a unitary operation or a measurement. A 2QCFA has a classical initial state q_0 and an initial quantum state $|\phi_0\rangle$.

The evolution of the classical part of the automaton and of the quantum state of the register is specified by a mapping Θ that assigns to each classical state q and a tape symbol σ an action $\Theta(q, \sigma)$.

One possibility is that $\Theta(q, \sigma) = (q', d, U)$, where q' is a new state, d is next movement of the head (to left, no movement or to right), and U is a unitary operator to be performed on the current quantum register state.

The second possibility is that $\Theta(q, \sigma) = (M, m_1, q_1, d_1, m_2, q_2, d_2, \ldots, m_k, q_k, d_k)$, where M is a measurement on the register state, $m_1, \ldots, m_k$ are its possible classical outcomes and for each measurement outcome new state and new movement of the head is determined. In such a case the state transmission and the head movement are probabilistic.

It has been shown in [4] that 2QCFA with only one qubit of quantum memory are already more powerful than 2FA. Such 2QCFA can accept the language of palindromes over the alphabet $\{0, 1\}$, which cannot be accepted by probabilistic 2FA at all, and also the language $\{0^i1^i \mid i \geq 0\}$, in polynomial time. This language can be accepted by probabilistic 2FA, but only in exponential time.

In the above model only projection measurements have been considered. It is not clear whether something, especially concerning the number of the classical states, could be obtained by considering also POVM.

2.13 Design methodologies

The first systematic methodology to design quantum finite automata has been due to Morita [24]. He has shown a method how to construct reversible classical (and therefore also quantum) automata using only Fredkin gates in such a way that the resulting circuits are "garbage-free".

A new type of universal reversible logic element, the so called *rotary memory*, (RM), has been introduced in [?]. It is a 4-input/output reversible element which has two states and it is a kind of switching element that changes the path of an input signal depending on its state. Rotary memory elements and Fredkin gates can simulate each other and therefore there is a methodology how to design RFA out of rotary elements. The main new features of RM is that when designing circuits with them one does not need synchronization elements for inputs, or a clock, as it is the case with Fredkin gates. (Moreover RM is a very useful element when designing reversible two-dimensional cellular automata.)

A variety of other computational is currently known and their use for design of QFA has to be explored.

RFA are, of course, a very simple model. However, unless their design methodology are well understood one can hardly expect to have more general design methodologies.

2.14 Quantum almost finite automata

The adjective *finite* in the case of classical finite automata refers to the fact that those automata needs only finite memory (to store states). This is actually true only for 1FA. For 2FA an additional memory of size $\lg n$ is needed to store the position of the head in the case of the input of length n. However, since 2FA can be simulated by 1FA, it is in a sense justified to use also for them the adjective "finite".

In the case of "finite quantum automata" situation is different and only 1QFA are actually true "finite state (memory) quantum automata". In the case of 2QFA a memory of the size $\mathcal{O}(n)$ is needed to store heads on any position of the tape. (This is true even for g1QFA.)

From the other models of almost finite automata of interest are **quantum counter automata** and **quantum push-down automata**. In the case of quantum counter automata an additional classical memory is added into which in one step the automaton can add or subtract 1, and can also check whether this additional memory is empty. In the case of quantum push-down automata a push-down tape is added.

For both of these types of automata the basic definitions for their quantum versions have been provided and also the well-formedness conditions [14, 21] to ensure unitarity of their evolution. In addition, the emptiness problem was shown to be undecidable for quantum counter automata and they have been shown to be more effective than their randomized classical versions [9].

2.15 Main open problems

In the area of QFA and QAFA the following problems seems to be of interest and importance:

- To explore in depth classical reversible finite automata of various types, their properties and design techniques for various sets of quantum computation primitives..
- To characterize class of the languages accepted by MM-1QFA by the bounded error acceptance.
- To explore the power of MO-1QFA with respect to unbounded-error acceptance.
- To find out out if it is decidable whether two given MM-1QFA are equivalent.
- To determine whether the increase in the size from n to $2^{\Omega(n)}$ is the worst possible when going from DFA to MM-1QFA.
- To determine whether g1QFA can recognize all regular languages.
- To determine whether g1QCFA are more powerful than g1QFA.
- To determine whether 2QCFA with n qubits are more powerful than those with $n - 1$ qubits.
- To determine whether ther is a regular language that cannot be recognized by 1QFA, but it can be recognized by quantum counter automata.
- To determine whether quantum pushdown-automata are more powerful than probabilistic push-down automata.

3 More Inherently Quantum Models

Let us explore some ways to go in order to have more inherently quantum models that could be more useful to investigate some other key problems of quantum information processing and communication, and whether researsch in quantum automata could help to deal with some of the key problems of quantum information processing and communication.

3.1 From the classical to quantum inputs

The common feature of all models of quantum automata considered so far is that they process de facto only classical input. Either a classical string of the input symbols or a classical string on a classical tape.

A natural alternative seems to be to consider also quantum inputs: either a sequence of quantum states (qubits), or a quantum register state. In the last case a head (heads) could be positioned on various qubits and they could move from one qubit to the next (previous) one.

In the case of input strings it seems to be reasonable to consider as inputs sequences of states from a set of nonorthogonal states.

3.2 Role of the entanglement and of the mixedness

Natural quantumization of the classical concepts of automata does not lead to bipartite quantum systems and therefore one does not see entanglement to play a role in it. Increased power of quantum (finite) automata (more exactly of some of the models considered so far), is therefore not due to the entanglement. A state of the automata can be seen as measuring the incoming state and transition is made, probabilistically, depending on the outcome of the measurement.

There is a space to investigate the power of entanglement in connection with the 2QCFA model, but this has not been done yet.

There may be, perhaps, a potential to make use of the entanglement of two heads positioned on the opposite ends of an input string to perform communication that classically requires $\mathcal{O}(n)$ steps.

It is also not clear yet how much one can obtain for quantum finite automata working with mixed states and if POVM are used instead of the projection measurements.

On the general level of quantum circuits, it has been shown in [1] that computation with mixed states and superoperators does not bring anything essential.

Pure and mixed states are not the only alternatives to explore. Computations with special pseudo-pure states are also of interest to investigate.

3.3 Quantum automata and decoherence

It is not quite clear how to use quantum automata models to get some insight into the problems decoherence causes and into the possibilities to deal with them

Of course it can be of interest to explore how quantum finite state transducers can be used to implement operations connected with quantum error-correcting codes, with encodings, decodings and error-correction processes.

3.4 Quantum automata and quantum processors

In the classical case the theory of finite automata, and finite state machines in general, contributed quite a bit to the development of techniques to design hardware and software. For example, it contributed much to the development of the hardware minimization and optimization techniques as well to the design of quite sophisticated and fast text processing systems.

It seems that a contribution to the hardware design and optimization could also be desirable in the case of quantum finite automata. (Of course, such a theory can be really useful only after larger scale quantum machines are feasible.) For developing such a theory much more insights are needed into quantum finite automata. many very basic things are still not clear enough in this area.

- How to get from a QFA that accepts a language with a certain probability to a QFA that accepts the same language, but with higher probability? When is such a construction possible and how much it costs in terms of the number of states and so?
- How to develop o theory of minimization of QFA. What is the price to pay for minimality in terms of the accepting probabilities for decreasing the number of states. (Such a minimization theory may not be easy, because already for RFA there is no unique, up to isomorphism, minimal RFA (Paschen, 2000)).

3.5 Getting beyond quantum

Once a new powerful model of computation starts to be considered, it is always of interest/importance to try to see how we can get beyond it, at least slightly.

Physically this would require to come with some "new physics". Mathematically there are usually far more easy ways of doing that (but of course it is less trivial to see which sense they really have).

All four main types of computation modes; deterministic, nondeterministic, probabilistic and quantum can be defined for basic types of automata by specifying types of transition matrices for going from one configuration to another, (here we assume that the corresponding Hilbert space is that over all configurations representing the standard basis states). A computation step is then represented by a matrix-vector multiplication. In the case of deterministic automata, entries of the matrix are zero's and one's and each column has at most one non-zero elements. In the case of nondeterministic computations the number of ones is not limited. In the case of probabilistic computations entries are nonnegative numbers smaller than one and the sum of numbers in each column has to be one. From that it follows that matrix multiplication preserves L_1-norm. In the case of quantum computations entries are complex numbers and transition matrices have to be unitary, what implies that they preserve L_2-norm (actually, it is sufficient to assume from the complexity point of view that entries are arbitrary real numbers, see [13]).

A natural generalization is to replace requirement of unitarity by a weaker one or to remove it at all. The second case was done in [22, 12], for the case of Turing machine computations. They introduce this way the class **AWPP** (Almost Wide Probabilistic Polynomial time) and showed various properties of it, For example

$$\mathbf{AWPP} \subseteq \mathbf{PP} \subseteq \mathbf{PSPACE}$$

and, in addition, that $\mathbf{BQP} \subseteq \mathbf{AWPP}$.

A natural and puzzling question may be why to consider such models? One pragmatic reasons is that they may be more easy to deal with. If, for example, we can show that something cannot be done with a more powerful model, at least mathematically more powerful, than it cannot be done for the weaker model either.

References

[1] Dorit Aharonov, Alexei Kitaev, and Noam Nisan. Quantum circuits with mixed states. In *Proceedings of* 30*th ACM STOC*, pages 20–30, 1998. quant-ph/9806029.

[2] Masami Amano and Kazuo Iwama. Undecidability on quantum finite automata. In *Proceedings of 31st ACM STOC*, pages 368–375, 1999.

[3] Andris Ambainis, Richard Bonner, Rūsiņš Freivalds, and Arnolds Ķikusts. Probabilities to accept languages by quantum finite automata. Technical report, quant-ph/9904066, 1999.

[4] Andris Ambainis and John Watrous. Two-way finite automata with quantum and classical states. Technical report, quant-ph/9911009, 1999.

[5] Andris Ambainis and Rūsiņš Freivalds. 1-way quantum finite automata: strengths, weaknesses and generalizations. In *Proceedings of* 39*th IEEE FOCS*, pages 332–341, 1998. quant-ph/9802062.

[6] Andris Ambainis, Ashwin Nayak, Amnon Ta-Shma, and Umesh Vazirani. Dense quantum coding and a lower bound for 1-way quantum finite automata. Technical report, quant-ph/9804043, 1998.

[7] Charles H. Bennett Logical reversibility of computation. *IBM Journal of Research and Development*, 17:525–532, 1973.

[8] Richard Bonner, Rūsiņš Freivalds, and Renars Gailis. Undecidability of 2-tape quantum finite automata. In *Proceedings of International Workshop on Quantum Computation and Learning, Sundbyholms, May 27-29, 2000*, pages 93–100, 2000.

[9] Richard Bonner, Rūsiņš Freivalds, and Maxim Kravtsev. Quantum versus probabilistic 1-way finite automata with counter. In *Proceedings of International Workshop on Quantum Computation and Learning, Sundbyholms, May 27-29, 2000*, pages 80–88, 2000a.

[10] Richard Bonner, Rūsiņš Freivalds, and Madars Rikards. Undecidability of quantum finite 1-counter automaton. In *Proceedings of International Workshop on Quantum Computation and Learning, Sundbyholms, May 27-29, 2000*, pages 65–71, 2000b.

[11] Alex Brodsky and Nicholas Pippenger. Characterization of 1-way quantum finite automata. Technical report, quant-ph/9903014, 1999.

[12] S. Fenner, L. Fortnow, S. Kurtz, and L. Li. An oracle builder's toolkit. In *Proceedings of 8th IEEE Structure on Complexity Theory Conference*, pages 120–131, 1993.

[13] Lance Fortnow. One complexity theorist's view of quantum computing. Technical report, quant-ph/0003035, 2000.

[14] Maratas Golovkins. On quantum pushdown automata. In *Proceedings of International Workshop on Quantum Computation and Learning, Sundbyholms, May 27-29, 2000*, pages 41–51, 2000.

[15] Jozef Gruska. *Foundations of computing*. International Thomson Computer Press, 1997.

[16] Jozef Gruska. *Quantum computing*. McGraw-Hill, 1999. See also additions and updatings of the book on http://www.mcgraw-hill.co.uk/gruska.

[17] Jozef Gruska. Descriptional complexity issues in quantum computing. *Journal of automata, languages and combinatorics*, page to be published, 2000.

[18] Stanley Gudder. Basic properties of quantum automata. Technical report, Department of Computer Science, University of Denver, 2000.

[19] Arnolds Kikusts and Zigmars Rasscevskis. On the accepting probabilities of 1-way quantum finite automata. In *Proceedings of International Workshop on Quantum Computation and Learning, Sundbyholms, May 27-29, 2000*, pages 72–79, 2000.

[20] Attila Kondacs and John Watrous. On the power of finite state automata. In *Proceedings of 36th IEEE FOCS*, pages 66–75, 1997.

[21] Maksim Kravtsev. Quantum finite one-counter automata. Technical report, quant-ph/9905092, 1999.

[22] L. Li. On the counting functions. Technical report, PhD Thesis, University of Chicago, Department of Computer Science, 1993.

[23] Cristopher Moore and James P. Crutchfield Quantum automata and quantum grammars. Technical report, Santa Fe, 1997.

[24] Kenichi Morita. A simple construction method of a reversible finite automaton out of Fredkin gates, and its related problems. *The transactions of the IEICE*, E73:978–984, 1999.

[25] Kenichi Morita, Yasuyuki Tojima, and Katsunobu Imai. A simple computer embedded in a reversible and number conserving two-dimensional cellular space. *Multiple-valued logic*, page to be published, 2000.

[26] Ashwin Nayak. Optimal lower bounds for quantum automata and random access codes. In *Proceedings of 40th IEEE FOCS*, pages 369–376. ACM, 1999.

[27] Jean-Erie Pin. On the languages accepted by finite reversible automata. In *Proceedings of 14th ICALP*, pages 237–249. LNCS 267, Springer-Verlag, 1987.

[28] Maris Valdats. The class of languages recognizable by 1-way quantum automata is not closed under union. Technical report, quant-ph/000115, 2000.

[29] John Watrous. On the power of 2-way quantum finite automata. Technical report, University of Wisconsin, 1997.

Decoherence and its Implications in Quantum Computation and Information Transfer
T. Gonis and P.E.A. Turchi (Eds.)
IOS Press, 2001

Unambiguous discrimination of linearly independent states as an eavesdropping strategy

Miloslav DUŠEK [1], Mika JAHMA[2], and Norbert LÜTKENHAUS [2]
[1] *Dept. of Optics, Palacký University, 17. listopadu 50, 772 00 Olomouc, Czech Rep.*
[2] *Helsinki Institute of Physics, P.O. Box 9, 00014 Helsingin yliopisto, Finland*

Abstract. Realistic implementations of quantum key distribution (QKD) mostly use signal states which are non-orthogonal but *linearly independent*. This fact enables an eavesdropper to perform unambiguous state discrimination and to get some information on the key without disturbing the transmission. In this paper, the limits for secure QKD, imposed by such an attack, are determined. It is also shown that security against beam-splitting attack does not necessarily imply security against the unambiguous-state-discrimination attack.

1. Introduction

The only *provably* secure way to communicate with guaranteed privacy is the so called one-time pad or Vernam cipher [1]. It requires both communicating parties share a secret key of the same length as the message. Quantum key distribution (QKD) is a technique to provide two parties with such a secure, secret and shared key. The first complete protocol for QKD was given by Bennett and Brassard [2] (BB84) following first ideas by Wiesner [3]. The essence of the protocol is that if non-orthogonal quantum states are used for communication and a channel transmit them perfectly then eavesdropping is detectable.

We consider the BB84 protocol in a typical quantum optical implementation. Ideally, Alice sends a sequence of single photons which are at random polarized in one of the following four states: right or left circular polarized, or vertically or horizontally polarized. Bob chooses at random between two polarization analyzers, one distinguishing the circular polarized states, and the other distinguishing the linear polarized states. Following a public discussion about the basis of the sent signals and the measurement apparatus applied to them, sender and receiver can obtain a shared key made up from those signals where the measurement device gives deterministic results. This is the *sifted key* [4]. Proofs of security of this scheme against the most general attack, even in the presence of noise, have been obtained [5, 6, 7]. In this article we follow another goal: we would like to illuminate to which extend even a very simple attack can render QKD impossible once realistic imperfections like lossy lines and non-ideal signal states are taken into account. The difficulties implied by the use of weak coherent states in combination with lossy lines has been pointed out earlier [8, 9, 10] and this subject has been illuminated in depth in [11], where bounds on coverable distances are given. Positive security proofs for sufficiently short distances considering realistic signals are given for individual attacks in [12]. The eavesdropping attacks which crack the secrecy of the key for setups exceeding this secure distances are still quite complicated (an eavesdropper needs to perform a QND measurement and store photons).

In this paper we deal with much simpler eavesdropping strategy that uses the opportunities arising from lossy lines and non-ideal signals. The attack has been first proposed by Bennett [13] and Yuen [9]. It is based on the fact that Eve can, with a finite probability, discriminate the four signal states unambiguously. Whenever such a discrimination is performed successfully, the eavesdropper knows immediately which of the four signal states was sent and can transmit this information via a classical channel to Bob's detector, in front of which she places a state preparation machine to prepare the identified state. This way this state does not experience the losses of the actual quantum channel without that Eve has to invest into a perfect quantum channel.

Our investigation of this scenario refines Bennett's and Yuen's analysis. It takes into account that the photon statistics of the signals arriving at Bob's detectors can be monitored to a certain extent. The results illuminate the restrictions placed on implementations of QKD on lines with strong losses. We show that, contrary to common belief, the use of unambiguous state discrimination can be a more efficient eavesdropping strategy than the beam-splitting attack [13], even for dim coherent states.

2. Unambiguous discrimination of signal states

Let us consider N different quantum states. Unambiguous state discrimination is possible whenever these N states are linearly independent. The problem can be described by a measurement which can give the results 'state 0', 'state 1', …'state $N-1$', and the result 'don't know'. The constraint is that the measurement results should never wrongly identify a state, and the goal is to keep the fraction of 'don't know' results as low as possible. This problem has been investigated by Ivanovic [14] for the case of two equally probable non-orthogonal states. Peres [15] solved this problem in a formulation with probability operator measures (POM). Later Jaeger and Shimony [16] extended the solution to arbitrary a priori probabilities. Peres's solution has been generalized to an arbitrary number of equally probable states which are generated from each other by a symmetry operator by Chefles and Barnett [17]. They used the fact that the symmetry allows to write the input states in the form

$$|\Psi_k\rangle = \sum_{j=0}^{N-1} c_j \exp\left(2\pi \mathrm{i}\frac{kj}{N}\right) |\phi_j\rangle\,, \tag{1}$$

where $|\phi_j\rangle$ represent a set of orthonormal states. Then they have shown that the maximum probability of the successful unambiguous state discrimination is given by the formula:

$$P_D = N \min_j |c_j|^2. \tag{2}$$

Practical implementations of QKD mostly employ attenuated laser pulses. Thus the reasonable description of realistic signal states seems to be that of a coherent state with a small amplitude α. The four BB84 signal states, carrying two linear and two circular polarizations, may be expressed in terms of two modes (corresponding to two linear orthogonal polarizations) as follows

$$\begin{aligned} |\Psi_0\rangle &= \left|\alpha/\sqrt{2}\right\rangle \left|\alpha/\sqrt{2}\right\rangle, \qquad & |\Psi_1\rangle &= \left|\alpha/\sqrt{2}\right\rangle \left|\mathrm{i}\alpha/\sqrt{2}\right\rangle, \\ |\Psi_2\rangle &= \left|\alpha/\sqrt{2}\right\rangle \left|-\alpha/\sqrt{2}\right\rangle, \qquad & |\Psi_3\rangle &= \left|\alpha/\sqrt{2}\right\rangle \left|-\mathrm{i}\alpha/\sqrt{2}\right\rangle. \end{aligned} \tag{3}$$

It turns out, however, that for realistic sources these states are not the correct description of the situation. The reason is that Eve does not have a phase reference. That means that for a given polarization she does not see the coherent state $|\alpha\rangle$ but the phase averaged density matrix $\frac{1}{2\pi}\int_\phi |e^{\mathrm{i}\phi}\alpha\rangle\langle e^{\mathrm{i}\phi}\alpha|\, \mathrm{d}\phi$. This results in signal states which are mixtures of Fock states with a Poissonian photon number distribution described by the density matrix

$$\rho = e^{-\mu} \sum_n \frac{\mu^n}{n!} |n\rangle\langle n|. \tag{4}$$

Here the state $|n\rangle$ denotes the Fock state with n photons in one of the four BB84 polarization states. The optimal strategy to discriminate between the four possible density matrices can be logically decomposed into a QND measurement on the *total* photon number and a following measurement which unambiguously discriminates between the four resulting conditional states for each total photon number. The justification for this is that the measurement of the total photon number "comes free", since the execution of this measurement does not change the signal states. Therefore the total probability of unambiguous state discrimination P_D is given in terms of the respective probabilities for each photon number subspace $P_D^{(n)}$ as

$$P_D = \sum_{n=0}^{\infty} e^{-\mu} \frac{\mu^n}{n!} P_D^{(n)}. \tag{5}$$

The conditional states resulting from the QND measurement and corresponding to n photons in total satisfy again the symmetry condition which allows to apply the results by Chefles and Barnett. The maximum probability of unambiguous state discrimination for fixed value of n is given by (for more details see [18])

$$P_D^{(n)} = \begin{cases} 0 & n \le 2 \\ 1 - 2^{1-n/2} & n \text{ even} \\ 1 - 2^{(1-n)/2} & n \text{ odd}. \end{cases} \tag{6}$$

It is possible to sum up the contributions from different photon numbers from the Poissonian distribution and we obtain the expression

$$P_D = \sum_{n=0}^{\infty} e^{-\mu} \frac{\mu^n}{n!} P_D^{(n)} = 1 - e^{-\mu} \left(\sqrt{2} \sinh \frac{\mu}{\sqrt{2}} + 2 \cosh \frac{\mu}{\sqrt{2}} - 1 \right). \tag{7}$$

This result is compared to the result for coherent states in Fig. 1. As expected, the probability for unambiguous state identification is lower for the mixture of Fock-states than for the coherent states. An expansion in terms of the photon number μ gives $P_D = \frac{1}{12}\mu^3 + O(\mu^4)$ for both situations. For lower than third order the signal states are not linearly independent, so that no unambiguous state discrimination is possible. Note that an actual implementation does not necessarily need to follow the decomposition into a QND and another measurement. Actually, Bennett et al. [13] and Yuen [9] gave a simple beam-splitter setup which obtains a discrimination probability of $P_D = \frac{1}{32}\mu^3 + O(\mu^4)$.

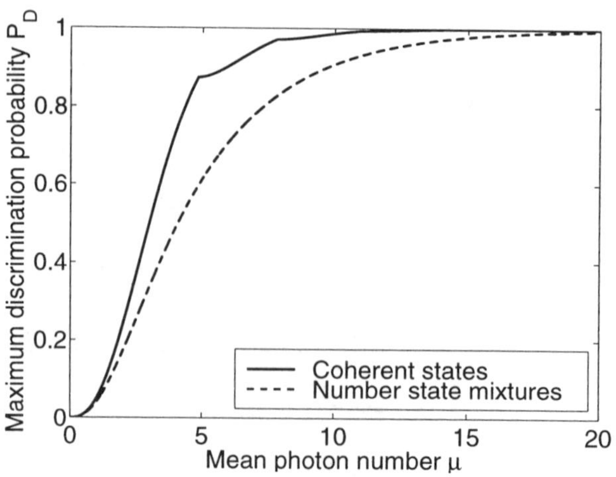

Figure 1. Comparison of the optimum probability of unambiguous states discrimination for coherent states and for the corresponding mixture of Fock states. Both have the same Poissonian photon number distribution with mean photon number μ.

3. Unambiguous state discrimination as eavesdropping strategy

Let us now consider the realistic situation when Alice uses the phase-averaged coherent states as signal states. These mixed states undergo Poissonian photon-number statistics with mean photon number μ. We fix our eavesdropping scenario, to which we refer as the *unambiguous state discrimination attack* (USD attack), as follows: The unambiguous state discrimination allows Eve to identify a fraction of the signals without error. For this fraction, she can prepare a corresponding state close to Bob's detectors such that no errors appear for these signals. Whenever the identification does not succeed, she sends the vacuum signal to Bob to avoid errors, which therefore will not be relevant in the considered scenario.

We will study this strategy under realistic constraint. Of course, any real quantum channel connecting both parties suffer by losses. Let its transmittance be characterized by the transmission efficiency η_L. Further, we will consider a detection setup where Bob monitors each polarization mode in the chosen basis by one detector. These detectors have a finite detection efficiency η_B, in which we include any additional loss on Bob's side. The detectors are modeled as "yes/no" detectors, which either fire, or they do not fire; they cannot distinguish the number of impinging photons.

Once Eve identified a signal she is interested to produce a signal in the corresponding polarization such that Bob will detect it despite his inefficient detectors. One strategy is to send a stronger signal than the original one in the correct polarization. This will work as long as Bob measures in the polarization basis which includes the signal polarization (sifted key), but it will lead to an increased coincidence rate of clicks in both of Bob's detectors otherwise. Our analysis includes the additional constraint put on the eavesdropping strategy by the fact that Bob observes not only the rate of clicks of one or the other detector, but also the rate of events when both detectors fire, each monitoring one of the orthogonal polarization modes. The latter event will be observed ideally only when Alice and Bob use different bases, independently of the presence or absence of an eavesdropper. Eve's aim is to reproduce these two observables with the minimum number of non-vacuum signals to make efficient use of the successfully identified signals.

In the absence of Eve, whenever Alice and Bob use the same polarization basis, Bob expects to find at most one detector clicks; the probability of a click is

$$\bar{P}_1 = 1 - \exp(-\eta_L \eta_B \mu), \tag{8}$$

as follows from the Poissonian photon-number statistics of coherent states. Whenever Alice and Bob use different bases a double-click may occur; its probability is

$$\bar{P}_2 = \left[1 - \exp\left(-\frac{\eta_L \eta_B \mu}{2}\right)\right]^2. \tag{9}$$

How the situation changes in the presence of Eve depends on the signals Eve sends for the successfully detected Alice's signals. It is clear that Eve can avoid the occurrence of double clicks when Alice and Bob measure in the same basis, since she unambiguously determined the signal. Therefore it is not useful to monitor the double click rate when Alice and Bob use the same basis.

Let us suppose now, that whenever Eve succeeds in the unambiguous state discrimination she sends a number state (with correct polarization) containing N photons to Bob. If she fails she simply sends no photon.

If Alice and Bob use the same basis, at most one of two Bob's detectors will click. The probability of this event is given by

$$P_1^{(N)} = P_D \left[1 - (1 - \eta_B)^N\right]. \tag{10}$$

This is the probability that one detector clicks if a state $|N\rangle$ comes, multiplied by the probability that Eve succeeds in USD (and sends $|N\rangle$).

If Alice and Bob use different polarization bases, we can think of the photons as being equally and independently distributed to both Bob's detectors. The probability of double click in Bob's "yes-no" detectors when Eve is active then reads (see [18])

$$P_2^{(N)} = P_D \left[1 - 2\left(1 - \frac{\eta_B}{2}\right)^N + (1 - \eta_B)^N\right]. \tag{11}$$

There is no reason to restrict Eve only to the use of number states. When she succeeds in state discrimination she can send to Bob any pure state or mixture. However, from Bob's point of view these signals are effectively mixtures of photon number states because of the nature of his detection. Therefore it is sufficient to analyze only a mixture of photon number states in the polarization of the identified signal, so that only the photon number statistics remains to be chosen by Eve.

As already mentioned, Bob is interested only in the number of single clicks (in case that his and Alice's bases coincide) and double clicks (if the bases differ). One can plot very illustrative diagram displaying relations between corresponding single-click and double-click probabilities (see Fig. 2). The situation where Eve sends number states to Bob is represented by a dot for each value of the photon number N. The positions of these dots have been calculated for fixed values of η_L and μ. Coordinates of a point corresponding to any mixture of number states can always be expressed as a linear convex combination of coordinates corresponding to individual number states. Because

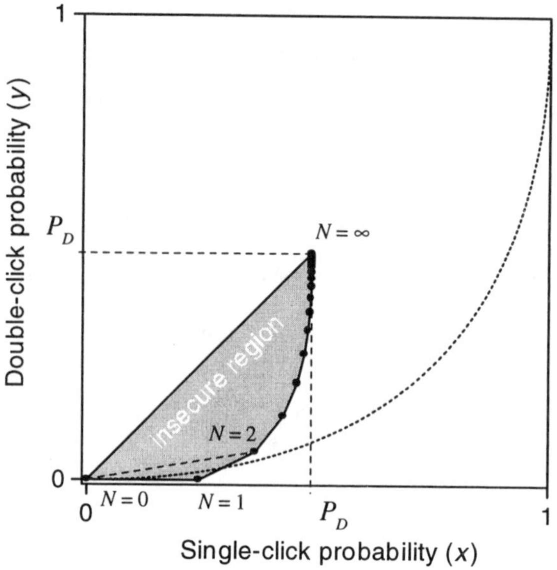

Figure 2. Diagram displaying relations between "single-click" and "double-click" probabilities. The highlighted area contains all possible combinations of Bob's detection probabilities stemming from Eve's activity for a given detection efficiency (here, particularly, $\eta_B = 0.5$) and a given mean photon number in states sent by Alice ($\mu = 4$). It is a region of *insecure* key generation. The shape of the area depends on η_B, the scaling on μ [through discrimination probability $P_D(\mu)$]. The value of $\mu = 4$ is chosen to make the diagram well readable. The structure is the same for lower, realistic values. The separate dotted curve represents a set of all possible "working points" without an eavesdropper, i.e. a set of all possible pairs of expected $\bar{P}_1$ and $\bar{P}_2$. Any particular position of a working point depends on the values of the line transmittance (η_L), the detection efficiency (η_B), and the mean photon number (μ).

of the convexity of the above mentioned curve all such points must lie inside (or on the boundary) of the polygon with vertices at the points corresponding to number states (i.e. in the area highlighted in Fig. 2). This area can be called a region of insecurity.

We define the working point of a setup as the point whose coordinates are given by expected values in the absence of an eavesdropper. If this working point falls into the region of insecurity, Eve can get complete information on the key without a risk of being disclosed. The set of all possible working points is represented by the dotted curve in the diagram. Expected single-click probability $\bar{P}_1$ represents x-coordinate, expected double-click probability $\bar{P}_2$ represents y-coordinate. Thus the explicit equation of the working point curve reads

$$y = \left[1 - (1-x)^{1/2}\right]^2 . \tag{12}$$

Now we have to find the values of parameters η_L, η_B, and μ for which the working point lies in the region of insecurity.

Working point	$x_w = 1 - \exp(-\eta_L\eta_B\mu)$	$y_w = \left[1 - \exp\left(-\frac{\eta_L\eta_B\mu}{2}\right)\right]^2$
Vertex $N = 1$	$x_1 = P_D\eta_B$	$y_1 = 0$
Vertex $N = 2$	$x_2 = P_D(2\eta_B - \eta_B^2)$	$y_2 = P_D\eta_B^2/2$

Table 1. Coordinates of selected points in the parameter space of "observables", which are the probabilities of single clicks (x) and double clicks (y) in Bob's detectors.

3.1. Necessary condition for insecurity

If the expected probability of single clicks satisfies inequality $\bar{P}_1 > P_1^{(N)}$ for all N then the working point will certainly not fall to the region of insecurity. This leads to the necessary condition for insecurity given by $\bar{P}_1 < P_D$. To evaluate the implication for the experimental parameters, we substitute Eq. (8). We get

$$\eta_L\eta_B < \frac{-\ln[1 - P_D(\mu)]}{\mu}. \tag{13}$$

It means that for a fixed expected photon number μ a system cannot be cracked by an USD attack if the total transmission efficiency $\eta_L\eta_B$ is higher than a certain threshold which depends on the the expected photon number μ. The surprising aspect is, that the threshold does not go to 1 as μ goes to infinity. Instead we find

$$(\eta_L\eta_B)^{(\infty)} = \lim_{\mu\to\infty} \frac{-\ln[1 - P_D(\mu)]}{\mu} = (1 - 2^{-1/2}) \approx 0.293. \tag{14}$$

This shows, that that the implementation of quantum cryptography with weak coherent states cannot be cracked completely by the USD attack for *all* values of the expected photon number μ as long as the total transmission satisfies $\eta_L\eta_B \geq 1 - 2^{-1/2}$.

3.2. Sufficient condition of insecurity

Now we will derive precise conditions determining when a working point falls into the region of insecurity. In a first step we will show that for parameters of practical applications it is sufficient to consider the scenario the working point falls below the straight line going through the origin and the vertex $N = 2$. This condition corresponds to

$$x_w \geq y_w \frac{x_2}{y_2}. \tag{15}$$

The coordinates of points used in this condition are defined in Tab. 1. In the second step we can then determine whether in this scenario the working point lies inside or outside the region of insecurity by checking on which side of the line connecting the vertices $N = 1$ and $N = 2$ it lies (see Fig. 2). If it is on the left, QKD is insecure. This corresponds to the inequality

$$x_w \leq y_w \frac{x_2 - x_1}{y_2} + x_1. \tag{16}$$

First, let us turn to the inequality (15). Substituting expressions for all coordinates according to Tab. 1 one obtains an inequality which is quadratic in the variable $R =$

$\exp(-\eta_L\eta_B\mu/2)$ with the parameter η_B. We find that the working point lies below the line connecting vertices $N=0$ and $N=2$ if $R \in \left(\frac{4-3\eta_B}{4-\eta_B}, 1\right)$. Thus the mean photon number in coherent states sent by Alice must be lower than a threshold μ_2 given by

$$\mu < \mu_2 = \frac{-2}{\eta_L\eta_B} \ln\left(\frac{4-3\eta_B}{4-\eta_B}\right). \tag{17}$$

We find that $\mu_2 \in [1/\eta_L, 2\ln 3/\eta_L]$ for any η_B and always $\mu_2 \geq 1$. As one can see, this condition is satisfied in all current experiments and does not pose a serious restriction to the validity of our analysis especially for non-negligible loss.

Now let us turn our attention to the condition (16) which, whenever condition (17) is fulfilled, determines whether the working point is in the region of insecurity. It can be expressed in the following form

$$F(\mu, \eta_L, \eta_B) := x_w\eta_B - 2y_w(1-\eta_B) - P_D\eta_B^2 \leq 0, \tag{18}$$

Due to the complicated dependence of P_D on μ we failed to find its analytical solution. The analytical statement we can do without any extra approximation is based on the observation that

$$\left.\frac{\partial F}{\partial \mu}\right|_{\mu=0} = \eta_L\eta_B^2 > 0 \quad \text{and} \quad F(0, \eta_L, \eta_B) = 0.$$

This implies that there exists always a range of values for μ starting from $\mu = 0$ for which we have $F > 0$, i.e. the security of the key distribution cannot be cracked completely by the USD attack.

Condition (18) can be easily evaluated numerically. E.g., for values of line transmittance $\eta_L = 0.1$ and detection efficiency $\eta_B = 0.5$ (so that $\mu_2 \approx 13.46$) function F is positive if $\mu \in (0, \mu_0)$ and negative (i.e. QKD is totally insecure) if $\mu \in (\mu_0, \mu_2)$ where the zero point lies at $\mu \approx 2.07$ photons.

3.3. Partly accessible loss in a system with large loss

Eve does not necessarily need to access the whole lossy quantum channel to be successful. (By *accessing* we mean, that she can avoid these losses either by replacing a quantum channel by a perfect, loss-free one, or by replacing it by classical communication and state preparation.) The formulas derived above still apply if we collect into the quantity η_B all those losses on the way to Bob's detector that are not accessible to Eve, while η_L denotes now only that loss that is accessible to her. It is instructive to look at the limit of high non-accessible losses ($\eta_B \ll 1$). In that case we can approximate function F of equation (18) by

$$F \approx \eta_B^2\left(\eta_L\mu - \frac{1}{2}\eta_L^2\mu^2 - P_D\right). \tag{19}$$

The insecurity criterion $F \leq 0$ in the region $\mu < \mu_2$ then leads to the condition

$$\eta_L \leq \eta_L^{\text{crit}} = \frac{1}{\mu}\left(1 - \sqrt{1 - 2P_D}\right) \approx \frac{P_D}{\mu} \approx \frac{1}{12}\mu^2, \tag{20}$$

which is independent of η_B. Dependence of η_L^{crit} on μ is shown as a solid line in Fig. 3. The additionally condition (17) can be approximated by $\mu < 1/\eta_L$ in leading order of η_B (a dashed line in Fig. 3).

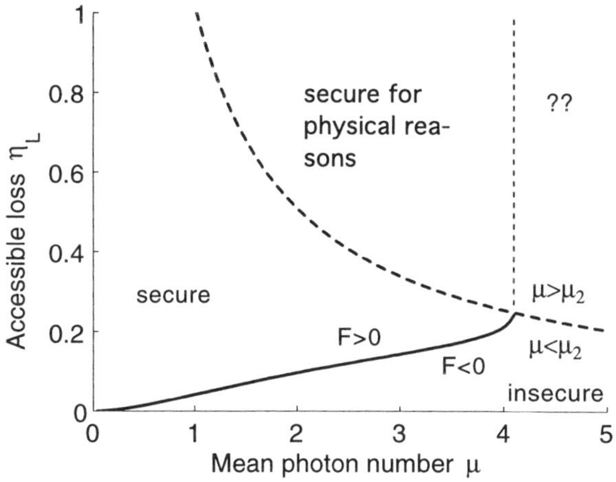

Figure 3. The secure parameter regime for the losses accessible to Eve for large Bob's losses ($\eta_B \ll 1$) is the region above the solid line ($F > 0$). In the region with $F \leq 0$ and $\mu < \mu_2$ the system is insecure. In the remaining region we have $F \leq 0$, but since $\mu > \mu_2$, we cannot make any definitive statements about security.

We can conclude that the system is secure against USD attacks in the regime of small detection efficiencies η_B if $F > 0$ and $\mu < \mu_2$. Furthermore, the system is insecure if $F \leq 0$ and $\mu < \mu_2$. For the parameter region with $\mu > \mu_2$ we can only make indirect statements. One is, that if the system is secure for a pair of values (μ, η_L), then it must be secure for all values (μ, η_L') with $\eta_L' > \eta_L$, otherwise Eve could gain an advantage by not accessing all the loss available to her. In the remaining region we have $F \leq 0$, but since $\mu > \mu_2$, we cannot make any definitive statements about security. Note that these considerations are valid for $\eta_B \ll 1$ and only in this limit η_B does no longer play any role. For higher values of η_B this changes.

4. Comparison of USD attack and beam-splitting attack

Traditionally, security against the beam-splitting attack [13] has been used as a *practical* level of security. In the beam-splitting attack the lossy line is replaced by an ideal lossless one complemented by a beam splitter such that the total loss of the original line is reproduced. The eavesdropper stores any photons coming out of the free arm of the beam splitter. If both the eavesdropper *and* the receiver obtain a photon (it is possible for multiphoton signals) Eve can measure her signal after she learns the polarization basis in the public announcement and she will learn thereby the bit value of these signals completely.

It is worth mentioning that security against beam-splitting attack allows that one can obtain a secure key even for large average photon numbers [12, 18]. It is clear from our analysis, however, that for large values of μ and typical loss rates, the USD attack will render the quantum key distribution protocol completely insecure.

In the USD attack the probability to identify a signal depends only on the average photon number μ, and once this probability is high enough to generate the expected number of signals for the receiver (which depends on the amount of loss) then the transmission becomes insecure. In the beam-splitting attack, on the other hand, the total probability of identified signals depend on μ *and* on the transmission coefficient η, and this probability goes *down* with increasing loss for fixed μ. In other words, the beam-splitting attack becomes less efficient with increasing loss. This is easy to see in a simple example of a two-photon signal. The probabilities $p(n, 2-n)$ that $n = 0, 1, 2$ photons arrive at Bob's detectors and $n-2$ photons go to Eve in the beam-splitting attack are given by

$$\begin{aligned} p(0,2) &= (1-\eta)^2, \\ p(1,1) &= 2\eta(1-\eta), \\ p(2,0) &= \eta^2. \end{aligned} \tag{21}$$

This means, that for high losses ($\eta \ll 1$) most likely both photons are sent to Eve. Probability of this event is $p(0,2) \approx 1 - 2\eta$, while the splitting probability goes down as $p(1,1) \approx 2\eta$. The respective probabilities for n-photon signals are of the same order of magnitude in η. Therefore, clearly, there is a crossover as a function of η where for fixed average photon number η the USD attack is more efficient than the beam-splitting attack.

We would like to stress again that from a technological point of view the USD attack seems to be easier to implement than the beam-splitting attack. There is no need of a quantum channel with reduced loss and no need of quantum memory, as required by the beam-splitting attack.

5. Conclusions

We have shown that unambiguous discrimination of linearly independent signal states can be used as an effective attack against realistic quantum crypto-systems. This attack enables eavesdropper to gain information on the key without causing any errors. It does not require the ability to store quantum states or to perform complicated quantum dynamics and it does not require to substitute the lossy quantum channel by a perfect one. We have derived a set of conditions which allow to judge whether a given system is insecure. In the limit of small detection efficiencies η_B we have obtained an analytic result that determines explicitly a set of parameters (line transmittances, detector efficiencies and mean photon numbers in coherent states sent by Alice) for which the transmission is secure with respect to the USD attack. We have also showed that security against beam-splitting attacks does not necessarily imply security against the USD attack.

Acknowledgments

This work was supported by the Academy of Finland (project 43336), the Czech Ministry of Education (research project "Wave and particle optics" and project VS 96028), and the Czech National Security Authority (19982003012).

References

[1] G. S. Vernam, Journal of the American Institute of Electrical Engineers **45**, 109 (1926).

[2] C. H. Bennett and G. Brassard, In: *Proceedings of IEEE International Conference on Computers, Systems, and Signal Processing, Bangalore, India*, IEEE, New York, 1984, pp. 175–179.

[3] S. Wiesner, SIGACT News **15**, 78 (1983).

[4] B. Huttner and A. K. Ekert, J. Mod. Opt. **41**, 2455 (1994).

[5] D. Mayers, In: *Advances in Cryptology – Proceedings of Crypto '96*, Springer, Berlin, 1996, pp. 343–357, also available as quant-ph/9606003.

[6] D. Mayers, Unconditional security in Quantum Cryptography, available as quant-ph/9802025v4.

[7] H.-K. Lo and H. F. Chau, Science **283**, 2050 (1999).

[8] B. Huttner, N. Imoto, N. Gisin, and T. Mor, Phys. Rev. A **51**, 1863 (1995).

[9] H. P. Yuen, Quantum Semiclass. Opt. **8**, 939 (1996).

[10] M. Dušek, O. Haderka, and M. Hendrych, Opt. Comm. **169**, 103 (1999).

[11] G. Brassard, N. Lütkenhaus, T. Mor, and B. C. Sanders, Security Aspects of Practical Quantum Cryptography, available as quant-ph/9911054.

[12] N. Lütkenhaus, Phys. Rev. A **61**, 052304 (2000), also available as quant-ph/9910093.

[13] C. H. Bennett, F. Bessette, G. Brassard, and L. Savail, J. Cryptology **5**, 3 (1992).

[14] I. D. Ivanovic, Phys. Lett. A **123**, 257 (1987).

[15] A. Peres, Phys. Lett. A **128**, 19 (1988).

[16] G. Jaeger and A. Shimony, Phys. Lett. A **197**, 83 (1995).

[17] A. Chefles and S. M. Barnett, Phys. Lett. A **250**, 223 (1998).

[18] M. Dušek, M. Jahma, N. Lütkenhaus, Unambiguous-state-discrimination attack in cryptography with weak coherent states, to appear in Physical Review A (August 2000), also available as quant-ph/9910106.

Decoherence and its Implications in Quantum Computation and Information Transfer
T. Gonis and P.E.A. Turchi (Eds.)
IOS Press, 2001

Molecular Classical Computer Basic Elements

A. Tamulis, Z. Rinkevicius*, J. Tamuliene

Instit. of Theoretical Physics and Astronomy, A. Gostauto 12, Vilnius, 2600, Lithuania
**Faculty of Fundamental Sciences, Kaunas Technology University., Studentu 50, Kaunas, 3000, Lithuania*

Abstract: The description of our designed molecular implementation of two variable logic functions: OR, AND is performed. The design of the basic elements of the molecular computers is based on the semiempirical quantum chemical calculations of organic photoactive electron donor, electron acceptor and insulator molecules, supermolecules, supramolecules and substituted fullerene molecules.

1. Introduction

Synthesis and construction of truly unimolecular devices which focuses on the properties of a single organic or biological molecule is in the initial stage now.[1-6] Investigation of basic elements for molecular neural networks is one of evolving area of neuromolecular computing science.[7-9] Networks constructed from nanoscale elements – organic molecules have advantage over silicon based networks[10] in their speed and learning capabilities. The projects of molecular shift register memory based on photoinduced electron transfer phenomenon in organic supermolecules,[11] molecular organic cellular automaton based on phthalocyanine molecules,[12] molecular electronic NAND gate,[13-14] and self-assembly computer based on fullerene C_{60} molecule dimmers[15] were proposed.
Some quantum chemical research of photoactive organic molecules, supermolecules and supramolecules were performed in our theoretical molecular electronics laboratory for the purpose of further design of molecular devices.[16-23] The quantum chemical calculations and investigations of three photo electron donor molecules: carbazole (abbreviation of this compound will be Cz and formula is $C_{12}H_9N$), N,N,N',N'-tetramethyl-1,4-phenylenediamine (abbreviation: TeMePhDA formula: $C_6H_4(N(CH_3)_2)_2$), 1,4-phenylenediamine (PhDA, $C_6H_4(NH_2)_2$); several organic electron acceptor molecules: 2,4,7-trinitro-9-fluorenone (TNF, $C_{12}H_5O(NO_2)_3$), 2,4,5,7-tetranitro-9-fluorenone (TeNF, $C_{12}H_4O(NO_2)_4$), 7,7,8,8-tetracyanoquinodimethane (TCNQ, $C_8H_4(CN)_4$), 2,4,7-trinitro-9-dicyanofluorene (TN9(CN)2F, $C_{12}H_5C(CN)_2(NO_2)_3$), 2,4,5,7-tetranitro-9-dicyano-fluorene (TeN9(CN)2F, $C_{12}H_4C(CN)_2(NO_2)_4$), 1,2,4,5-tetracyano-benzene (TCNB, $C_6H_2(CN)_4$), 2,3,5,6-tetrachlor-1,4-benzoquinone (TeClBQ, $C_6Cl_4O_2$), 2,3,5,6-tetrafluoro-7,7,8,8-tetracyanoquinodimethane (TeFTCNQ, $C_8F_4(CN)_4$) and others; carbazole-containing supermolecules (carbazole is connected via the alkane or alkene electron insulator bridges with electron acceptor molecules) and carbazole-containing supramolecules (carbazole and electron acceptor molecules are stabilized by the van der Waals forces) were performed using semiempirical MNDO, AMI, PM3[24-25] and CNDO/S-Configuration Interaction methods.[26] The results of our investigations show that

photoactive charge transfer carbazole-containing supermolecules and supramolecules should be molecular implementation of photodiodes.[17-21] Taking into account 120 single excited configurations, calculations of singlet state bands of two C_{60} molecules in softer region are in good agreement with experimental values of absorption spectrum of C_{60} clusters in water solution or in the Langmuir-Blodgett films. The CNDO/S-CI calculation results show three new additional bands in the regions 342.90-364.46 nm, 427.28-437.08 nm and 481.58-495.88 nm.[22] Quantum chemical investigations of a charge transfer singlet state band of the proposed system consisting of two C_{60} molecules and two small charges are performed in the framework of CNDO/S-CI method taking into account 60 single excited configurations. Results of calculations show the existence of a charge transfer band in this system of two C_{60} molecules. This enables to design unimolecular and bulk photoactive molecular devices.[27] The projects of photodiodes and radiation solar energy conversion cells based on photoactive polymers and C_{60} layer or other C_{20+2n} molecular films are proposed.[27] The quantum mechanical investigations of fullerene C_{24}, C_{26}, C_{28} molecule conformers are performed in the framework of point set group theory and semiempirical PM3 configuration interaction, MNDO, AMI. The main criterion of stability of calculated fullerene molecules we stated the lowest total energy of various isomers and conformers that appears due to the Jahn-Teller distortion.[23] The most stable occurs C_{24} D_6 symmetry conformation with term $^1A_1$ and open shell C_{26} D_{3h} symmetry conformation with term $^5A_1$.

Some preliminary results of our design of molecular implementation (MI) of basic elements of digital and postdigital computers have been published,[27-28] and reported in conferences.[29-30] They are as follows.

MI of two, three, four variable logic functions, summators of neuromolecular networks, cells of molecular cellular automata, molecular trigger, devices of electronically genome regulation were designed based on results of semiempirical quantum chemical calculations of electron donor, electron insulator, electron acceptor and fullerene molecules.[16-23, 27, 28] Complete set of sixteen MIs of two variable logic functions (for example: OR, AND, Implication, Equivalence, Difference, etc.) was designed and also proposed using MIs of two variable molecular logic function initial basic sets: {OR, AND, Negation} or {NOR} and, or {NAND}. We have described in more detail the designed MIs of: a) two variable logic functions OR, NOR, AND, NAND (from fullerene molecules) , Converse Unitary Negation-1, Converse Unitary Negation-0, Unitary Negation-1, Unitary Negation-0, "0" and "1 " Matrix Constants; b) three variable logic functions AND, NAND, OR, NOR analogs; c) four variable logic functions OR, NOR, AND, NAND analogs. This was done based on quantum chemical investigations of organic photo-induced electron donor molecules: a) carbazole, 3,6-dibromcarbazole, TeMePhDA, PhDA; b) electron acceptor molecules: TCNQ, TCNB, TeClBQ, small empty and endohedral fullerene molecules: C_{60}, C_{28}, $C_{28}H_4$, $C_{24}H_2$, C_{20}, $C_{20}H_6$, A@C_{60}, A=Be, Zn, Cd and c) electron insulator molecules by using semiempirical MNDO, AMI and PM3 methods in MOPAC-7 programs package. We have investigated in more detail the electronic structure of above mentioned planar electron donor and electron acceptor molecules and series of fullerene C_{60} substituted derivatives: $C_{60}CH_2$, $C_{60}C_2H_4$, $C_{60}C_3H_6$, $C_{60}C_4H_8$ using *ab initio* Density Functional Theory [31] (DFT) and Hartree-Fock[32] (HF) methods and designed new series of more correctly MIs of two variable logic functions OR, NOR, AND, NAND (two sets: one designed from planar molecules and another - from fullerene molecules) , Converse Unitary Negation-1, Converse Unitary Negation-0, Unitary Negation-1, Unitary Negation-0, "0" and "1" Matrix Constants based on geometry optimization procedure of above mentioned molecular devices. The electron hoping via the insulator bridges in the supermolecules: electron donor-bridge-electron acceptor

phenomenon was investigated by using CNDO/S-Configuration Interaction method.
In this paper we shall concentrate on more detailed explanations of designed Mis of two variable logic functions: OR, AND.

2. Molecular implementation of disjunction logical function

Four various MI of disjunction were designed: 1) from planar PhDA, TeMePhDA, TeClBQ, TCNQ molecules[33] 2) from PhDA, carbazole and fullerene C_{60}, $C_{28}H_4$ molecules[28] 3) from PhDA, carbazole and fullerene derivative $C_{60}C_2H_4$ (see Figure 1) 4) from PhDA, carbazole and TeClBQ (see Figure 2).

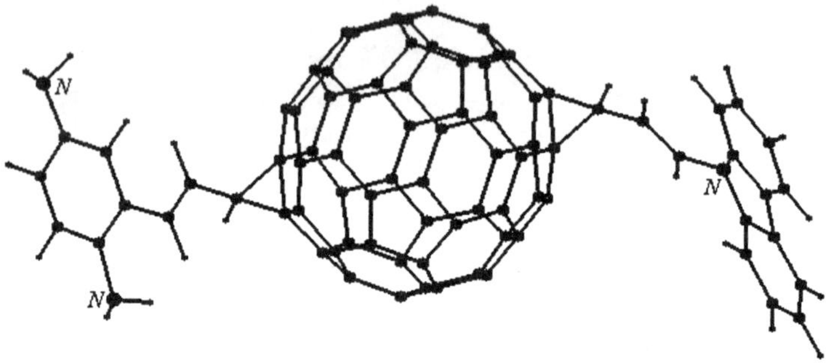

Figure 1. Molecular implementation of two variable disjunction (OR) is designed from PhDA, carbazole and $C_{60}C_2H_4$ molecules and the electron insulator -CH=CH- bridges.

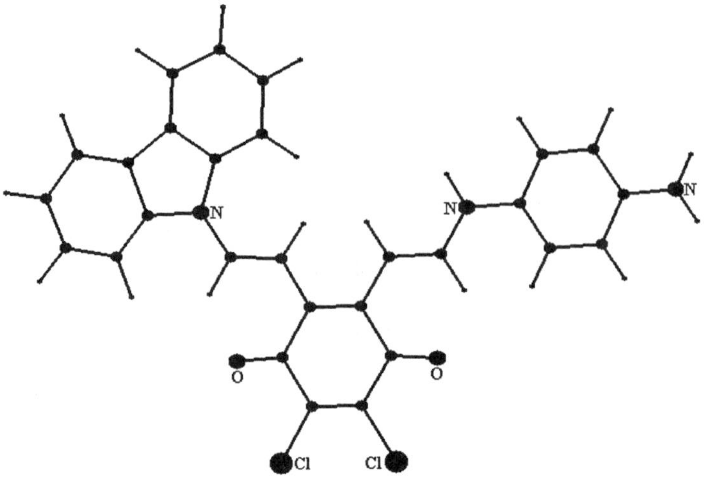

Figure 2. Molecular implementation of two variable disjunction (OR) is designed from PhDA, carbazole and TeClBQ molecules and the electron insulator -CH=CH- bridges

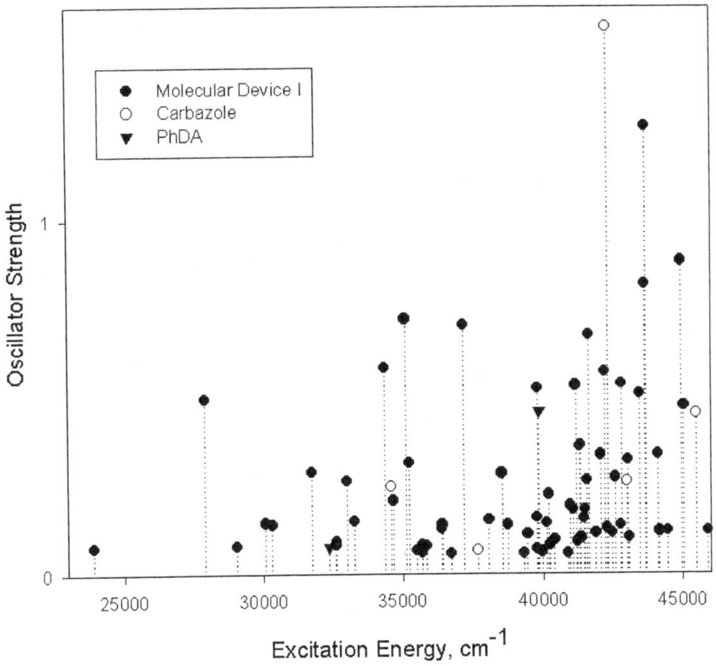

Figure 3. Molecular Device I and PhDA, carbazole molecules UV/Vis absorption spectrum calculated using ZINDO1/S method.

Two photoelectron donor molecules: PhDA and carbazole (two inputs of this logic gate) that is possible in principle to excite by two different wavelengths are connected via an electron insulator -CH=CH- bridges to electron acceptor molecule (output of this logic gate): $C_{60}C_2H_4$ (Device I) or TeClBQ (Device II). The present state of experimental physics allow to measure the single electron charge[34-37] therefore in principle it possible to measure the additional electron charge on single electron acceptor $C_{60}C_2H_4$ (in Device I) or TeClBQ (in Device II). Let us imagine that we do not excite by light no one photo electron donor molecule (not send signals on the input) of this MI OR gates. Then we do not measure the additional electron charge on the $C_{60}C_2H_4$ (in Device I) or TeClBQ (in Device II) molecule (on the output of this gate) - we should have response 0. In the case if we should excite by light one photo electron donor molecule (one of two inputs) electron should jump from photo electron donor molecule to the fragment of $C_{60}C_2H_4$ (in Device I) or TeClBQ (in Device II) molecule and we should measure the additional electron charge on $C_{60}C_2H_4$ (in Device I) or TeClBQ (in Device II) molecule. Therefore we should have responses 1 on the output of the gate in both cases: if excite PhDA or carbazole molecule. In the case if we excite both photo electron donor molecules PhDA and carbazole on the input of the gate at the same time we should measure additional electron on the $C_{60}C_2H_4$ (in Device I) or TeClBQ (in Device II) molecule as well - we should have response 1 on the output of the gate. Summarizing in all four possible input cases of this two variable logic gate we should have sequence of responses on the output - $C_{60}C_2H_4$ (in Device I) or TeClBQ (in Device II) molecule: 0, 1, 1, 1. It means that this our MI logic function is disjunction.

Device I constructed from PhDA, carbazole and $C_{60}C_2H_4$ molecules connected via electron insulators -CH=CH- bridges (see Fig. 1) geometry optimization performed using semiempirical PM3 method with GAMESS-US[38] package. UV/Vis absorption spectrum

of Device I calculated with ZINDO1/S[39-40] method taking into account 430 single configurations on PM3 optimized geometry (see Figure 3).
Two wavelength 228.44 nm and 251.26 nm corresponding to carbazole and PhDA excitation in Molecular Device I can be used to send input signal to this logic gate (see Figure 3). From Mulliken atomic charges analysis we obtained that charge transfer for these excitation wavelengths from photoelectron donors (PhDA or carbazole) to electrons acceptor $C_{60}C_2H_4$ in Molecular Device I exist: -0.3562 electron charge (e) and -0.361 e.

3. Molecular implementation of conjunction logic function

Three various MI conjunctions were designed: 1) from planar PhDA, TeMePhDA, TeClBQ, TCNB molecules[33] 2) from PhDA, carbazole, empty fullerene C_{60} and endohedral metallofullerene Me@C_{60} molecules[28] 3) from PhDA, carbazole, indane derivative (abbreviation of this compound will be Ind and formula is $C_9H_4(CH_2)(C(CN)_2)_2$) and $C_{60}CH_2$ (see Figure 4).

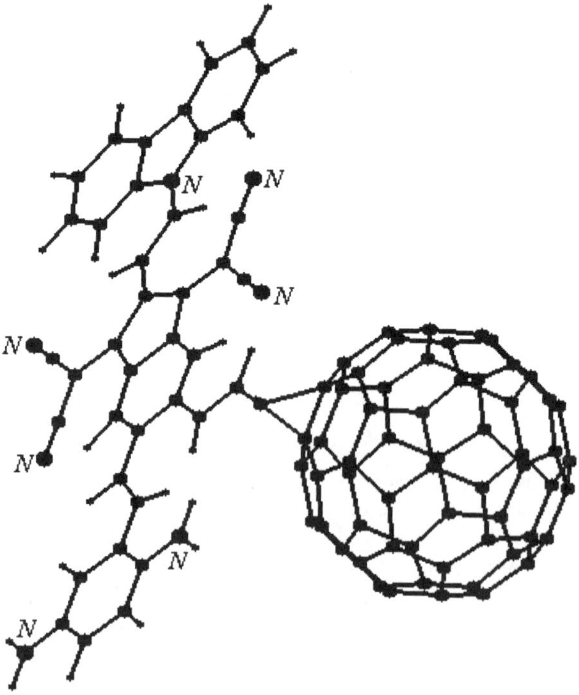

Figure 4. Molecular implementation of two variable conjunction (AND) is designed from PhDA, carbazole, Ind and $C_{60}CH_2$ molecules and the electron insulator -CH=CH- bridges.

Two photo electron donor molecules: PhDA, carbazole (two inputs of this gate) are connected via the electron insulator -CH=CH- bridges with the fragment of electron acceptor molecule Ind and further via the -CH=CH- bridge with the second electron acceptor molecule $C_{60}CH_2$ - output (see Figure 4). Let us imagine that we measure additional electron charge in the ground state $C_{60}CH_2$ molecule. In the case if we should not excite no one of two photo electron donor molecules on the input of this gate we should not measure the additional electron charge on the output - $C_{60}CH_2$ molecule. Therefore in this case we should have response on the output: 0. In the case if we should excite by light one of two photo electron donor molecules we also should not measure the additional electron charge on the output of this gate - $C_{60}CH_2$ molecule (because according to our quantum chemical calculations Ind molecule is better electron acceptor than $C_{60}CH_2$ and electron hopping from the fragment of Ind molecule to $C_{60}CH_2$ molecule should not exists). Therefore in all these two cases we should have responses on the output; 0. In the case if we should excite at the same time both photo electron donor molecules (two inputs) we should measure electron on the output - $C_{60}CH_2$ molecule because one additional electron prefer to jump from twice charged fragment of Ind molecule to $C_{60}CH_2$ molecule. Therefore we should have response on the output: 1. Summarizing in all four different cases of this two variable logic gate we should have sequence of responses on output - $C_{60}CH_2$ molecule: 0, 0, 0, 1. It means that this our MI logic function is conjunction.

Acknowledgements

This research was mainly financed by EOARD contract F61775-00-WE050
Authors thanks Prof. A. Graja and Poznan's Supercomputing and Networking Center for possibility to use CRAY and GAUSSIAN 98 DFT-TD method. We are grateful to Computer Centre at Oklahoma State University, USA with installed the GAUSSIAN 98 package and Dr. N. A. Kotov for cooperation in computations.
A. Tamulis is grateful to Lithuanian Science and Education foundation supported by two grants or possessing EOARD contract F61775-00-WE050 and National Research Council Twinning Program 2000-2001 with Lithuania grant.

References

1. M. Pomerantz, A. Aviram, R. A. McCorckle, L. Li, and A. G. Schrott, "Reactification of STM current to graphite covered with Phthalocyanine molecules". *Science* **255**, pp. 1115-1118, 1992.
2. A. S. Martin, J. R. Sambles and G. J. Ashwell, "Molecular rectifier", *Phys. Rev. Lett.* **70,** pp. 218-221, 1993.
3. J.-W. Choi, J. Min, s. Y. Oh, D. M. Shin, and W. H. Lee, "Molecule photodiode consisted of TCNQ-Pyrene-Ferrocene hetero-Langmuir-Blodgett films" *Book of Extended Abstracts of the 5th Intern. Symp. on Bioelectronic and Molecular Electronic Devices,* R&D Assoc. for Future Electron Devices, Okinava, Japan, pp. 189-190, Nov. 28-30, 1995.
4. I. Lee, W. Lee, R. J. Warmack, D. P. Allison, and E. Greenboum, "Molecular electronics of a single photosystem I reaction center: studies with scaning tunneling microscopy and spectroscopy:. *Book of Extended Abstracts of the 5th Intern. Symp. on Bioelectronic and Molecular Electronic Devices,* R&D Assoc. for Future Electron Devices, Okinava, Japan, pp. 19-23, Nov. 28-30, 1995.
5. Masuo Aizawa, "Molecular assemblies for bioelectronic devices and molecular electronic devices". *Book of Extended Abstracts of the 5th Intern. Symp. on Bioelectronic and Molecular Electronic Devices,* R&D Assoc. for Future Electron Devices, Okinava, Japan, pp. 3-6, Nov. 28-30, 1995.
6. M. Garzon, P. Neathery, R. Deaton, R. C. Murphy, D. Franceschetti, "A new metric for DNA computing", *Proc. 2nd Genetic Programming Conference,].* R. Koza, K. Deb, M. Dorigo, D. B. Fogel, H. Iba, R. L. Riolo, eds., Morgan Kaufinann, pp. 472-478, 1997.
7. H. M. Hastings and T. W. Siegel, "Molecular neural networks". *Proceedings of the Twelfth Annual International Conference of the IEEE - Engineering in Medicine and Biology Society,* Philadelphia, Pennsylvania, USA, pp. 1768-1769, November 1-4, 1990.
8. M. Conrad, "Emergent computation through self-assembly" *Nanobiology* **2**, pp. 5-30, 1993.

9. A. Miyata, H. Aral and Y. Unuma, "Development of optical neural devices using photochromic molecular assemblies" *Book of Extended Abstracts of the 5th Intern. Symp. on Bioelectronic and Molecular Electronic Devices,* R&D Assoc. for Future Electron Devices, Nov. 28-30, 1995, Okinava, Japan, pp. 201-202.

10. J. A. McDonald, "Neural nets are starting to make sense". *Biosensors & Bioelectronics,* **7**, pp. 621-626, 1992.

1l. J. J. Hopfield, J. N. Onuchie, and D. N. Beratan, "A molecular shift register based on electron transfer". *Science,* pp. 817-820, 12 August 1988.

12. Kikukazu Sakurai and Shigemasa Takano, "Organic quantum computing systems for electronic brain", *Proceedings of the Twelfth Annual International Conference of the IEEE - Engineering in Medicine and Biology Society,* Philadelphia, Pennsylvania, USA, pp. 1756-1757, November 1-4, 1990.

13. R. R. Birge, A. F. Lawrence, and L. A. Findsen, "Current research and future potential of the application of molecular electronics to computer architecture". *Proceedings of the 1986 International Congress on Technology and Technology Exchange,* Pittsburgh, Pennsylvania USA, October 6-8, 1986.

14. R. R. Birge, B. R. Ware, P. A. Dowben, and A. F. Lawrence, "Quantum statistics and reliability of molecular electronic devices". *Proceedings of the Engineering Foundation Conference on Molecular Electronics - Science and Technology,* Keauhou-Kona, Hawaii, p.p. 1-9, February 19-24, 1989.

15. D. L. Koruga, S. R. Hameroff, J. Simic-Krstic, S. Jankovic, M. Trifiinovic, and M. Sundareshan, "Self-assembled molecular computer based on fullerene C_{60}: from nanobiology to nanotechnology". *Book of Extended Abstracts of the 4th Intern. Symp. on Bioelectronic and Molecular Electronic Devices,* R&D Assoc. for Future Electron Devices, Miyazaki, Japan, pp. 1-4, Nov. 30-Dec. 2, 1992.

16. A. Tamulis, A. Undzenas, and L. Bazan, "The electronic processes in carbazolyl containing polymers sensitized by pyrylium and Rhodamine 6G", *The Proceedings of the Second international symposium "Organic Materials for Non-Linear Optics",* held 4th-6th September, 1990, in Oxford. Edited by R.A. Harm and D. Bloor. Published by The Royal Society of Chemistry, Thomas Graham House, Science Park, Cambridge CB4, 4WF, pp. 267-272, 1991.

17. A. Tamulis, S. Janusonis, and L. Bazhan, "Selection rules for self-formation in the molecular nanotechnology", *Makromolecular. Chem., Macromol. Symp.* **46,** pp. 181-185, 1991.

18. A. Tamulis and L. Bazan, "Charge photogeneration in carbazole-containing compounds and valence bands of Oligomers", *Progress in Colloid & Polymer Science* **90** , pp. 186-193, 1992.

19. A. Tamulis and L. Bazan, "Electronic structure of the supermolecules and supramolecules constructed from carbazole and acceptor molecules", *Mol. Cryst. Liq. Cryst.* 216pp. 49-54, 1992.

20. A. Tamulis and L. Bazan, "Quantum chemical investigations of photoactive supermolecules and supramolecules, their self assembly and design of molecular devices", *Synthetic Metals* 55-57, pp. 4685-4690, 1993.

21. A. Tamulis, L. Bazhan, "Quantum chemical investigations of photoactive supermolecules formation, supramolecules self assembly and solar cells design". *The Proceedings of the Fifth National Conference with International Participation: "Self-Formation Physics, Technology and Application",* held in Vilnius, The Republic of Lithuania, pp. 48-50, 12th-13th May, 1992.

22. A. Tamulis, M. Braga, A. Klimkans, "Quantum Chemical Investigation of Two Fullerene C_{60} Molecules", *Fullerene Science and Technology* **3**, pp. 603-610, 1995.

23. L. M. Balevicius, E. Stumbrys, A. Tamulis, "Conformations and electronic structure of fullerene C_{24} and C_{26} molecules", *Fullerene Science and Technology* 5, pp. 85-96, 1997.

24. J. J. P. Stewart, "MOPAC - A semiempirical molecular orbital program". Journal of Computer-Aided Molecular Design **4,** pp. 1-105, 1990.

25. J. J. P. Stewart, *J. Am. Chem. Soc.* **10,** p. 221, 1989.

26. J. del Bene and H H. Jaffe, *J. Chem. Phys.* **49**, pp. 1221-1229, 1968.

27. A. Tamulis V. Tamulis, "Molecular electronics - advanced technology". *Science and Arts of Lithuania* **2**, pp. 40-47, (in Lithuanian), 1994.

28. Tamulis A., Stumbrys E., Tamulis V. and Tamuliene J., "Quantum Mechanical Investigations of Photoactive Molecules, Supermolecules, Supramolecules and Design of Basic Elements Molecular Computers", in NATO ASI series. High Technology; Vol. 9, Ed. by F. Kajzar, V.M. Agranovich and C.Y.-C. Lee, *"Photoactive Organic Materials: Science and Applications",* June 25-30, 1995, Avignon, France, Kluver Academic Publishers, Doderecht/Boston/London, p.p. 53-66, 1996.

29. A. Tamulis and V. Tamulis, "Quantum chemical design of molecular cellular automata, neuromolecular networks and solar cells based on advanced molecules". *Book of abstracts ofpapers of International Workshop on "Quantum Communications and Measurement",* University of Nottingham, Nottingham, U. K., pp. 81-82, 11-16 July 1994.

30. A. Tamulis, E. Stumbrys, V. Tamulis, J. Giceviciute-Tamuliene and A. Nakas, "Stability investigations of small empty and endohedral fullerene molecules, disc-like supramolecules and design of basic elements of molecular computers". *Book of abstracts ofpapers of NATO ASI on "Localized and Itinerant Molecular Magnetism: From molecular Assemblies to the Devices",* Tenerife, Spain, p. 520, 22 April-03 May, 1995.

31. R.G. Par and W. Yang, Density-functional theory of atoms and molecules, Oxford Univ. Press: Oxford, 1989.

32. J. A. Pople, and R. K. Nesbeth, "Self-consistent orbitals for radicals", *J. Chem. Phys.* 22, pp. 571-574, 1959, R. McWeeny and G. Dierksen, *J. Chem.Phys.* **49,** p. 4852, 1968.

33. A. Tamulis, V. Tamulis, "Design of Basic Elements of Molecular Computers Based on Quantum Chemical Investigations of Photoactive Organic Molecules", Proceedings of the SPIE Photonics WEST_ Conference on Optoelectronic Integrated Circuits II, held on 24-30 January, 1998, San Jose Convention Center, San Jose, California, U.S.A., Volume 3290, p.p. 315-324.

34. P. Delsing, K. K. Likharev, LI S. Kuzmin, and T. Claeson, *Phys. Rev. Lett.* **63,** pp. 1862-1864, 1989.

35. H. Pothier, P. Lafarge, C. Urbina, D. Esteve and M. H. Devoret, *Europhys. Lett.* **17**, pp. 249-254, 1992.

36. E. R. Williams, R. N. Ghosh and J. M. Martinis, *J. Res. Natl. Inst. Stand. Technol.* **97**, pp. 299-304, 1992.

37. A. Gourdon, "Toward single molecule addressing". *Book of Extended Abstracts of the 5th Intern. Symp. on Bioelectronic and Molecular Electronic Devices,* R&D Assoc. for Future Electron Devices, Okinava, Japan, pp. 207-208, Nov. 28-30, 1995.

38. M. W. Schmidt, K. K. Baldridge, J. A. Boatz, S. T. Elbert, M. S. Gordon, J. H. Jensen, S. Koseki, N. Matsunaga, K. A. Nguyen, S. J. Su, T. L. Windus, M. Dupuis, J. A. Montgomery, *J. Comput. Chem.* **14**, pp. 1347-1363, 1993.

39. J. Ridley and M. Zerner, TCA (Berl.) **32**, pp.111-134, 1973.

40. Semiempirical Molecular Orbital Methods Michael C. Zerner in Reviews in Computational Chemistry II Ed. Kennith B. Libkowitz and Donald B. Boyd VCS Publishers Inc. 1991.

Author Index